The Mathematical Method

"YOU WANT PROOF? I'LL GIVE YOU PROOF!"

THE MATHEMATICAL METHOD

A Transition to Advanced Mathematics

Murray Eisenberg

University of Massachusetts, Amherst

Prentice Hall
Upper Saddle River, New Jersey 07458

Library of Congress Cataloging-in-Publication Data

Eisenberg, Murray
    The mathematical method : a transition to advanced mathematics /
Murray Eisenberg.
       p.   cm.
    Includes index.
    ISBN 0-13-127002-8
    1. Proof theory.   I. Title.
QA9.4.E47   1996
511.3—dc20   95-650
                                        CIP

Composed by the author using the Y&YTeX System with the $\LaTeX 2_\varepsilon$
Document Preparation System and the $\mathcal{A}_{\mathcal{M}}\mathcal{S}$-$\LaTeX$ and $P\!\!|\,C\!T_{\!E}X$ macros.
Typeset in Lucida® Bright, Lucida New Math, and LucidaBright
Expert fonts designed by Bigelow & Holmes, Inc.
$T_{\!E}X$ is a trademark of the American Mathematical Society.

Acquisitions Editor: George Lobell
Editorial Production/Supervision: Rachel J. Witty, Letter Perfect, Inc.
Manufacturing Buyer: Alan Fischer
Marketing Manager: Frank Nicolazzo
Cover Designer: Jayne Conte
Cover Photo: Rotunda, Worrell Center for Law and Management
                Wake Forest Univ., N. C. © 1996 Timothy Hursley
Photo Editor: Lorinda Morris-Nantz
Photo Researcher: Joelle Burrows
Editorial Assistant: Gale Epps

© 1996 Prentice-Hall, Inc.
Simon & Schuster/A Viacom Company
Upper Saddle River, NJ 07458

Printed in the United States of America
10 9 8 7 6 5 4 3 2 1

ISBN 0-13-127002-8

Prentice-Hall International (UK) Limited, *London*
Prentice-Hall of Australia Pty. Limited, *Sydney*
Prentice-Hall Canada, Inc., *Toronto*
Prentice-Hall Hispanoamericano, S.A., *Mexico*
Prentice-Hall of India Private Limited, *New Delhi*
Prentice-Hall of Japan, Inc., *Tokyo*
Simon & Schuster Asia Pte. Ltd., *Singapore*
Editora Prentice-Hall do Brasil, Ltda., *Rio de Janeiro*

For Phyllis

Contents

Preface

This text is a gentle introduction to mathematical rigor and proof, intended especially for those expecting to major in mathematics. Its aims are to help prepare students for the demands of standard junior-senior courses in analysis and algebra and, at the same time, to give students an opportunity to see whether they really like engaging in mathematical reasoning.

Many topics could have been chosen for this text that would contribute toward fulfilling the stated aims. My choice of topics was, of course, influenced by my own interests and was guided by the following principles:

1. The topics should not depend heavily upon specific content of prior math courses; in particular, they should have no prerequisites beyond one semester of calculus and should require very little even of that (one exception: some experience with limits of sequences—*not* with epsilon-delta formalities—is useful in parts of Chapter 3).

2. No topic should significantly overlap the content of a standard upper-division course taken subsequently by most math majors.

3. When a topic does overlap the content of another course, the approach should differ somewhat from the one normally expected in that course (so, for example, here the geometrically appealing Archimedean Ordering Property and Nested Interval Property are taken as axioms for the real numbers, and the Order Completeness Property is derived as a theorem).

4. Each topic should start with fairly simple ideas but build toward at least a few meaty ones.

5. No topic should unduly dwell upon proving what readers already know.

6. Each topic should be intrinsically interesting and include some
   results that will be new to all readers and many that will be new
   to nearly all.

I have *not* attempted to present systematically the fundamentals of
mathematics from the ground up—covering in order logic, sets, relations, functions, etc. In fact, my experience is that such a presentation
at this level can be too difficult for students because of its very generality, and sterile because of its superficiality. Rather, I prefer to introduce
these fundamentals *in context*, with "real" mathematical topics. (These
contexts are not isolated little extracts of mathematics but instead are
actual developments of such topics.)

For example, rather than introduce the general notions of equivalence relation and quotient set at the outset, I allow these notions to
emerge while treating congruence in number theory. (The general notions are defined and examined in Appendix C.) Students who had used
a preliminary version of this book have told me later, when taking advanced math courses, how much better a grasp they felt they had of
the general notions than did students whose first exposure to them
had been more abstract.

My experience is that students will learn the fundamentals about
logic, sets, relations, functions, etc., by working through the "real" topics in the chapters and by referring, when necessary, to the appendices.
Of course, the individual instructor does have the option of explicitly
covering all or parts of an appendix early, whether before, in parallel
with, or immediately after covering a topic in the chapters. Indeed, in
my own course I typically spend a little time on equivalence relations
in general after Sections 2.4 and 2.5. And I interpose material from Appendix A on set algebra and from Appendix B on function composition
and inversion when covering cardinality in Chapter 4.

When asked to prove things about natural numbers or integers, students are sometimes confused as to which properties they may legitimately assume and which not. Therefore, in Sections D.1 and D.3 I
collect and explicitly state these properties. I recommend that these
two sections be used only for reference rather than be covered explicitly.

Should the instructor wish, the properties of the natural numbers
collected in Section D.1 can be justified on the basis of the Peano Postulates by covering Section D.2; addition and multiplication then can
be defined recursively, and their properties established inductively, as
in Section 1.2.

A first course in mathematical rigor and abstraction is probably not
the place to construct the various fundamental number systems from
the ground up. For this reason, the integers are taken as being known,
and in Chapter 3 the real numbers are treated axiomatically. Should

the instructor nonetheless prefer a constructive approach, this can be followed. See Section D.4 and Exercise 3.4.12 for two alternative constructions of the integers from the natural numbers; Example 3.4.10 for a construction of the rational numbers from the integers; and Exercises 3.3.18 and 3.5.46 for two alternative constructions of the real numbers from the rational numbers.

A distinctive feature of this book is the arrangement of the exercises. They are interspersed throughout the text rather than segregated into sets at the ends of sections. This arrangement is designed to engage students more fully in the development and to make sure they understand one idea before moving on to the next.

The exercises range from the routine to the challenging. However, I have deliberately refrained from rating them by difficulty: I do not want to discourage students from attempting harder ones (and I want to make the point that in reality one often does not know at the outset how hard a problem might be).

At several places, especially in Chapters 1 and 2, a few examples and exercises are included that involve writing a computer program or, better yet, defining a function in a computer algebra system such as MATHEMATICA. Although these computer excursions can be omitted, I recommend including them. They will surely spice up the course—and at times perhaps provide some welcome relief from the ever-present commands to "prove", "prove", "prove." More significantly, using the computer as I suggest—to "teach the computer" how to do something— is, I believe, a valuable way to learn some otherwise abstract ideas. For the computer work, I particularly recommend MATHEMATICA or other software with equivalent power and expressiveness.

Here and there a few results are mentioned whose proofs are quite beyond the scope of this book but which are so significant or interesting it would have been a sin to omit. (See, for example, Section 2.3 on prime numbers.)

One of the aims of a course for which this book is intended is to learn to read as well as to write mathematics. With enough preparation by the students, lecturing by the instructor will ordinarily be superfluous. Class time can therefore be devoted primarily to students' presenting their proofs and other solutions to the class, perhaps themselves lecturing on some of the material, and reacting to their colleagues' presentations.

Acknowledgments

This book evolved from notes developed and used over several years in my version of Math 300, Fundamental Concepts of Mathematics, at the University of Massachusetts, Amherst. During that evolution, many people helped me. My colleague Berthold Schweizer aided me in digging out some historical information and pointed me to a few of the epigraphs.

The following reviewers provided useful feedback on early and late drafts of the manuscript.

Michael Albert
Carnegie Mellon University

Peter Colwell
Iowa State University

Dennis Garity
Oregon State University

Juan Gatica
University of Iowa, Iowa City

Daniel Sweet
University of Maryland, College Park

Thomas Tucker
Colgate University

Using LATEX not just to prepare the manuscript but also to convert it into camera-ready pages has been an adventure. Many TEXperts generously offered me help over the Internet when I could not convince LATEX to do what I wanted. My colleague George Avrunin resolved TEX problems as fast as I threw them at him. The technical staff at Y&Y, Inc., who supplied both the TEX system and the fonts used, provided prompt assistance when I ran into difficulties.

Michael Barr kindly granted me permission to use his TEX diagram macros for the several commutative diagrams that appear in the book.

The editorial staff at Prentice Hall has been helpful throughout this project. I thank my editor, George Lobell, not only for causing the project to reach completion but also for calmly keeping me on track when various crises, real or imagined, arose as the manuscript approached going to press.

Most especially I thank the many students whose reactions have helped me eradicate errors in and improve the notes that became this book. Should any errors remain, I will appreciate hearing about them—although I do not promise to reward you with the dimes and quarters my students earned when they found errors. More generally, I welcome your suggestions for further improvements.

Amherst, Massachusetts   Murray Eisenberg
June, 1995   murray@math.umass.edu

To the Student

Modern mathematics is nearly characterized by the use of rigorous proofs. This practice, the result of literally thousands of years of refinement, has brought to mathematics a clarity and reliability unmatched by any other science.

— *Arthur Jaffe and Frank Quinn*

In earlier mathematics courses, especially in calculus, the emphasis is usually upon completing calculations and solving problems without worrying too much about how to justify the methods used. In advanced mathematics, by contrast, there is considerable attention paid to carefully formulated definitions, clearly stated assumptions, and logically rigorous proofs.

As the words of Jaffe and Quinn[1] state, it is this rigor and proof—the mathematical method—that characterizes mathematics among fields of intellectual endeavor. Yet doing mathematics involves many activities, among them: calculating, exploring and experimenting, discovering, visualizing, applying, analogizing, generalizing, abstracting, and proving. Although this book certainly affords the opportunity to engage in all these activities, nonetheless the emphasis in it is decidedly upon proving. Try not to forget, though, that proving is just one aspect of doing mathematics.

Students in advanced math courses are often surprised to discover how much greater the emphasis there is upon words, and how much less upon numbers and formulas and graphs, than in their previous mathematical studies. Indeed, mathematically rigorous reasoning requires being adept at manipulating abstractions through words. I hope that this book will help you to acquire the requisite adeptness and thereby make your transition to advanced mathematics easier.

---

[1] Arthur Jaffe and Frank Quinn, "Theoretical Mathematics": Toward a Cultural Synthesis of Mathematics and Theoretical Physics, *Bull. Amer. Math. Soc.* **29** (1993), 1–13.

To a considerable extent, this is a "do-it-yourself" book. Only the bare minimum of proofs is included—just enough to give some models (plus a few proofs somewhat harder or less obvious than the sort I hope you will be concocting yourself by the end of the course). Look in these proofs for both general logical patterns and particular kinds of reasoning appropriate to particular contexts. However, do not expect the supplied proofs to serve as templates in which you make only minor modifications so as to fit each new problem that arises.

Here are a few suggestions on how to proceed:

- As you read, read actively—with paper and pencil ready to assist you in filling in a gap or completing an argument.

- Take your time when reading. Actively reading a math book— closely following the reasoning as you go—is just not a speedy business. So don't panic if you find yourself taking half an hour, an hour, or even longer to read some pages.

- Answer the parenthetical "why?"s that often (especially in the earlier sections) punctuate proofs; get into the habit of asking your own "why?"s as you read.

- After you have completed reading a proof of mine (or one of a fellow student, or even one of your own), cover it up and try writing it out yourself.

- If you find a definition difficult to grasp, look for objects that are instances of it as well as other objects that are not. (Many examples and exercises in the text are intended to guide you in doing that.)

- When trying to prove something, talk to yourself. Hold an internal dialog in which you ask yourself, "What is the hypothesis? What is the conclusion? What do all the terms in each mean? What previous results do I know that are related to what appears in the hypothesis and in the conclusion?"

- Especially if you get stuck in devising a requested proof, bounce your ideas—however vague and ill-formed they might seem to you—off another student in the class. And be ready to reciprocate. (Of course, do this only in so far as it is consistent with your instructor's policies about collaboration).

- Don't give up too quickly: if you have some idea of how to proceed, keep working at it. (Sometimes it helps to put a problem aside and come back to it later.) But when you find yourself stuck in a blind alley—and surely you will do so often—be ready to abandon your approach and try something else.

- Whatever you write, read it critically to see if it makes sense. Preferably, have someone else read it, too.

- Write in complete sentences (English or mathematical or a mixture of the two), with all the needed "for all"s, "for some"s, "if"s, "then"s, "and"s, and "or"s (as words or symbols).

- After you have written a proof or other solution to an exercise, rewrite it! Improve it in form and substance.

- Just because you have solved an exercise and even written the solution in a really nice form, don't stop there. Look for alternate solutions, too.

Unfortunately, there is no magic set of directions for how to prove all theorems. The road to mastering the art of mathematical proof may be steep and tortuous, but the journey is rewarding.

Ordinarily a textbook includes a bibliography to identify sources and to suggest further reading. This one does not. You will not learn what I would like you to learn here by consulting other books but only by working through everything yourself and with colleagues and an instructor. Please resist the temptation just this once to look elsewhere for help. But if you master a subject here and do want to pursue it further before you have a chance for a full-fledged course, then your instructor surely can suggest what to read.

My students' reactions to preliminary versions have been of great help in preparing this book. I welcome your own reactions; if you wish, you may correspond with me at the e-mail address shown below.

Amherst, Massachusetts   Murray Eisenberg
June, 1995   murray@math.umass.edu

Chapter 1

Induction

Great fleas have little fleas upon their backs to bite 'em,
And little fleas have lesser fleas, and so ad infinitum.
And the great fleas themselves, in turn, have greater fleas to
go on;
While these again have greater still, and greater still, and so
on.

— Augustus De Morgan

In this book we shall examine several number systems—integers, rational numbers, and real numbers—with which you are doubtless familiar, at least in an informal way. We shall also take a look at the system of complex numbers, with which you may be less familiar.

Chapter 2 is devoted to number theory, which concerns the system of integers. The present chapter, concerning mathematical induction, lays some of the groundwork for that discussion.

1.1 Ordinary Induction

To begin, we study the Principle of Mathematical Induction, which is a method for proving things about the set of all positive integers. We shall use the following standard notation.

Notation 1.1.1.

$$\mathbb{N} = \{0, 1, 2, 3, \dots\}$$
$$= \textit{the set of all } \textbf{natural numbers}$$
$$= \textit{the set of all } \textbf{nonnegative integers}$$

1

and

$$\mathbb{N}^* = \{1, 2, 3, \dots\}$$
$$= \textit{the set of all \textbf{positive integers}.}$$

If you look closely at the printed symbol $\mathbb{N}$ magnified—

$$\mathbb{N}$$

—you will see that its diagonal part has two strokes. That is how you should form the character when writing with pen, pencil, or chalk. The two diagonal strokes make the symbol readily recognizable and distinguish it from an ordinary mathematical N.

We have, for example, $19 \in \mathbb{N}^*$, but the negative integer $-3 \notin \mathbb{N}$ and the rational number $7/5 \notin \mathbb{N}$. Here we have used the set membership relation $\in$ ("element of") and its negation $\notin$.

By definition, $0 \in \mathbb{N}$ whereas $0 \notin \mathbb{N}^*$. Hence $\mathbb{N}^* \neq \mathbb{N}$.[1] Of course, $n \in \mathbb{N}$ whenever $n \in \mathbb{N}^*$—that is, in terms of logical implication $\Rightarrow$ ("implies," "if ... then"), $n \in \mathbb{N}^* \Rightarrow n \in \mathbb{N}$. Thus $\mathbb{N}^* \subset \mathbb{N}$, that is, $\mathbb{N}^*$ is a subset of $\mathbb{N}$. In summary, $\mathbb{N}^*$ is a *proper* subset of $\mathbb{N}$.

You are doubtless familiar with many basic properties of the natural numbers that concern the operations of adding and multiplying them and the order relation $<$ between them. Here are a few of these properties.

Examples 1.1.2.   (1) The natural number 1 is an identity element for multiplication—for each natural number n, the product $1 \cdot n = n$. In symbols, using the **universal quantifier** $\forall$ ("for every," "for each," "for all"), this property is

$$(\forall n \in \mathbb{N})\,(1 \cdot n = n),$$

that is, in terms of logical implication,

$$(\forall n)\,(n \in \mathbb{N} \;\Rightarrow\; 1 \cdot n = n).$$

In such a statement, we understand the implication sign $\Rightarrow$ to have a low precedence, and so this statement means the same thing as its fully parenthesized form

$$(\forall n)\,((n \in \mathbb{N}) \;\Rightarrow\; (1 \cdot n = n)).$$

---

[1]Two sets are equal precisely when they have the same elements. See Section A.1.

(2) The *distributive law* for multiplication over addition is

$$(\forall k, m, n \in \mathbb{N})\, (k(m + n) = k \cdot m + k \cdot n).$$

(3) Each natural number other than 0 is the *successor*—one more than—some natural number. In symbols,

$$(\forall n \in \mathbb{N})\, (n \neq 0 \implies (\exists k \in \mathbb{N})(n = k + 1)).$$

Here we used the **existential quantifier** $\exists$ ("there exists some," "there is a"). Thus this property is, in words: For each natural number n, if $n \neq 0$, then there exists some natural number k such that $n = k + 1$.

Later, when we examine the larger system of real numbers, we shall list all such basic properties. For now, we shall use them freely as needed, usually without special mention.

Exercises 1.1.3.   (1) Tell what is meant in Example 1.1.2 (2) by the phrase $\forall k, m, n \in \mathbb{N}$.

(2) State the commutative law for addition in words and then in symbols (using quantifiers).

(3) Symbolize the fact that 1 is the *only* multiplicative identity.

(4) State (in words) the meaning of

$$(\forall n \in \mathbb{N})(\exists m \in \mathbb{N})(n < m).$$

Then show that this statement is true.

One of the basic properties of the positive integers—the *Principle of Mathematical Induction*—may not be as familiar as the simple properties above. It allows us to prove that something is true about *all* the positive integers by showing that it is true about the first positive integer 1 and that, whenever it is true about a positive integer n, it must be true about the next positive integer $n + 1$. Here is a precise statement of this principle in terms of subsets of $\mathbb{N}^*$.

Axiom 1.1.4 (Principle of Mathematical Induction). *Let I be a set of positive integers with the following properties:*

(a) *The first positive integer $1 \in I$.*

(b) *For each n, if $n \in I$, then its successor $n + 1 \in I$.*

Then every positive integer belongs to I.

In terser language,[2] the Principle of Mathematical Induction is:

If $I \subset \mathbb{N}^*$ and

$$1 \in I \ \& \ (\forall n)(n \in I \implies n + 1 \in I),$$

then $I = \mathbb{N}^*$.

The Principle of Mathematical Induction expresses an intuitively obvious idea: Imagine all the positive integers written in order on a very, very long list. Suppose that you cross off the first number in the list and suppose further that whenever you cross off a given number in the list, you can then cross off the next one, too. Then eventually you can cross off all of them. After all, since you can cross off 1, then you can cross off $2 = 1 + 1$. But then you can also cross off $3 = 2 + 1$. And then you can also cross off 4, 5, etc. The magic, of course, is in the "etc."

Or, imagine a very long line (infinitely long, in fact) of people all standing facing the same direction. You, who are not in line, come along and tap on the shoulder the person at the end of the line; whenever a person in line is tapped on the shoulder, he or she likewise taps on the shoulder the person immediately in front of him or her. Then, eventually, everyone in the line will be tapped on the shoulder.

The Principle of Mathematical Induction is designated here as an **axiom**, that is, a statement we *assume* as a fundamental property—not as something to be deduced from more basic truths about positive integers.[3]

Exercise 1.1.5. Write the Principle of Mathematical Induction as a single symbolic statement, with all its variables fully quantified, beginning $(\forall I \subset \mathbb{N}^*) \ldots$.

To use induction to prove that something[4] is true about every positive integer n, we may form the set I of all positive integers n for which the thing is true and then use the Principle of Mathematical Induction to show that $I = \mathbb{N}^*$.

Example 1.1.6. Observe that $2^1 = 2 > 1$, $2^2 = 4 > 2$, $2^3 = 8 > 3$, etc. There seems to be a general formula here: $2^n > n$.

We use mathematical induction to prove

$$2^n > n \qquad (n = 1, 2, \ldots).$$

---

[2]We use the ampersand & to denote logical conjunction ("and"). Feel free, however, to use instead the word "and" when writing such symbolic formulations.

[3]Later, when we axiomatize the real number system, we shall define the set of natural numbers in such a way that the Principle of Mathematical Induction is a consequence of other properties!

[4]Technically, by "something" we mean a "predicate" $P(n)$ in n. See page 6.

[The parenthesized phrase $(n = 1, 2, \ldots)$ to the right is shorthand for "for every integer $n \geq 1$," that is, "for every $n \in \mathbb{N}^*$," and indicates that the formula to the left is being asserted for every positive integer n.]

Proof. Let $I = \{ n \in \mathbb{N}^* : 2^n > n \}$. We shall use induction to show $I = \mathbb{N}^*$.

First, $1 \in I$ because $2^1 = 2 > 1$.

Next, we show that $n \in I \implies n + 1 \in I$. Assume $n \in I$. We deduce that $n + 1 \in I$. From the inductive assumption (that $n \in I$),

$$2^n > n.$$

Multiply by 2:

$$2^{n+1} = 2 \cdot 2^n > 2 \cdot n \geq n + 1.$$

The final inequality here holds because $2 \cdot n = n + n$, and certainly $n + n \geq n + 1$ for $n \geq 1$. $\square$

Notice the pattern here to prove $I = \mathbb{N}^*$ by induction:

- First, $1 \in I$ because $\ldots$.

- Next, $n \in I \implies n + 1 \in I$.

  To prove this:

  - Assume $n \in I$.
  - Then $n + 1 \in I$ because $\ldots$.

In every such proof by induction there must be a **base step** (here $1 \in I$) to start things off and then an **inductive step** (here $n \in I \implies n + 1 \in I$) to establish the logical link between each n and its successor. To establish the inductive step, which is an implication, we typically make an **inductive assumption** of the implication's hypothesis (here $n \in I$) and deduce from it the implication's conclusion (here $n + 1 \in I$).

It is essential to establish both the base step and the inductive step, for otherwise you might "prove" something truly absurd. For example, the statement $(\forall n \in \mathbb{N}^*)(n > 99)$ is false, yet the inductive step $n > 99 \implies n + 1 > 99$ is correct! (Can you think of a similar example where the base step is correct but the inductive step is wrong?)

Exercises 1.1.7. Use mathematical induction to prove the following. Identify the base step, the inductive step, and the inductive assumption.

(1) For all $j \in \mathbb{N}^*$, we have $j \leq 2^{j-1}$.

(2) The sum $n^3 + 2n$ is divisible by 3 for each positive integer n. (By definition, integer m is *divisible by* integer d when there is some integer k for which $m = k \cdot d$. Try to use this formulation of divisibility rather than actually dividing to obtain quotients.)

In induction examples such as those shown above, the set I typically has the form

$$I = \{ n \in \mathbb{N}^* : P(n) \}$$

for a suitable predicate $P(n)$. By a **predicate** $P(n)$ in the variable n we mean a mathematical formula in which the variable n is *not* quantified by either the universal quantifier $\forall$ or the existential quantifier $\exists$. For example, $2 \cdot n > n$ is such a predicate. However, neither $(\exists n)(2 \cdot n > 100)$ nor $(\forall n)(2 \cdot n > n)$ is a predicate. In Example 1.1.6, the predicate $P(n)$ is the formula $2^n > n$.

A predicate $P(n)$ used in an induction may involve another variable, as in the next example, and then we require that every other such variable *is* quantified.

Example 1.1.8. Prove:

$$\frac{d\,x^n}{dx} = nx^{n-1} \qquad (n = 1, 2, \ldots).$$

What is to be proved, for each $n \in \mathbb{N}^*$, is actually that $d(x^n)/dx = nx^{n-1}$ for all real x, that is,

$$(\forall x \in \mathbb{R}) \left(\frac{d\,x^n}{dx} = nx^{n-1} \right),$$

where $\mathbb{R}$ denotes the set of all real numbers. The preceding formula (in which the variable x is quantified by $\forall$) is the predicate $P(n)$. The set I to be formed is

$$I = \left\{ n \in \mathbb{N}^* : (\forall x \in \mathbb{R}) \left(\frac{d\,x^n}{dx} = nx^{n-1} \right) \right\}.$$

Exercises 1.1.9.   (1) Finish Example 1.1.8 by using the Principle of Mathematical Induction to prove the formula for the derivative of x^n. [You may use formulas from calculus for the derivative of a product of two functions and the derivative of a constant function.]

(2) Prove: The nth derivative $d^n(xe^x)/dx^n = e^x(x + n)$ for $n = 1, 2, \ldots$. [You may use formulas from calculus for the derivative of a product of two functions and the derivative of e^x.]

It is often simpler to write a proof by induction directly in terms of $P(n)$ rather than in terms of the corresponding set $I = \{ n \in \mathbb{N}^* : P(n) \}$. For this, we use the following alternate form of the Principle of Mathematical Induction.

Theorem 1.1.10. (Principle of Mathematical Induction— Predicate Form) *Let $P(n)$ be a predicate in the variable n. Suppose:*

(a) *The statement $P(1)$ is true.*

(b) *The statement $(\forall n \in \mathbb{N}^*)(P(n) \implies P(n+1))$ is true.*

Then the statement $(\forall n \in \mathbb{N}^*)(P(n))$ is true.

Proof. Define $I = \{ n \in \mathbb{N}^* : P(n) \}$. Now apply the original, set form of the Principle of Mathematical Induction to show that $I = \mathbb{N}^*$. $\square$

To illustrate this form of induction, we redo our first example using predicates.

Example 1.1.11. (Reprise of Example 1.1.6) Prove:

$$2^n > n \qquad (n = 1, 2, \dots).$$

Proof. We use induction on n. Let $P(n)$ be the predicate $2^n > n$. First, $P(1)$ is true because $2^1 = 2 > 1$.

Next, let $n \geq 1$ and assume $P(n)$ is true. We want to deduce that $P(n+1)$ is true, that is, $2^{n+1} > n + 1$. The inductive assumption is

$$2^n > n.$$

Multiply this by 2:

$$2^{n+1} = 2 \cdot 2^n > 2 \cdot n \geq n + 1.$$

The final inequality here holds because $2 \cdot n = n + n \geq n + 1$ (since $n \geq 1$). $\square$

Once you become comfortable enough using the Principle of Mathematical Induction, you may dispense with explicit mention of both the set I and the corresponding predicate $P(n)$. Here's our original example a third and last time, written now in the more informal way mathematicians ordinarily formulate proofs by induction.

Example 1.1.12. (Second reprise of Example 1.1.6) Prove:

$$2^n > n \qquad (n = 1, 2, \dots).$$

Proof. We use induction on n. First, $2^1 = 2 > 1$.
   Next, let $n \geq 1$ and assume

$$2^n > n.$$

(We want to deduce that $2^{n+1} > n + 1$.) Multiply the inductive assumption $2^n > n$ by 2:

$$2^{n+1} = 2 \cdot 2^n > 2 \cdot n \geq n + 1.$$

The final inequality here holds because $2 \cdot n = n + n \geq n + 1$ (since $n \geq 1$).   □

Exercise 1.1.13. In each of the following variants of the proof in Example 1.1.12, something is wrong—somehow the wording is incorrect (or at least undesirable) or the logic is defective. Indicate what is wrong.

   (a) We use induction on n. First, $2^1 = 2 > 1$. Next, assume $2^n > n$ for all $n \geq 1$. (We want to deduce that $2^{n+1} > n + 1$ for all $n \geq 1$.) Let $n \geq 1$. From the inductive assumption, $2^{n+1} = 2 \cdot 2^n > 2 \cdot n = n + n \geq n + 1$.

   (b) We use induction on n. First, $2^1 = 2 > 1$. Next, assume $2^n > n$. (We want to deduce that $2^{n+1} > n + 1$.) From the inductive assumption, $2^{n+1} = 2 \cdot 2^n > 2 \cdot n = n + n \geq n + 1$.

   (c) We use induction on n. First, $2^1 = 2 > 1$. Next, let $n \geq 1$. (We want to deduce that $2^{n+1} > n + 1$.) Multiply $2^n > n$ by 2 to obtain $2^{n+1} = 2 \cdot 2^n > 2 \cdot n = n + n \geq n + 1$.

   (d) We use induction on n. First, $2^1 = 2 > 1$. Next, let $n \geq 1$ and assume $2^n > n$. We want to deduce $2^{n+1} > n + 1$. Now $2 \cdot n = n + n \geq n + 1$, and so it is enough to show that $2^{n+1} > 2n$. Divide this by 2 to get $2^n > n$. But the latter inequality is true because it is the inductive assumption.

   For practice, rewrite your inductive proofs from previous exercises, first using predicates explicitly and then implicitly.

Exercise 1.1.14.   (a) Justify the following version of induction. Let $I \subset \mathbb{N}^*$ such that (i) $1 \in I$ and (ii) for every integer $n > 1$, if $n - 1 \in I$, then $n \in I$. Then $I = \mathbb{N}^*$.

   (b) Restate (a) in predicate form.

   (c) Use this version of induction—in either set form or predicate form—to reprove Example 1.1.6.

If you have met mathematical induction before this course, you may
have seen it used to prove summation formulas like

$$1 + 2 + 3 + \cdots + n = \frac{n(n+1)}{2}$$

for the sum of the first n positive integers. The "etc." indicated by the
ellipsis $(\cdots)$ in such a formula is succinctly and precisely expressed by
using *sigma notation:*

$$\sum_{j=1}^{n} j = \frac{n(n+1)}{2}.$$

Recall that, in general, $\sum_{j=1}^{n} a_j$ is shorthand for $a_1 + a_2 + \cdots + a_n$. For
example,

$$\sum_{j=1}^{5} 8j^2 = 8 \cdot 1^2 + 8 \cdot 2^2 + 8 \cdot 3^2 + 8 \cdot 4^2 + 8 \cdot 5^2.$$

When either the upper or the lower limit of summation (like the 5 or
the 1 in the preceding display) is a variable n, then the question arises
of how to express the sum in terms of n with a *closed-form* formula—
that is, a formula involving no summation sign $\sum$.

Example 1.1.15. Use induction to prove that, for every integer $n \geq 1$,

$$\sum_{j=1}^{n} j = \frac{n(n+1)}{2}.$$

Proof. Let I denote the set of positive integers n for which the above
formula holds. We use induction to show $I = \mathbb{N}^*$.
  Since $\sum_{j=1}^{1} j = 1 = 1(1+1)/2$, certainly $1 \in I$.
  Now assume $n \in I$. We deduce that $n + 1 \in I$. The inductive assumption is

$$\sum_{j=1}^{n} j = \frac{n(n+1)}{2}.$$

What we want to deduce is

$$\sum_{j=1}^{n+1} j = \frac{(n+1)[(n+1)+1]}{2}.$$

Split up the sum of $n + 1$ terms into the sum of the first n terms and
the last, $(n + 1)$st, term:

$$\sum_{j=1}^{n+1} j = \sum_{j=1}^{n} j + (n+1).$$

Now use the inductive assumption on the sum of n terms on the right:

$$\sum_{j=1}^{n+1} j = \frac{n(n+1)}{2} + (n+1).$$

Factor out $n + 1$ from both terms on the right and simplify:

$$\sum_{j=1}^{n+1} j = (n+1)\left(\frac{n}{2} + 1\right)$$

$$= (n+1)\frac{n+2}{2}$$

$$= \frac{(n+1)[(n+1)+1]}{2}. \quad \square$$

The inductive step here was given in great detail. To put it more succinctly: Assume $n \in I$, that is,

$$\sum_{j=1}^{n} j = \frac{n \cdot (n+1)}{2}.$$

Then

$$\sum_{j=1}^{n+1} j = \sum_{j=1}^{n} j + (n+1)$$

$$= \frac{n(n+1)}{2} + (n+1) \qquad \text{(by the inductive assumption)}$$

$$= (n+1)\left(\frac{n}{2} + 1\right)$$

$$= (n+1)\frac{n+2}{2}$$

$$= \frac{(n+1)[(n+1)+1]}{2}.$$

How did we know what formula to prove for the sum? Perhaps we saw the pattern from a few cases, such as $1 + 2 = 3 = 2 \cdot 3/2$ and $1 + 2 + 3 = 6 = 3 \cdot 4/2$. Perhaps, even, we confirmed the $n(n + 1)/2$ pattern with the special case $1 = 1 \cdot 2/2$ or with a few further cases such as $1 + 2 + 3 + 4 = 10 = 4 \cdot 5/2$ and $1 + 2 + 3 + 4 + 5 = 15 = 5 \cdot 6/2$.

Perhaps, instead, we saw the pattern geometrically, as in Figure 1.1. Make a triangular stack of $\sum_{j=1}^{n} j$ dots with n rows by putting one dot in the first row at the top, two dots in the second row, three in the third, and so forth, ending with n dots in the bottom row. Next, make another such stack of $\sum_{j=1}^{n} j$ dots, but upside down. Finally, put the

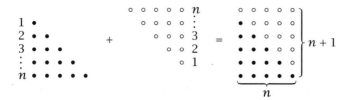

Figure 1.1: "Visual proof" of a summation formula.

two stacks together to form a rectangle consisting of $n(n + 1)$ dots in all. Then $2\sum_{j=1}^{n} j = n(n + 1)$, and so $\sum_{j=1}^{n} j = n(n + 1)/2$. (Of course, this "visual proof" is merely suggestive, because the figure is for a specific n.)

Before being in a position to establish the formula for $\sum_{j=1}^{n} j$ by mathematical induction, we had to have that formula at hand. We had to discover the formula somehow—to perceive the pattern. And this points out a significant limitation of mathematical induction: Induction merely allows us to prove rigorously what we already surmise to be true.[5]

For summation formulas such as Example 1.1.15, there are often systematic methods for discovering the correct answer and even for proving the answer correct as you discover it. We will return to such matters later, in Section 1.3. For now, we shall usually continue to tell you the answer and ask you merely to prove it by induction (but if you think you see a way to prove it without induction, try that, too).

Exercise 1.1.16. Calculate $1 + 2 + \cdots + 1000$.

Exercise 1.1.17. Prove by mathematical induction the formulas for:

(a) the sum of the squares of the first n positive integers: $\sum_{j=1}^{n} j^2 = n(n + 1)(2n + 1)/6$.

(b) the sum of the first n even positive integers: $2 + 4 + 6 + \cdots + 2n = n(n + 1)$.

Exercise 1.1.18. Discover the flaw in the following "proof" that 1 is the only positive integer.

Proof. Let n be an integer with $n \geq 2$. We shall show that $n = 1$.

---

[5]The situation is not unlike that in calculus, where you use the rigorous (ϵ, δ) definition of a limit to prove that $\lim_{x \to a} kx = ka$. There, too, you already have to "know" the answer before you can prove it correct.

Previously we established by induction the formula

$$\sum_{j=1}^{m} j = \frac{m(m+1)}{2}$$

for all integers $m \geq 1$. Apply it for $m = n - 1$:

$$\sum_{j=1}^{n-1} j = \frac{(n-1)n}{2},$$

that is,

$$1 + 2 + 3 + \cdots + (n-1) = \frac{(n-1)n}{2}.$$

Add 1 to both sides of the last equation:

$$1 + 2 + 3 + \cdots + n = \frac{(n-1)n}{2} + 1.$$

But we also know

$$\sum_{j=1}^{n} j = \frac{n(n+1)}{2}.$$

Comparing the last two identities, we have

$$\frac{(n-1)n}{2} + 1 = \frac{n(n+1)}{2}.$$

Solve this for n to get $n = 1$. □

Be careful not to jump to an erroneous conclusion about all positive integers from insufficient evidence concerning just some special cases. The following exercise shows how misleading such evidence can be.

Exercise 1.1.19.   (a) By direct computation, show that $n^2 - n + 41$ is prime for $n = 1, 2, \ldots, 20$. (Recall that an integer $p > 1$ is said to be *prime* if it cannot be written as the product of two smaller positive integers. For example, 3 and 29 are prime, whereas $21 = 3 \cdot 7$ and $16 = 4 \cdot 4$ are not prime.)

(b) Determine whether $n^2 - n + 41$ is prime for some higher values of n.

(c) Prove or disprove that $n^2 - n + 41$ is prime for every positive integer n.

The preceding exercise also highlights the difference between mathematical induction, on the one hand, which gives a rigorous proof of a general mathematical proposition, and scientific induction, on the other hand, which suggests a generalization about the empirical world on the basis of many individual instances. Mathematical induction is, of course, a form of logical deduction. Scientific induction, to the contrary, is not. The biologist might reach the generalization that "All crows are black" after observing many crows and noting that each and every one of them is black.[6]

Insufficient evidence from some special cases is not the only trap in using induction. As we said earlier, you must also be sure that the inductive step really works. Try the following exercise.

Exercises 1.1.20. (1) Find the flaw in the following "proof" that all cars are red. Then explain mathematical induction to your friends and amaze them with this proof (but don't give away the secret flaw!).

Proof. There is some car that is red. Hence it suffices to prove that, for each positive integer n, every n cars are of the same color as each other. For that we use induction on n.

Base step: All the cars in a set of 1 car are surely of the same color as each other.

Inductive step: Let n be a positive integer and assume that every n cars are of the same color as each other. Let S be a set of $n + 1$ cars. Number the cars in S as $1, 2, \ldots, n, n + 1$. By the inductive assumption, the first n cars $1, 2, \ldots, n$ in S are of the same color as each other. Likewise, by the same inductive assumption (which applies to *every* set of n cars), the last n cars $2, 3, \ldots, n + 1$ in S are of the same color as each other. The situation is this:

$$\overbrace{1, \quad \underbrace{2, \quad 3, \quad \ldots, \quad n, \quad n + 1}_{\text{same color}}}^{\text{same color}}$$

Hence all the $n + 1$ cars in S are of the same color. □

---

[6] The more black crows one observes, without ever seeing a nonblack crow, the more confident one is supposed to become of the truth of "All crows are black." Now the statement "All crows are black" means, of course, "For every x, if x is a crow, then x is black." Logically, that is equivalent to, "For every x, if x is not black, then x is not a crow." Hence another way of confirming that all crows are black is to observe anything that is not black and note that it is not a crow. So it seems that we do not have to go out into the fields looking for crows after all. Instead, sitting at home we can become more confident that all crows are black by observing a red apple, a brown sock, or a green rug! (We leave the reader to ponder this conundrum.)

By the way, not all crows are black: there are some nonblack crows in Australia!

(2) Construct a similarly fallacious proof that all blondes have blue eyes.

(3) Concoct and "prove" another fallacious result similar to those in (1) and (2).

(4) Construct a fallacious proof that all positive integers are equal by using induction on n to "prove" that, for each positive integer n, if the maximum of any two positive integers x and y is n, then $x = y$.

All mathematical proof is proof by deduction. But deduction comes in many varieties, just one of which is proof by induction. (Then there is "proof by intimidation"—see the frontispiece.)

Induction may start at 0 instead of 1. Here is the formal statement:

Proposition 1.1.21. (Principle of Mathematical Induction—Zero Base Form) *Let I be a subset of* $\mathbb{N}$ *such that:*

(a) $0 \in I$.

(b) *For each natural number n, if $n \in I$, then $n + 1 \in I$.*

Then $I = \mathbb{N}$.

Proof. Let $J = \{ j \in \mathbb{N}^* : j - 1 \in I \}$. Then $1 \in J$ because $1 - 1 = 0 \in I$; furthermore, $j \in J$ implies $j + 1 \in J$ (why?). By ordinary induction, $J = \mathbb{N}^*$. Hence $I = \mathbb{N}$ (why?). □

Or course, we usually apply this form of induction, too, in terms of predicates.

Example 1.1.22. If $r \neq 1$, prove the following formula for the sum of a **geometric progression** having $n + 1$ terms, first term 1, and ratio r:

$$1 + r + r^2 + \cdots + r^n = \frac{1 - r^{n+1}}{1 - r}.$$

Proof. The possible values of n were not specified, but both sides of the formula are meaningful for $n = 0, 1, 2, \ldots$, and the formula is, in fact, correct for such n. And so we use induction on n, starting with $n = 0$, to prove the formula.

When $n = 0$, the formula reduces to $1 = (1 - r)/(1 - r)$, which is certainly true.

Let $n \geq 0$ and assume

$$1 + r + r^2 + \cdots + r^n = \frac{1 - r^{n+1}}{1 - r}. \tag{*}$$

We wish to deduce that

$$1 + r + r^2 + \cdots + r^{n+1} = \frac{1 - r^{(n+1)+1}}{1 - r}.$$

To do this, add r^{n+1} to both sides of (*) and perform a bit of algebraic simplification. □

The preceding formula is, of course, interesting only for $n \geq 1$. Nonetheless, we find it more convenient to prove for $n \geq 0$. Establishing the induction's base step—that $1 = (1 - r)/(1 - r)$—is slightly easier starting with 0 than starting with 1: if we started with $n = 1$, the base step would be that $1 + r = (1 - r^2)/(1 - r)$.

Exercises 1.1.23.   (1) What was the predicate $P(n)$ used implicitly in the preceding proof? Redo the proof, but explicitly using a subset of $\mathbb{N}$ instead of this predicate.

(2) Evaluate $\sum_{j=0}^{n} 2^j$.

Induction may also start with some integer $s > 1$. The following proposition covers that case as well as the base 0 case and the original, base 1, case.

Proposition 1.1.24. (Principle of Mathematical Induction—Generalized Form) *Let $s \in \mathbb{N}$ and let*

$$\mathbb{N}_s = \{ n \in \mathbb{N} : n \geq s \}.$$

Suppose I is a subset of $\mathbb{N}_s$ such that:

(a) $s \in I$.

(b) *For each integer $n \geq s$, if $n \in I$, then $n + 1 \in I$.*

Then $I = \mathbb{N}_s$.

Proof. Exercise. □

Again, we most often apply this generalized form of induction in terms of predicates. You should write out its predicate formulation yourself.

Example 1.1.25. Prove: $n^2 > n + 1$ for all integers $n > 1$.

Proof. The inequality is to be proved, in other words, for $n \geq 2$. We use induction on n, starting with $n = 2$.
    For $n = 2$, we have $2^2 = 4 > 3 = 2 + 1$.

Let $n \geq 2$ and assume $n^2 > n + 1$. We must deduce $(n + 1)^2 > (n + 1) + 1$. By squaring $n + 1$ and applying the inductive assumption, we get

$$
\begin{aligned}
(n + 1)^2 &= n^2 + 2n + 1 \\
&> (n + 1) + 2n + 1 \\
&= 3n + 2 \\
&> n + 2 = (n + 1) + 1. \quad \square
\end{aligned}
$$

Exercise 1.1.26. Prove or disprove each of the following for the indicated range of values of the integer n. If, however, no range of values of n is stated, first determine an appropriate range (some experimentation may be required) and prove it for that range. [*Note:* In some exercises, you may want to prove, instead, a simpler result equivalent to the stated one. In some, there may be an easier way than induction to prove the result, and then other exercises (established by induction) may furnish the easier way.]

(a) $2^n > 2n + 1$.

(b) $2^n > n^2$.

(c) $2^n < n!$. Recall that the factorial function ! is defined by $0! = 1$, $1! = 1$, $2! = 2 \cdot 1$, $3! = 3 \cdot 2 \cdot 1 = 6$, and, in general, $(n + 1)! = (n + 1)(n!)$.

(d) $n^2 \leq 4n!$

(e) For $x \geq 0$, $(1 + x)^n \geq 1 + nx$ for $n = 0, 1, 2, \ldots$.

(f) $\dfrac{d^{n+1}}{dx^{n+1}} x^n = 0$.

(g) $n^3 + (n + 1)^3 + (n + 2)^3$ is divisible by 9.

(h) For integers a and b with $a > b$, $a^n - b^n$ is divisible by $a - b$.

(i) $\log_2 n \leq \sqrt{n}$.

(j) $2^n < n^{10} + 2$.

(k) $2^{2^n} + 1$ is prime.

(l) $\sqrt{n} \leq \sqrt[n]{n!}$ for $n = 1, 2, \ldots$.

Here are some additional exercises where you first have to discover the correct formula.

Exercise 1.1.27. Discover and prove a formula for each of the following.

(a) The sum of the interior angles of a convex n-gon (that is, a convex polygon with n sides). A polygon is *convex* when each line segment whose ends are inside the polygon lies completely inside the polygon; equivalently, a polygon is convex when each of its interior angles is less than π. [*Hint:* Of course, start with $n = 3$, that is, a triangle. For $n = 4$, divide the polygon (a quadrilateral) into two triangles.]

(b) The number of possible handshakes among n people, each shaking hands with each other exactly once.

(c) The number of lines joining n points in the plane no three of which are collinear.

(d) The number of subsets of a set with n elements. Recall that the empty set $\varnothing$ (which has no elements) and the entire set A are both subsets of a set A. (*Hint:* Start with $n = 0$, that is, $A = \varnothing$.)

(e) The number of strictly increasing lists of integers starting at 1 and ending at n. We call a list of integers $a_1, a_2, \ldots, a_k$ *strictly increasing* when $a_1 < a_2 < \cdots < a_k$. (Note that, for a given n, such lists may have different lengths k. For example, when $n = 5$, both $1, 2, 5$ and $1, 3, 4, 5$ are such strictly increasing lists starting at 1 and ending at n.)

(f) The smallest number $k(n)$ such that each list of $k(n)$ distinct integers has a length n sublist (in the given order of appearance) that is strictly increasing or a sublist that is strictly decreasing.

   For example, take $n = 3$. Then $k(3) > 3$ because a list a_1, a_2, a_3 of distinct numbers with $a_1 < a_2 > a_3$ has neither a strictly increasing nor a strictly decreasing sublist of length 3 (although it has the strictly increasing sublist a_1, a_2 of length 2 and the strictly decreasing sublist a_2, a_3 of length 3). Likewise, $k(3) > 4$ because a list a_1, a_2, a_3, a_4 of distinct numbers with $a_1 < a_2 > a_3 < a_4$ has neither a strictly increasing nor a strictly decreasing sublist of length 3.

1.2 Recursion

In a temple at Benares, so it is told, there are 64 circular gold disks of different sizes, each with a hole bored straight through its center, and a brass plate in which are mounted upright three diamond needles. At the beginning of the world, all the disks were stacked from largest (on the bottom) to smallest (on the top) on the first of the three needles. The priests attending the temple have the sacred obligation to move all the disks from one needle to another, but always moving only one disk at a time and never placing a larger disk on top of a smaller disk. The priests work day and night at this task. When they have finally moved the entire stack to the third of the three needles, then the world will come to an end.

The preceding is a fictional account of the origin of the *Tower of Hanoi* puzzle invented in 1883 by Edouard Lucas. Lucas' puzzle consists of a base with three vertical pegs A, B, and C and only 8 disks—see Figure 1.2. The disks are stacked, bottom to top, in order of size from largest to smallest on the first peg (peg A) of the three pegs. The object

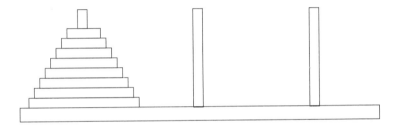

Figure 1.2: The Tower of Hanoi Puzzle.

of the puzzle is to move the disks so as to place the entire stack on the third peg, peg C. The rules are as follows.

- Only one disk can be moved at a time, from the top of a (possibly empty) stack on any one of the pegs to the top of a (possibly empty) stack on any other peg.

- A disk may not be moved on top of a smaller disk.

Is this puzzle solvable? If so, how many moves does it take? (By a *move* we mean the movement of one disk from one peg to another in accordance with the two rules above.) Rather than answer that question straight off, let us generalize it instead to the situation of n disks—but still three pegs.

When $n = 1$, clearly the puzzle is solvable, in one move: Move the one and only disk from peg A to peg C. And when $n = 2$, it is readily solvable, in three moves: Move the top disk (the smaller of the two) from peg A to the middle peg, peg B; move the remaining (larger) disk from peg A to peg C; and, finally, move the disk from peg B onto the top of the larger disk on peg C.

Exercises 1.2.1.   (1) Show that the Tower of Hanoi is solvable for $n =$ 3 disks by listing the moves; try to do it in the fewest moves possible. Tell how many moves it takes. (Did the solution for $n = 2$ disks help to see how to solve it for $n = 3$?)

(2) Show that the puzzle is solvable for $n = 0$ disks. How many moves does it take? (Yes, this seems silly, but there's a reason for considering this degenerate case, too.)

(3) Prove that the Tower of Hanoi puzzle is solvable for an arbitrary number n of disks. (*Hint:* Use mathematical induction on n to prove the slightly more general statement that a stack of n disks on *any* one of the three pegs can be moved to either one of the other two pegs.)

(4) Write a program in your favorite programming language (such as Pascal) or use a mathematical software package (such as MATH-EMATICA)—one that allows recursion—that will solve the n-disk Tower of Hanoi puzzle. The input should be n, and the output should be the sequence of moves needed to solve the puzzle for that n. You might describe a typical move of the (top) disk from peg X to peg Y by the output X -> Y. Then the output for input 2 would be as follows.

```
A  ->  B
A  ->  C
B  ->  C
```

Please read no further until you have done (3)!

According to the penultimate exercise above, the n-disk Tower of Hanoi puzzle is solvable for every positive integer n. (In fact, it is solvable no matter which of the three pegs the stack starts on and which of the other two it ends up on.) If we include the degenerate case of no disks, we even know the puzzle is solvable for every nonnegative integer n. But how many moves, exactly, does it take? We mean, of course, the *least* number of moves needed. Let us denote this number by T_n (T as in "Tower").

We have

$$T_0 = 0, \qquad T_1 = 1, \qquad T_2 = 3.$$

If we know T_n, then we can determine T_{n+1} as follows. The stack of the top n disks on peg A can be transferred (one by one, according to the rules) to the middle peg, B, in T_n moves; the remaining, largest, disk on peg A can be transferred to peg C in 1 move; and, finally, the stack of n disks temporarily resting on peg B can be transferred to peg C, on top of the largest disk just placed there, in T_n moves. So, in all, we can transfer all $n + 1$ disks from peg A to peg C in $T_n + 1 + T_n = 2T_n + 1$ moves. Perhaps there is some shorter way to do it, and so all this shows is that the *least* number T_{n+1} of moves for the $n + 1$ disks satisfies

$$T_{n+1} \le 2T_n + 1.$$

Exercise 1.2.2. Show that $T_{n+1} \ge 2T_n + 1$ also. Thus, $T_{n+1} = 2T_n + 1$, exactly.

Altogether, the sequence $(T_n)_{n=0,1,2,...} = (T_0, T_1, T_2, ...)$ of integers satisfies the pair of formulas

$$\begin{cases} T_0 & = 0, \\ T_{n+1} & = 2T_n + 1. \end{cases}$$

The first of this pair of formulas is referred to as the **initial condition**, and the second as the **recurrence relation**. From the two formulas we can, in principle, calculate the value of T_n for any particular n we choose, however large (providing we have enough time and patience). For example,

$$T_1 = 2T_0 + 1 = 2 \cdot 0 + 1 = 1,$$
$$T_2 = 2T_1 + 1 = 2 \cdot 1 + 1 = 3,$$
$$T_3 = 2T_2 + 1 = 2 \cdot 3 + 1 = 7,$$
$$T_4 = 2T_3 + 1 = 2 \cdot 7 + 1 = 15.$$

Exercises 1.2.3.   (1) Continue by calculating T_5 and T_6.

(2) Do you see a general pattern suggesting a closed-form formula for T_n as a function of n? Check your formula for all the values previously calculated. Does it fit the special case $n = 0$, too? Use it to predict the value of T_7; then confirm your prediction by using the recurrence relation to calculate T_7.

(3) Use induction to prove your formula for T_n.

(4) How many moves does it take to solve Lucas' original 8-disk puzzle? If you move one disk per second, how long will it take?

(5) How many moves does it take to transfer the Tower of Hanoi's 64 disks from the first needle to the third? If the priests could move one disk each second, how long would it take? If a computer were programmed to simulate the transfer and could make one move per microsecond, how long would it take?

Suppose you did not actually see the pattern to T_n (although we hope and assume that eventually you did). Is there some systematic way to derive a formula for T_n in terms of n? Here is one, which involves using the recurrence relation $T_k = 2T_{k-1} + 1$ repeatedly n times:

$$
\begin{aligned}
T_n &= 2T_{n-1} + 1 \\
&= 2\left(2T_{n-2} + 1\right) + 1 = 2^2 T_{n-2} + 2 + 1 \\
&= 2^2\left(2T_{n-3} + 1\right) + 2 + 1 = 2^3 T_{n-3} + 2^2 + 2 + 1 \\
&\;\;\vdots \\
&= 2^n T_0 + 2^{n-1} + \cdots + 2^2 + 2 + 1.
\end{aligned}
$$

But $T_0 = 0$, and so the result is

$$
T_n = 2^{n-1} + \cdots + 2^2 + 2 + 1 = \sum_{j=0}^{n-1} 2^j.
$$

We recognize the latter sum as that of a geometric progression (see Example 1.1.22). Hence the sum T_n is $\left(1 - 2^{(n-1)+1}\right)/(1-2) = 2^n - 1$.

There is something troublesome about the preceding calculations— the "etc." implied by the vertical dots $\vdots$. The *idea* of what's going on is clear enough, but what *exactly* is going on for arbitrary n? Further, the calculations seem overly complicated, made so by the constant term 1 on the right-hand side of the recurrence relation $T_k = 2T_{k-1} + 1$. *If* the recurrence relation had been simply $T_k = 2T_{k-1}$, then the calculations would have been drastically simpler: $T_n = 2T_{n-1} = 2\left(2T_{n-2}\right) = 2^2 T_{n-2}$, etc.

Here is a cleaner solution, accomplished by reducing the actual recurrence relation to such a simplified one that is **homogeneous** in the sense of having no constant term on its right-hand side. The crux of the reduction is to realize that $(2T_{k-1} + 1) + 1 = 2\left(T_{k-1} + 1\right)$. Thus, add 1 to both sides of the initial condition and both sides of the recurrence relation:

$$
\begin{cases}
T_0 & + 1 = 1, \\
T_{n+1} + 1 = 2T_n + 2 = 2(T_n + 1).
\end{cases}
$$

This suggests defining

$$
U_n = T_n + 1
$$

for all $n \in \mathbb{N}$, so that

$$
\begin{cases}
U_0 & = 1, \\
U_{n+1} = 2U_n.
\end{cases}
$$

Thus $U_0 = 1$, $U_1 = 2U_0 = 2 \cdot 1 = 2$, $U_2 = 2U_1 = 2 \cdot 2 = 4$, $U_3 = 2U_2 = 2 \cdot 4 = 8$, etc. The recurrence relation for the sequence of numbers $(U_n)_{n \in \mathbb{N}}$ is simpler than the one for the original sequence $(T_n)_{n \in \mathbb{N}}$, and now the pattern should be clear:

$$U_n = 2^n.$$

Exercise 1.2.4.   (a) Use induction to prove the preceding formula for U_n.

(b) Use the formula for U_n to get again the formula for T_n.

The recurrence relation $U_n = 2U_{n-1}$ is a special case of the more general homogeneous linear first-order recurrence relation

$$s_n = \alpha s_{n-1}. \tag{*}$$

It is easy to solve any recurrence relation of this form—to find an explicit formula for s_n as a function of n. The method is to see what geometric sequences $c, cr, cr^2, cr^3, \ldots$ satisfy the recurrence relation.

Exercise 1.2.5. For a given constant $\alpha \neq 0$, consider a first-order recurrence relation of the form (*), above. Of course, the "trivial" sequence $0, 0, 0, \ldots$ satisfies (*), but we are interested in other, nontrivial solutions.

(a) Let r be a nonzero constant and let $s_n = r^n$ for all $n \in \mathbb{N}$. Show that the sequence $(s_n)_{n \in \mathbb{N}}$ satisfies the recurrence relation (*) if and only if $r = \alpha$.

(b) If $(s_n)_{n \in \mathbb{N}}$ satisfies (*), show that for arbitrary constant c the sequence $(c s_n)_{n \in \mathbb{N}}$ also satisfies it.

(c) Use induction to show that, for a given number a, there is at most one sequence $(s_n)_{n \in \mathbb{N}}$ satisfying both the recurrence relation (*) and the initial condition $s_0 = a$.

(d) Conclude that, for a given a, there is exactly one sequence $(s_n)_{n \in \mathbb{N}}$ satisfying both the recurrence relation (*) and the initial condition $s_0 = a$, namely, the sequence defined by $s_n = a\alpha^n$.

Exercise 1.2.6. Solve the n-disk Tower of Hanoi puzzle with the additional rule:

- Direct moves from peg A to peg C, or from peg C to peg A, are not allowed; that is, each move must be from or to the middle peg B.

That is, prove that the puzzle can be solved with this additional restriction, and, if possible, determine the number of moves needed.

Example 1.2.7. How many pieces of pizza at most can you get by making n straight cuts going all the way across it? In somewhat more mathematical terms: What is the maximum number of regions into which n lines separate the plane?

Evidently we get fewer regions if two of the lines are parallel (including if they are coincident) than if they are not, and fewer regions if three of the lines are concurrent (pass through the same point) than if they do not. Hence we stipulate that no two of the lines are parallel (or coincident) and that no three are concurrent—the lines are, as is said, "in general position."

Let R_n be the maximum number of regions possible for n lines. Certainly

$$R_0 = 1$$

(the one region being the entire plane) and

$$R_1 = 2$$

(the two half-planes). What about R_2? Look at the rather obvious "general" picture in Figure 1.3. We see that

$$R_2 = 4.$$

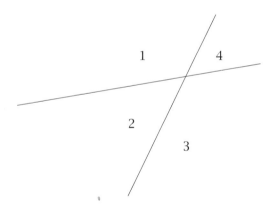

Figure 1.3: Two lines divide the plane into 4 regions.

Do you see a pattern here yet? Do you think $R_n = 2^n$ in general? Then you have been too rash. Add a third line to the preceding picture— see Figure 1.4. The way we have numbered the regions, each of the first three regions 1, 2, and 3 is subdivided into 2 regions 1a and 1b, 2a and 2b, and 3a and 3b, respectively, and so the total number of regions for $n = 3$ lines is 7:

$$R_3 = 4 + 3 = R_2 + 3.$$

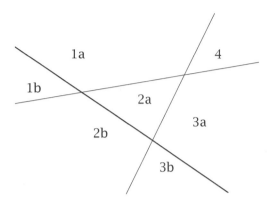

Figure 1.4: Three lines divide the plane into 7 regions.

This suggests a recurrence relation:

$$R_n = R_{n-1} + n.$$

The next exercise asks you to prove this relation and to solve it.

Exercise 1.2.8.   (a) Explain why $R_n \le R_{n-1} + n$ for all $n \ge 1$. (*Hint:* The nth line increases the old number of regions for $n - 1$ lines by k if and only if it splits exactly k old regions. It splits exactly k old regions if and only if it intersects the old lines at exactly $k - 1$ points.)

(b) Show that $R_n \ge R_{n-1} + n$.

(c) Solve the recursion $R_0 = 1$, $R_n = R_{n-1} + n$ for $n \ge 1$.

Our aim in this section is not so much to discuss methods for solving a recurrence relation with given initial condition(s)—that is, finding an explicit formula for the sequence so described—as to explore the very idea of such a sequence and its connection with induction.

First, let us clarify our notation. A sequence $(s_0, s_1, s_2, \dots)$ whose indices $0, 1, 2, \dots$ start with 0 is denoted more succinctly by $(s_n)_{n=0,1,\dots}$ or $(s_n)_{n \in \mathbb{N}}$. A sequence $(s_1, s_2, s_3, \dots)$ whose indices $1, 2, 3, \dots$ start instead with 1 is denoted by $(s_n)_{n=1,2,\dots}$ or $(s_n)_{n \in \mathbb{N}^*}$. We may even have a starting index other than 0 or 1, and the notation for that situation should be obvious. Also, for simplicity we often denote a sequence by (s_n) when it is clear from the context (or is irrelevant) what the starting index is.

How do we know that there really *is* a sequence $(T_n)_{n \in \mathbb{N}}$ satisfying the conditions $T_0 = 0$ and $T_{n+1} = 2T_n + 1$? You could argue that, for each n, the number T_n is, by definition, the least number of moves needed to solve the n-disk Tower of Hanoi puzzle, and such a number exists because, as you have proved, the puzzle has a solution for each n.[7] Although that involves the route we took to obtain the recurrence relation, still you would not be answering the question we are actually asking. Forget all about the Tower of Hanoi. Suppose *all* you are given is the pair of conditions

$$\begin{cases} T_0 & = 0, \\ T_{n+1} & = 2T_n + 1 \qquad (n \in \mathbb{N}). \end{cases}$$

Why must a sequence $(T_n)_{n \in \mathbb{N}}$ exist that satisfies these conditions?

What we have again here is an example of **recursion**—where a sequence $(s_n)_{n \in \mathbb{N}}$ is described by an **initial condition**

$$s_0 = z$$

specifying its starting term together with a **recurrence relation**

$$s_{n+1} = G(s_n)$$

telling how each term is related to the preceding term. Here G is a rule producing a value as output given the input s_n, that is, a function whose domain and range both consist (at least) of all the terms of the sequence.

Roughly speaking, then, such a **recursive definition** "defines a sequence in terms of itself." What we need is a principle guaranteeing that such a sequence really exists.

It is tempting to offer a "proof" like the following that the sequence exists: Let I be the set of all nonnegative integers n for which s_n is defined. We use induction to show that $I = \mathbb{N}$. First, $0 \in I$ because we are given the value $s_0 = z$. Now assume $n \in I$, that is, s_n is defined. But by the recurrence relation $s_{n+1} = G(s_n)$, we have that s_{n+1} is also defined so that $n + 1 \in I$, too.

Does this "proof" convince you? Reread it several times. The flaw is rather subtle. Look at the set I we allegedly formed:

$$I = \{ n \in \mathbb{N} : s_n \text{ is defined} \}.$$

What, exactly, does the predicate "s_n is defined" mean? Evidently, it is supposed to mean that n is in the domain of the sequence s. But *what* sequence s are we talking about? We do not have any sequence

---

[7] Of course, with this answer you would be mixing up the formal world of mathematics with a "real-world" puzzle, but never mind about that: you could always formulate a mathematical model of that puzzle in terms of sets, numbers, and functions.

s yet; the whole point of what we are doing is to try to prove that the sequence *s* actually exists!

It turns out that a rigorous proof that recursion works is technically a bit complicated, depending as it does on a careful definition of functions in terms of sets and relations. Such a proof is outlined in Exercise B.4.2. Here we merely state the result formally.

Theorem 1.2.9 (Ordinary Recursion Principle). *Let X be a set, let $z \in X$ be a given element, and let*

$$G: X \to X$$

be a given function. Then there exists one and only one sequence $(s_n)_{n \in \mathbb{N}}$ such that

$$\begin{cases} s_0 & = z, \\ s_{n+1} = G(s_n) & (n \in \mathbb{N}). \end{cases}$$

There is nothing sacred about 0 as the starting point for such a recursion, and indeed 0 may be replaced by 1 and $\mathbb{N}$ by $\mathbb{N}^*$ to get a sequence $(s_n)_{n=1,2,\dots}$ indexed by the positive integers.

Examples 1.2.10.   (1) For the sequence (T_n) arising in the Tower of Hanoi puzzle, we could take $X = \mathbb{N}$, and then $z = 0$ and $G: X \to X$ is the function given by $G(k) = 2k + 1$ for all $k \in X$.

(2) Suppose we did not know how to add arbitrary pairs of natural numbers but only knew how to add 1 to an arbitrary natural number. Then we could define addition of natural numbers recursively. The idea is to say

$$\begin{cases} m + 0 & = m, \\ m + (n + 1) = (m + n) + 1. \end{cases}$$

The details are as follows.

Let $\sigma: \mathbb{N} \to \mathbb{N}$ be the function defined by $\sigma(k) = k + 1$ for all $k \in \mathbb{N}$. For each $k \in \mathbb{N}$, the number $\sigma(k)$ is the immediate successor of k, that is, the least natural number that is greater than k. Accordingly, σ is sometimes called the **successor function** of $\mathbb{N}$.

Temporarily fix $m \in \mathbb{N}$. By ordinary recursion, there is a unique sequence $\left(s_n^{(m)}\right)_{n \in \mathbb{N}}$—the parenthesized superscript m reminds us that the sequence involves the number m—such that

$$\begin{cases} s_0^{(m)} = m, \\ s_{n+1}^{(m)} = \sigma\left(s_n^{(m)}\right). \end{cases}$$

Now define $m + n$ to mean $s_n^{(m)}$.

There is a significant point to the preceding example where we recursively defined something—addition of natural numbers—with which you already are quite familiar. The point is that this definition allows us to *prove* various laws of addition that you have previously taken for granted. The proofs, typically, involve induction.

Exercises 1.2.11.   (1) In each part of this exercise, the only facts you should use concerning addition of natural numbers are the preceding recursive definition and results established in earlier parts of the exercise.

(a) Use induction on k to prove the *associative law*

$$(m + n) + k = m + (n + k)$$

for addition of natural numbers.

(b) By the recursive definition, we know that $n + 0 = n$ for every $n \in \mathbb{N}$. Now use induction to prove that $0 + n = n$ for every $n \in \mathbb{N}$.

(c) Prove the *commutative law*

$$m + n = n + m$$

for all natural numbers m, n.

(d) Starting from the definitions $2 = 1+1, 3 = 2+1$, and $4 = 3+1$, deduce that $2 + 2 = 4$.

(2) Suppose you did not already know what it means to say that one natural number is less than another. For $m, n \in \mathbb{N}$, define $m < n$ to mean there exists some $d \in \mathbb{N}$ such that $m + d = n$. Prove that $m < n \implies m + k < n + k$ for all $k \in \mathbb{N}$.

(3) Define multiplication of natural numbers recursively. Then use your definition to prove several properties of multiplication such as the associative law and the commutative law.

The program, begun above, for defining operations and relations of the system of natural numbers, can be continued to reconstruct all the standard operations and relations and their familiar properties. In fact, just such a program was carried out by Giuseppe Peano in 1891. His assumptions, known as the **Peano Postulates**, can be stated in modern terminology as follows:

1. There is a particular element $0 \in \mathbb{N}$.

2. There is a function $\sigma : \mathbb{N} \to \mathbb{N}$. [What we denote by $\sigma(n)$, Peano referred to as the *successor* of n. Thus this postulate states that each natural number has exactly one successor.]

3. $0 \notin \text{range}(\sigma)$; that is, $0 \neq \sigma(n)$ for all $n \in \mathbb{N}$. (In other words, 0 is not the successor of any natural number.)

4. The function σ is one-to-one; that is, if $m \neq n$, then $\sigma(m) \neq \sigma(n)$. (In other words, distinct natural numbers have distinct successors.)

5. If $I \subset \mathbb{N}$ such that $0 \in I$ and $\sigma(n) \in I$ whenever $n \in I$, then $I = \mathbb{N}$. (This is, of course, the zero base form of the Principle of Mathematical Induction.)

Exercise 1.2.12. Strictly speaking, a sequence $(s_n)_{n \in \mathbb{N}}$ in a set X is a function $s \colon \mathbb{N} \to X$, with s_n being simply the value $s(n)$ of this function at n (see Section B.4). Restate the conclusion of the Ordinary Recursion Principle (Theorem 1.2.9) in a strictly functional way—in terms of such a function s and the successor function σ.

Let $n > 1$ be an integer. What does it mean to form the nth power a^n of a real number a? You might answer, "Multiply a by itself n times." What, though, does it mean to multiply n numbers together when $n > 1$? After all, the operation of multiplication produces the product of a *pair* of numbers, not of three or four or more. What we need (short of a definition of the product of any n numbers) is a recursive definition of a^n.

Example 1.2.13. Let a be a fixed real number. Then the nonnegative powers a^n of a may be defined recursively by

$$\begin{cases} a^0 & = 1, \\ a^{n+1} = a \cdot a^n & (n \in \mathbb{N}). \end{cases}$$

A function in the Pascal programming language for implementing that recursive definition might look like this:

```
function Power (a :  real; n :  real) :  real;
   begin
      if n = 0 then
         Power := 1.0
      else
         Power := a*Power (a, n - 1)
   end;
```

Of course, this is just a program fragment. A complete program would include statements to read values of a and n as input and to print the value of the nth power of a as output.

In MATHEMATICA, the corresponding recursive definition would be as follows.

```
power[a_Real, n_Integer] :=
   If[n == 0, 1., a*power[a, n - 1]] /; n >= 0
```

The same thing can be accomplished in MATHEMATICA by the following pair of definitions.

```
power[a_Real, 0] := 1.;
power[a_Real, n_Integer] :=
   a*power[a, n - 1] /; n > 0
```

Exercises 1.2.14.   (1) In the preceding example, identify explicitly the set X, the element $z \in X$, and the function $G: X \to X$ used for the recursion, as well as the sequence obtained.

(2) Write and run a complete computer program that implements the recursive definition of powers.

(3) From the recursive definition of powers, use induction to derive the law of exponents

$$a^{m+n} = a^m \cdot a^n.$$

You may assume that multiplication is commutative, that is, $xy = yx$ for all real x, y.

(4) Let A be a given set. Use ordinary recursion to justify the existence, for each positive integer n, of the set, usually denoted by A^n, that consists of all ordered n-tuples $(a_1, a_2, \ldots, a_n)$ of elements from A. [*Hint:* Your problem is to figure out what, exactly, should be meant by such n-tuples for arbitrary n. Of course, for $n = 2$, the set A^2 consists of all ordered pairs (a_1, a_2) with both coordinates belonging to A; that is, $A^2 = A \times A$, the cartesian product of A with itself. What is A^1? How can you define ordered triples (3-tuples) in terms of ordered pairs (2-tuples)?]

Look again at the Ordinary Recursion Principle (Theorem 1.2.9). For a given $z \in X$ and $G: X \to X$, the sequence $(s_n)_{n \in \mathbb{N}}$ whose existence is guaranteed by this theorem satisfies

$$
\begin{aligned}
s_0 &= z, \\
s_1 &= G(s_0) = G(z), \\
s_2 &= G(s_1) = G(G(z)), \\
s_3 &= G(s_2) = G(G(G(z))), \text{ etc.}
\end{aligned}
$$

In terms of composition of functions (see Section B.2), the sequence is given by

$$s_0 = z = G^0(z),$$
$$s_1 = G(z) = G^1(z),$$
$$s_2 = (G \circ G)(z) = G^2(z),$$
$$s_3 = (G \circ G \circ G)(z) = G^3(z), \text{ etc.}$$

Here G^0, G^1, G^2, G^3, etc., are powers—under iterated composition—of the function G. Of course, for the "etc." to make sense, we have to establish the existence of such powers G^n for *all* integers n. Needless to say, that requires ordinary recursion again.

Exercises 1.2.15.   (1) Let X be a set and $f : X \to X$ be a function.

    (a) Prove the existence of a unique sequence $(f^n)_{n \in \mathbb{N}}$ of functions such that f^0 is the identity function of X and $f^{n+1} = f \circ f^n$ for every $n \in \mathbb{N}$. (*Hint:* What are the appropriate set X, the element z of X, and function G needed to apply ordinary recursion? The set X to be used in Theorem 1.2.9 will *not* be the same as the given set X of this exercise. To get started and avoid confusion, use different names for the two Xs!)

    (b) Prove that $f^{m+n} = f^m \circ f^n = f^n \circ f^m$ for all $m, n \in \mathbb{N}$.

(2) Prove that $m + n = \sigma^m(n)$, where $\sigma : \mathbb{N} \to \mathbb{N}$ is the successor function given by $\sigma(k) = k + 1$ and addition in $\mathbb{N}$ is defined recursively as in Example 1.2.10 (2).

The Ordinary Recursion Principle (Theorem 1.2.9) also furnishes the underpinning for *discrete dynamical systems*. Such a system may model the evolution of the state of some physical, biological, or other real-world system at discrete times $n = 0, 1, 2, \ldots$ (each second, hour, year, or whatever time unit is appropriate) with $n = 0$ being the initial time.

Briefly, here is the setup for a discrete dynamical system. Start with a set X and a function $f : X \to X$. Elements of the set X are the system's possible states; the function f is the dynamical rule that tells how the system changes from a state x at any one time to another state $f(x)$ one unit of time later. (We are tacitly assuming that the rule does not depend upon when the change occurs, only upon the state from which it is changing.) Then for an initial state $z \in X$, the sequence $(s_n)_{n \in \mathbb{N}}$ satisfying $s_0 = z$ and $s_{n+1} = f(s_n)$ for $n = 0, 1, \ldots$ gives the successive states to which that initial state evolves as time passes.

Some recursive definitions take more complicated forms than that in the Ordinary Recursion Principle. One of the simplest is the definition of the factorial function.[8] The idea behind the definition is, of course, to be able to say

$$n! = 1 \cdot 2 \cdots n.$$

The reason a recursive definition is needed is the "etc." implied by the ellipsis dots.

Example 1.2.16. The **factorial function** ! is defined by

$$\begin{cases} 0! & = 1, \\ (n+1)! = (n!) \cdot (n+1) & (n \in \mathbb{N}). \end{cases}$$

If we denote $n!$ by $f(n)$, then the recurrence relation here is

$$f(n+1) = f(n) \cdot (n+1).$$

This kind of recurrence relation is *not* covered by the Ordinary Recursion Principle: The right-hand side of the recurrence relation does not have the form $G(f(n))$ for any function $G\colon \mathbb{N} \to \mathbb{N}$. To compute the right-hand side, you must use both n itself and the value of $f(n)$. In other words, the right-hand side has the form $H(f(n), n)$, where H is the function of *two* variables given by $H(k, n) = k \cdot (n+1)$.

The definition of the factorial function can be justified by the theorem that follows. In it, we use the notion of the **(cartesian) product** $X \times Y$ of two sets X and Y, defined as

$$X \times Y = \{\, (x, y) : x \in X, \ y \in Y \,\},$$

that is, the set of all ordered pairs (x, y) whose first coordinate x belongs to X and whose second coordinate y belongs to Y. In our use of this notation, $Y = \mathbb{N}$.

Theorem 1.2.17 (Primitive Recursion Principle). *Let X be a set, let $z \in X$ be a given element, and let*

$$H\colon X \times \mathbb{N} \to X$$

be a given function. Then there exists one and only one sequence $(s_n)_{n \in \mathbb{N}}$ such that

$$\begin{cases} s_0 & = z, \\ s_{n+1} = H(s_n, n) & (n \in \mathbb{N}). \end{cases}$$

Proof. Omitted. □

---

[8]Curiously, unlike the case for most functions, the symbol ! for the factorial function traditionally *follows* its argument, as in 5!, rather than precedes it, as in !(5).

Exercises 1.2.18.   (1) Everybody knows that $3! = 1 \cdot 2 \cdot 3 = 6$ and $2! = 1 \cdot 2 = 2$ and hence finds it reasonable to accept that $1! = 1$. But why should we define $0! = 1$ rather than, say, $0! = 0$?

(2) Write two computer programs[9] for computing factorials:

    (a) A recursive program—one that directly uses the recursive definition above.

    (b) An iterative program, which successively multiplies the numbers $1, 2, \ldots, n$.

(3) Assuming the preceding Primitive Recursion Principle, deduce the Ordinary Recursion Principle as a consequence.

In some forms of recursion, the next term of the sequence depends not just on the preceding term but also on the two (or more) preceding terms. Perhaps the most famous of such sequences is described in the next example.

Example 1.2.19. In his thirteenth-century book *Liber abaci*, the mathematician Leonardo of Pisa, also known as Fibonacci, posed the following problem:

> A newborn pair of rabbits, one of each sex, is placed on a desert island. A male-female pair of rabbits does not breed until they are mature at two months. Each month thereafter, they breed and produce exactly one new male-female pair. Assuming no rabbits ever die or hop off the island, how many pairs of rabbits are there after n months?

Denote by F_n the number of *pairs* of rabbits on the island at the end of n months. Because the original pair does not breed until after the first 2 months, evidently

$$F_1 = 1, \qquad F_2 = 1.$$

Then they breed, producing 1 new pair that month, and so after 3 months there are 2 pairs:

$$F_3 = 2.$$

During the fourth month, the newly born pair is still immature, and so only the original pair breeds, producing 1 new pair, giving

$$F_4 = 3.$$

---

[9]In a "typed" language such as Pascal, you will doubtless want to declare the result (output) variables to be of type Real rather than Integer. The reason is that a variable of the latter type can hold only fairly small integers, whereas values of $n!$ grow quite rapidly with n.

During the fifth month, though, the first pair born on the island reaches maturity and breeds, as does the original pair, and so during the fifth month 2 new pairs are added, for a total of

$$F_5 = 5.$$

And so on. Table 1.1 shows what is happening.

Table 1.1: Fibonacci's rabbits.

Month	Mature Pairs	Immature Pairs	Total Pairs
1	0	1	1
2	0	1	1
3	1	1	2
4	1	2	3
5	2	3	5
6	3	5	8
7	5	8	13

Evidently $F_{n+1} = F_n + F_{n-1}$ for each n. Is there some closed-form formula for F_n?

Definition 1.2.20. The **Fibonacci numbers** are the entries in the sequence $(F_n)_{n=1,2,...}$ defined recursively by

$$\begin{cases} F_1 & = 1, \\ F_2 & = 1, \\ F_{n+1} = F_n + F_{n-1} & (n \geq 2). \end{cases}$$

Note that the right-hand side of the recurrence relation $F_{n+1} = F_n + F_{n-1}$ involves the two integers preceding $n+1$, and so of course we need *two* initial conditions $F_1 = 1$ and $F_2 = 1$ in order to get the recursion going.

Exercises 1.2.21. (1) Use the recursive definition of the Fibonacci numbers to check all the totals in Table 1.1 and to calculate F_8 and F_9 as well.

(2) Draw a diagram (like a family tree) that traces the lineages of Fibonacci's rabbits for several generations.

(3) Suppose we use the same recurrence relation as for the Fibonacci numbers but initial conditions $F_1 = 1$ and $F_2 = 3$. Calculate F_n for $n = 3, 4, 5, 6$.

(4) Formulate a general version of recursion which guarantees that a definition such as Definition 1.2.20 really establishes the existence of a sequence.

The Fibonacci numbers have many fascinating properties, two of which appear in the next exercises. Now the Fibonacci sequence is defined by a **second-order** recursion—the recurrence relation expresses each term as a function of *two* preceding terms. (By way of contrast, the numbers T_n for the Tower of Hanoi are defined by a **first-order** recursion.) Therefore it should not be surprising that ordinary induction does not always suffice to prove properties of this sequence. After all, in ordinary induction, it is an assumption about only one term that allows us to deduce something about the next term. For second-order (or higher-order) recursion, we need a form of induction that allows us to deduce something about a term from an assumption about *two* (or more) preceding terms. This sort of "strong" induction is discussed in Section 1.4, where additional properties of Fibonacci numbers appear.

Exercises 1.2.22. Using induction, or otherwise, prove the following properties of the Fibonacci numbers:

(1) $F_{n+2} = 1 + \sum_{j=1}^{n} F_j$ for all n.

(2) No two consecutive Fibonacci numbers are divisible by any integer greater than 1.

The second-order recursion relation $F_{n+1} = F_n + F_{n-1}$ is an instance of the class of second-order *linear* recursions that are *homogeneous* in the sense that the recurrence relation has no constant term on its right-hand side. Such recursions take the general form

$$\begin{cases} s_0 = a, \\ s_1 = b, \\ s_n = \alpha s_{n-1} + \beta s_{n-2} \quad (n \geq 2), \end{cases}$$

for given initial values a and b and given coefficients α and β. The following exercises let you try to solve one such recursion and then derive a general method for solving all such recursions. The method is similar to that for the homogeneous linear first-order recursions considered in Exercise 1.2.5.

Exercises 1.2.23. (1) Solve the linear second-order recursion $s_0 = 4$, $s_1 = 5$, $s_n = 2s_{n-1} + 3s_{n-2}$ for $n = 2, 3, \ldots$. That is, find a closed-form formula for s_n. (*Hint:* Try using the recurrence relation repeatedly.)

(2) For given constants α and β, consider the second-order recurrence relation

$$s_n = \alpha s_{n-1} + \beta s_{n-2}. \tag{*}$$

(a) For a nonzero constant r, show that the sequence $s_n = r^n$, $n = 0, 1, 2, \ldots$, satisfies (*) if and only if r is a root of the quadratic polynomial $r^2 - \alpha r - \beta$.

(b) Prove: If the sequences $(t_n)_{n \in \mathbb{N}}$ and $(u_n)_{n \in \mathbb{N}}$ both satisfy (*), then for any constants c_1 and c_2 so does the sequence $(s_n)_{n \in \mathbb{N}}$ defined by $s_n = c_1 t_n + c_2 u_n$.

(c) Show that then, for any constants c_1 and c_2 and any roots r_1 and r_2 of $r^2 - \alpha r - \beta$, the sequence $s_n = c_1 r_1^n + c_2 r_2^n$ satisfies the recurrence relation (*).

(d) Suppose the quadratic polynomial $r^2 - \alpha r - \beta$ has the two distinct roots r_1 and r_2. Show that a sequence $(s_n)_n$ satisfies the recurrence relation (*) *only if* it has the form $s_n = c_1 r_1^n + c_2 r_2^n$ for some coefficients c_1 and c_2.

(3) Apply (2) to re-solve the recursion in (1).

(4) Apply (2) to solve the recursion for the Fibonacci sequence.

(5) What can you say about solving the recurrence relation (*) when the quadratic polynomial $r^2 - \alpha r - \beta$ has a double root r_1 [that is, when this polynomial factors as $(r - r_1)^2$]?

1.3   Summation

Earlier, we used mathematical induction to establish the closed-form formula

$$\sum_{j=1}^{n} j = \frac{n(n+1)}{2}$$

for the sum of the first n positive integers (see Example 1.1.15). Actually, induction is not needed at all in order to prove this formula. In fact, you do not even have to know the answer ahead of time. The following argument is a special case of an insight by the 10-year-old Carl Friedrich Gauss (1777–1855).

Example 1.3.1. Denote the sum $\sum_{j=1}^{n} j$ by S. Then S is also the sum of the same numbers but in reverse order, and so we have

$$
\begin{aligned}
S &= 1 &+& 2 &+& \cdots &+& (n-1) &+& n, \\
S &= n &+& (n-1) &+& \cdots &+& 2 &+& 1.
\end{aligned}
$$

Add these two equations, but group terms in vertical pairs on the right-hand sides:

$$2S = (1 + n) + [2 + (n - 1)] + \cdots + [(n - 1) + 2] + (n + 1)$$
$$= \underbrace{(n + 1) + (n + 1) + \cdots + (n + 1) + (n + 1)}_{n \text{ terms}}$$
$$= n(n + 1)$$

so that

$$S = \frac{n(n + 1)}{2},$$

as claimed.

So now we have a second, noninductive proof of the formula for the sum of the first n positive integers. What is the formula if we add, instead, the *squares* of the first n positive integers? Or their cubes? We shall return to these questions shortly.

Before that, we need some general rules for manipulating sums. (The second rule in Proposition 1.3.2, below, was already used implicitly in Example 1.3.1.)

Proposition 1.3.2. *Let n be a positive integer. Then for all n-tuples $(a_j)_{j=1,2,\ldots,n}$ and $(b_j)_{j=1,2,\ldots,n}$ of real numbers, and for all real constants c,*

$$\sum_{j=1}^{n} ca_j = c \sum_{j=1}^{n} a_j, \tag{a}$$

$$\sum_{j=1}^{n} (a_j + b_j) = \sum_{j=1}^{n} a_j + \sum_{j=1}^{n} b_j. \tag{b}$$

How does one prove such formulas? You might be satisfied with the "etc." approach for, say, the first of the preceding formulas:

$$\sum_{j=1}^{n} ca_j = (ca_1) + (ca_2) + \cdots + (ca_n)$$
$$= c(a_1 + a_2 + \cdots + a_n)$$
$$= c \sum_{j=1}^{n} a_j.$$

But what is lurking behind those dots? In fact, what is lurking behind the dots in the very definition,

$$\sum_{j=1}^{n} a_j = a_1 + a_2 + \cdots + a_n,$$

of summation? After all, addition is a function that operates only on pairs of numbers, not arbitrary n-tuples. Recursion to the rescue!

Definition 1.3.3. The **sum** $\sum_{j=1}^{n} a_j$ of ordered n-tuples $(a_1, a_2, \ldots, a_n)$ of numbers is defined recursively by

$$\begin{cases} \displaystyle\sum_{j=1}^{1} a_j = a_1, \\ \displaystyle\sum_{j=1}^{n+1} a_j = \sum_{j=1}^{n} a_j + a_{n+1} \qquad (n = 0, 1, 2, \ldots). \end{cases}$$

Although it is correct, even this definition is a bit sloppy and incomplete. First of all, we did not specify in what number system we are doing the adding. Let us do it in the set $\mathbb{R}$ of all real numbers, which also contains the set $\mathbb{Q}$ of all rational numbers and the set $\mathbb{Z}$ of all integers; we could equally well do the adding in the still larger set $\mathbb{C}$ of all complex numbers.

Second, what we really did in Definition 1.3.3 was to use the Ordinary Recursion Principle to justify the existence of a sequence $(A_n)_{n \in \mathbb{N}^*}$ of *functions*

$$A_n \colon \mathbb{R}^n \to \mathbb{R},$$

where $\mathbb{R}^n$ denotes the set of all ordered n-tuples of real numbers. (This set was defined recursively earlier.) In the language of these functions A_n, the recursion takes the form

$$\begin{cases} A_1(a_1) = a_1 \qquad \text{for each } a_1 \in \mathbb{R}, \\ A_{n+1}(a_1, a_2, \ldots, a_n, a_{n+1}) = A_n(a_1, a_2, \ldots, a_n) + a_{n+1} \\ \qquad\qquad \text{for all } (a_1, a_2, \ldots, a_n, a_{n+1}) \in \mathbb{R}^{n+1}. \end{cases}$$

Then we let

$$\sum_{j=1}^{n} a_j = A_n(a_1, a_2, \ldots, a_n).$$

Exercises 1.3.4. (1) What, according to the informal "etc." meaning of summation (the form with dots), is the value of $\sum_{j=1}^{n} 1$? Use induction and the recursive definition of summation to prove your answer.

(2) Gauss had actually discovered a closed-form formula for the sum

$$a + (a + d) + (a + 2d) + \cdots + (a + nd)$$

of an *arithmetic progression.* (The case $a = 0$, $d = 1$ reduces the arithmetic progression to the sum $1 + 2 + \cdots + n$ previously considered.) State and prove such a formula.

(3) Derive a closed-form formula for the sum of the first n *even* positive integers.

(4) Derive a closed-form formula for the sum of the first n *odd* positive integers.

(5) Give rigorous proofs of the identities in Proposition 1.3.2 based upon the recursive definition of summation.

Exercise 1.3.5.   (a) Informally, the product notation $\prod_{j=1}^{n} a_j$ means

$$\prod_{j=1}^{n} a_j = a_1 \cdot a_2 \cdots a_n.$$

Give a recursive definition for the product $\prod_{j=1}^{n} a_j$.

(b) State and prove identities for products analogous to those in Proposition 1.3.2 for sums.

Here are a few more exercises involving summation for you to try.

Exercises 1.3.6.   (1) Discover and prove a closed-form formula for the sum $\sum_{j=1}^{n} j(j!)$.

(2) In general, is the following formula true or false?

$$\left(\sum_{j=1}^{n} a_j \right) \left(\sum_{j=1}^{n} b_j \right) = \sum_{j=1}^{n} a_j b_j.$$

If it is true, prove it; if it is false, give a specific counterexample and then state and prove a correct formula for the product on the left.

(3) Everybody knows the factorization

$$x^2 - y^2 = (x - y)(x + y),$$

and perhaps you know

$$x^3 - y^3 = (x - y)(x^2 + xy + y^2).$$

Generalize to obtain a formula for factoring $x^n - y^n$ (with first factor $x - y$). Prove your formula.

(4) Without using induction, reprove the formula

$$1 + r + r^2 + \cdots + r^n = \frac{1 - r^{n+1}}{1 - r}$$

for the sum of a geometric progression with ratio $r \neq 1$. (An inductive proof appeared in Example 1.1.22.) (*Hint:* Call the sum G_n. What happens if you multiply G_n by r?)

(5) Most tables of integrals include the recurrence relation

$$\int x^n e^x \, dx = x^n e^x - n \int x^{n-1} e^x \, dx.$$

  (a) Use integration by parts to derive this recurrence relation.

  (b) Find an expression for $\int x^n e^x \, dx$ that does not involve an integral sign. (You *may*, however, use summation in your expression.)

Number (3) of the preceding exercises requested a formula for factoring a difference $x^n - y^n$ into two factors, one being $x - y$. An application of this formula is the following version of the **Factor Theorem:**

Let $p(x) = \sum_{j=0}^{n} c_j x^j$ be a polynomial having integers as its coefficients $c_0, c_1, \ldots, c_n$. Then an integer a is a root of $p(x)$ if and only if $x - a$ is a factor of $p(x)$ over the integers.

To say that a is a root of $p(x)$ means, of course, that $p(a) = 0$. To say that $x - a$ is a factor of $p(x)$ *over the integers* means that $p(x) = (x - a)q(x)$ for some polynomial $q(x)$ whose coefficients are also integers.

Exercise 1.3.7.   (a) Prove the Factor Theorem. [*Hint:* If a is a root of $p(x)$, then $p(x) = p(x) - p(a)$.]

  (b) Deduce that a polynomial $p(x) = \sum_{j=0}^{n} c_j x^j$ of degree n has at most n roots that are integers.

The Factor Theorem is also true, more generally, when rational, real, or complex numbers are allowed instead of just integers. In these cases it can also be deduced from the *Remainder Theorem* (see page 65). At the start of this section, we gave Gauss' proof of the formula

$$\sum_{j=1}^{n} j = \frac{n(n + 1)}{2}$$

for the sum of the first n positive integers. As promised, we now return to the question of generalizing this to the sum of squares (or higher powers). We could, of course, do some experimentation with low values of n, guess at the general rule, and then prove that rule by induction. There is an easier, more systematic way, however. The following example gives one of the "tricks" needed.

Example 1.3.8. For any positive integer n and numbers $a_1, a_2, \ldots, a_n$, we find a closed-form expression for the sum

$$\sum_{j=1}^{n} (a_{j+1} - a_j).$$

Using the identities in Proposition 1.3.2, we have

$$\sum_{j=1}^{n} (a_{j+1} - a_j) = \sum_{j=1}^{n} a_{j+1} - \sum_{j=1}^{n} a_j \tag{1}$$

$$= \sum_{j=2}^{n+1} a_j - \sum_{j=1}^{n} a_j \tag{2}$$

$$= \left\{ \sum_{j=2}^{n} a_j + a_{n+1} \right\} - \left\{ a_1 + \sum_{j=2}^{n} a_j \right\} \tag{3}$$

$$= a_{n+1} - a_1 + 0 \tag{4}$$

$$= a_{n+1} - a_1. \tag{5}$$

In step (2) we used

$$\sum_{j=1}^{n} a_{j+1} = \sum_{j=2}^{n+1} a_j.$$

This cute trick is known as *index shifting*. Formally, we decremented by 1 the dummy index j inside the sum and, to balance that, incremented by 1 the upper and lower limits of summation 1 and n. Do you see that we are still adding exactly the same terms but just indexing them differently? (In other situations, you might need to increment inside and decrement the limits, rather than the reverse; or, you might need to increment or decrement by some amount larger than 1.)

The purpose in shifting indices was to be able eventually to cancel the sum $\sum_{j=2}^{n} a_j$ that appeared twice, with opposite signs, in step (3). But to do that, we needed the two sums involving a_j to begin and end with the same indices. That is why, in going from step (2) to step (3), we *peeled off* the last term a_{n+1} from $\sum_{j=2}^{n+1} a_j$ and the first term a_1 from $\sum_{j=1}^{n} a_j$.

For fairly obvious reasons, a sum of the kind evaluated in this example is known as a *collapsing sum*.

Exercise 1.3.9. Find (and prove) a closed-form formula for the sum

$$\frac{1}{1 \cdot 2} + \frac{1}{2 \cdot 3} + \cdots + \frac{1}{n(n+1)}.$$

[*Hint:* Break the typical term $\dfrac{1}{j(j+1)}$ into partial fractions with denominators j and $j+1$.]

Next, we use collapsing sums to give yet a third way to find the sum of the first n positive integers.

Example 1.3.10. We evaluate the sum

$$\sum_{j=1}^{n} [(j+1)^2 - j^2]$$

in two different ways. First, as a collapsing sum:

$$\sum_{j=1}^{n} [(j+1)^2 - j^2] = (n+1)^2 - 1 = n^2 + 2n.$$

Second, by squaring the $j+1$ inside the sum and splitting up the sum into several sums:

$$\sum_{j=1}^{n} [(j+1)^2 - j^2] = \sum_{j=1}^{n} (j^2 + 2j + 1 - j^2)$$

$$= \sum_{j=1}^{n} (2j + 1)$$

$$= 2\sum_{j=1}^{n} j + \sum_{j=1}^{n} 1$$

$$= 2\sum_{j=1}^{n} j + n.$$

Comparing the two expressions for the collapsing sum, we have

$$2\sum_{j=1}^{n} j + n = n^2 + 2n.$$

Now we solve this equation for $\sum_{j=1}^{n} j$ to obtain the by-now-familiar result

$$\sum_{j=1}^{n} j = \frac{n^2 + 2n - n}{2} = \frac{n^2 + n}{2} = \frac{n(n+1)}{2}.$$

Exercises 1.3.11.   (1) Using the technique of the preceding example, derive the formula

$$\sum_{j=1}^{n} j^2 = \frac{n(n+1)(2n+1)}{6}$$

for the sum of the squares of the first n positive integers.

(2) Reprove the preceding formula by induction.

(3) Derive a formula for the sum of the cubes of the first n positive integers.

(4) Check your answer to (3) by proving the formula by induction.

(5) Derive a formula for the sum of the cubes of the first n *odd* positive integers.

1.4 Well-Ordering and Strong Induction

In this section, we are going to prove a form of induction—the Well-Ordering Principle—that, at first glance, does not look like induction at all. For it, we need the following preliminary result which asserts that the next natural number after a natural number m is $m + 1$, in other words, that there is no natural number strictly between m and $m + 1$. You might regard this as so completely obvious as to require no proof whatsoever, but in fact it can be deduced from more fundamental properties of natural numbers (such as are listed in Appendix D). Hence we state it as a lemma.[10]

Lemma 1.4.1 (Gap Lemma). *For each natural number m, there exists no natural number n for which $m < n < m + 1$.*

Proof. We use induction on m.

Base step ($m = 0$):   We show there is no natural number n with $0 < n < 1$, that is, for every $n \in \mathbb{N}$, it is *not* the case that $0 < n < 1$. To do this, we use induction on n. (Yes, this is an induction within an induction!)

First, it is not the case that $0 < 0 < 1$, because $0 = 0$.

Now let $n \in \mathbb{N}$ and assume it is not the case that $0 < n < 1$. (We want to deduce that it is not the case that $0 < n + 1 < 1$.) By the inductive assumption, either $n \leq 0$ or else $n \geq 1$. In the former case

---

[10]A **lemma** is a proposition of subsidiary interest, perhaps of no real interest in itself—one whose primary purpose is to facilitate proving other, more interesting propositions.

$(n \leq 0)$, we have $n = 0$, whence $n + 1 = 1$, and so it is not the case that $n + 1 < 1$. In the latter case $(n \geq 1)$, we have $n + 1 \geq 1 + 1 = 2 > 1$, and so it is not the case that $n + 1 < 1$.

This completes the induction on n that establishes the base step $(m = 0)$ of the induction on m.

Inductive step: Now let $m \in \mathbb{N}$ and assume there is no natural number n with $m < n < m + 1$. (We want to deduce that there is no natural number n with $m + 1 < n < m + 2$.) Just suppose $m + 1 < n < m + 2$ for some natural number n. Now $n \neq 0$ because $1 \leq m + 1 < n$. Then $n > 0$, and so we can write $n = k + 1$ for a natural number k (namely, $k = n - 1$). Then $m + 1 < k + 1 < m + 2 = (m + 1) + 1$ whence $m < k < m + 1$. This contradicts the inductive assumption about m. $\square$

Exercises 1.4.2.   (1) Use quantifiers, logical negation signs, etc., to formalize and completely explain the logic of the proof of the Gap Lemma.

(2) Precisely what properties of addition and ordering of natural numbers were used in the preceding proof of the Gap Lemma?

(3) The proof, above, that no natural number n satisfies $0 < n < 1$ was by induction on n. Is the following is a valid alternate proof of the inductive step? (Why or why not?) Where is the inductive assumption actually used?

Let $n \in \mathbb{N}$ and assume it is not the case that $0 < n < 1$. Then $n + 1 \not< 1$, for otherwise $n < 0$, which is impossible. Hence it is not the case that $0 < n + 1 < 1$.

(4) Carry out an alternate proof of the Gap Lemma in which the assertion about arbitrary m is reduced to the case where $m = 0$. What additional properties of natural numbers are needed?

(5) Show that the Gap Lemma holds for arbitrary integers, not just for natural numbers.

Below we speak of a least element of a subset of $\mathbb{N}$. The meaning of "least element" is probably perfectly obvious, but it is nonetheless worthwhile to write it out precisely. Let $A \subset \mathbb{N}$. We call an $m \in \mathbb{N}$ a **least element** of A if:

(a) $m \in A$; and

(b) $m \leq n$ for each $n \in A$; that is, $m < n$ for each $n \in A$ with $n \neq m$.

For example, 0 is a least element of $\mathbb{N}$ itself, 1 is a least element of $\mathbb{N}^*$, and 5 is the least element of $\{\, n \in \mathbb{N} : n \geq 5 \,\}$.

Exercises 1.4.3. (1) Prove the preceding assertion about 5 directly from the definition of "least element."

(2) Let $A \subset \mathbb{N}$ and let $k \in \mathbb{N}$.

    (a) If $k \in A$, state carefully what it means to say that k is *not* a least element of A.

    (b) If k is not necessarily in A, state carefully what it means to say that k is *not* a least element of A.

(3) Give an example of a subset of $\mathbb{N}$ that does not have a least element.

(4) We said "*a* least element" of A rather than "*the* least element" of A. Prove that, in fact, a subset A of $\mathbb{N}$ has at most one least element. (Thus, if a subset of $\mathbb{N}$ has a least element, it has exactly one, and then we may speak of *the* least element.)

(5) What properties of the order relation $<$ in $\mathbb{N}$ allowed you to prove uniqueness in (4)?

(6) The notion of a least element is also meaningful for subsets of the set $\mathbb{Z}$ of all integers and even for subsets of the set $\mathbb{R}$ of all real numbers. Show that the uniqueness of a least element also holds for these situations.

(7) Does each nonempty subset of $\mathbb{R}$ have a least element?

(8) Give a definition of a "greatest element" of a subset A of $\mathbb{N}$. Give an example. Show that such a greatest element, if it exists, must be unique (so that then we may speak of *the* greatest element of A).

(9) Does each nonempty subset of $\mathbb{N}$ have a greatest element?

(10) What is wrong with the following "proof" that 1 is the largest positive integer? Let n be the largest positive integer. Since $k^2 \geq k$ for every positive integer k, in particular $n^2 \geq n$. Since n is the largest positive integer and n^2 is a positive integer, then $n^2 = n$. Division by n gives $n = 1$.

If you were very, very observant when we were investigating the Tower of Hanoi puzzle, you may have noticed that we originally defined T_n as the *least* number of moves needed to solve the n-disk puzzle. All that was proved, however, was that there is *some* number of moves in which the n-disk puzzle can be solved, that is, that the set of all natural numbers k such that the puzzle is solvable in k moves is a nonempty set. That this set of k actually must have a least element is guaranteed by the following theorem, for which we have been preparing.

Theorem 1.4.4 (Well-Ordering Principle). *Each nonempty subset of* $\mathbb{N}$ *(or of* $\mathbb{N}^*$*) has a least element.*

Proof. We give the proof for subsets of $\mathbb{N}$ and leave as an exercise modification of the proof for subsets of $\mathbb{N}^*$.

Let $A \subset \mathbb{N}$. Just suppose A has no least element. We use induction on n to show

$$(\forall n \in \mathbb{N}) \left((\forall m \in A)\, (n < m) \right)$$

from which it will follow that $A = \varnothing$. (Why will that follow?)

Base step: If $m \in A$, then $m \neq 0$ (for otherwise 0 would be a least element of A), and so $0 < m$.

Inductive step: Let $n \in \mathbb{N}$ and assume

$$(\forall m \in A)\, (n < m).$$

We need to deduce that $n + 1 < m$ for all $m \in A$. Just suppose that

$$n + 1 \not< m$$

for *some* $m \in A$. Then $m \leq n+1$. By the inductive assumption, $n < m$. Thus

$$n < m \leq n + 1.$$

From the lemma,

$$m = n + 1.$$

We now claim that $n+1$ is a least element of A. Already, $n+1 \in A$ since $m \in A$. Moreover, if $k \in A$, then $n < k$ by the inductive assumption, and so $n + 1 \leq k$.

Thus, A has $n + 1$ as a least element. This contradicts the original assumption that A has no least element. $\quad\square$

Well-ordering is often a viable alternative to induction; the choice between the two may be a matter of taste. Let us take up yet once again this chapter's first example of induction.

Example 1.4.5. (Third reprise of Example 1.1.6) For each positive integer n, the inequality $2^n > n$ holds.

Proof. Just suppose $2^n \not> n$ for some $n \in \mathbb{N}^*$. Let

$$A = \left\{ n \in \mathbb{N}^* : 2^n \leq n \right\},$$

so that A is nonempty. By well-ordering, A has a least element n_1.

We know that $n_1 \geq 2$, because $2^1 = 2 > 1$ means that $1 \notin A$. Let $n = n_1 - 1$. Then $n \in \mathbb{N}^*$, and

$$2^n = 2^{n_1 - 1} = \frac{2^{n_1}}{2} \leq \frac{n_1}{2} \leq n_1 - 1 = n.$$

Thus $n \in A$ with $n < n_1$, the least element of A. This is absurd. □

As is the case with induction, when applying well-ordering one often uses a predicate $P(n)$, explicitly or implicitly, instead of a subset A of $\mathbb{N}$. Thus the preceding proof might be phrased: Just suppose $2^n \leq n$ for some positive integer n. By well-ordering, there is a least such positive integer n_1. Let $n = n_1 - 1$. Then (as above) $2^n \leq n$. This contradicts the choice of n_1 as the *least* such n.

The strategy used in Example 1.4.5 is typical of many—but not all— proofs that use well-ordering. Given a predicate $P(n)$, instead of forming the set

$$I = \{ n \in \mathbb{N}^* : P(n) \}$$

and using ordinary induction to show that $I = \mathbb{N}^*$, one forms the complementary set

$$A = \mathbb{N}^* \setminus I = \{ n \in \mathbb{N}^* : \text{not } P(n) \}$$

and uses well-ordering to show that A is empty (by daring to suppose it is nonempty and obtaining a contradiction).

The next example is a surprising application of well-ordering to prove that $\sqrt{2}$ is irrational. (Another proof, not involving well-ordering or any overt use of induction, appears later, in Example 2.2.20.) Recall that a real number x is said to be **rational** when $x = m/n$ for some integers m and n with $n \neq 0$, and **irrational** otherwise.

Example 1.4.6. The number $\sqrt{2}$ is irrational, that is, is not the quotient of two integers.

Proof. First, note that $1 < \sqrt{2} < 2$ because $1 < \left(\sqrt{2} \right)^2 = 2 < 4$.

Just suppose that $\sqrt{2}$ is rational. Then $n\sqrt{2}$ is an integer for some positive integer n (why?). By well-ordering, we may let n_1 be the least such n. Then the number k defined by

$$k = n_1 \left(\sqrt{2} - 1 \right)$$

is also a positive integer (why?) such that $k\sqrt{2}$ is an integer (why?). But $k < n_1$ (why?). This contradicts the choice of n_1 as the *least* n. □

Exercises 1.4.7.   (1) In the proof of the preceding example, what is the subset of $\mathbb{N}^*$ to which the Well-Ordering Principle is applied?

(2) Can you prove that $\sqrt{3}$ is irrational by using the same argument as in Example 1.4.6 except that $\sqrt{2}$ is replaced everywhere by $\sqrt{3}$? What about $\sqrt{5}$?

(3) Modify the proof in Example 1.4.6 in order to show that $\sqrt{s}$ is irrational whenever the positive integer s is not already a perfect square (the square of an integer).

(4) Supply the missing details in the following variant of the proof in Example 1.4.6: Just suppose that $\sqrt{2}$ is rational. Then the subset A of $\mathbb{N}^*$ defined by

$$A = \left\{ n \in \mathbb{N}^* : \sqrt{2} = \frac{m}{n} \text{ for some } m \in \mathbb{N}^* \right\}$$

is nonempty. By well-ordering, A must have a least element n_1. Since $n_1 \in A$, there is some $m \in \mathbb{N}^*$ with

$$\frac{m}{n_1} = \sqrt{2}. \tag{*}$$

Note that $n_1 < m < 2n_1$ (why?), and so both the integers $m - n_1$ and $2n_1 - m$ are positive.

In turn subtract 1 from both sides of equation (*), write the left-hand side as a single fraction, invert both sides, and rationalize the denominator in the resulting right-hand side to get

$$\frac{n_1}{m - n_1} = \sqrt{2} + 1.$$

Next, subtract 1 and write the left-hand side as a single fraction to get

$$\frac{2n_1 - m}{m - n_1} = \sqrt{2}.$$

Now both the numerator and the denominator are positive integers, and so $m - n_1 \in A$. But $m - n_1 < n_1$ (why?), and so we have an element of A that is less than the least element of A. This is impossible.

(5) Use well-ordering to prove that $n^3 + 2n$ is divisible by 3 for each positive integer n.

(6) Use well-ordering to prove that $n^2 > n + 1$ for all integers $n \geq 2$.

(7) Use well-ordering to prove that each integer greater than 1 is either a prime or else is a product of primes. (Recall that an integer $p > 1$ is said to be *prime* if it cannot be written as the product of two smaller positive integers. For example, 3 and 29 are prime, whereas $21 = 3 \cdot 7$ and $16 = 4 \cdot 4$ are not prime.)

(8) Finish and discuss the following "proof" that all natural numbers are interesting. The number 0 is certainly interesting since it is the very first natural number. Just suppose there is some natural number that is *not* interesting. By well-ordering, there must be a smallest natural number n that is not interesting. Now $n > 0$ (why?), and so $n - 1$ is also a natural number ....

The next set of exercises develops a generalization of well-ordering to certain subsets of the set

$$\mathbb{Z} = \{\ldots, -3, -2, -1, 0, 1, 2, 3, \ldots\}$$

of all integers. Of course, an arbitrary nonempty subset A of $\mathbb{Z}$ need not have a least element—for example, $\mathbb{Z}$ itself has no least element.

Exercises 1.4.8. Given a subset A of $\mathbb{Z}$, an integer b is called a **lower bound** of A in $\mathbb{Z}$ when $b \le a$ for every $a \in A$. For example, -9 is a lower bound of $A = \{n \in \mathbb{Z} : n \ge -2\}$; so are -5 and -2, but 3 is not. Thus a lower bound of a set A may belong to A but need not.

A subset A of $\mathbb{Z}$ is said to be **bounded below** in $\mathbb{Z}$ if it has some lower bound in $\mathbb{Z}$. For example, $\{n \in \mathbb{Z} : n \ge -2\}$ is bounded below in $\mathbb{Z}$, whereas (see below) $\mathbb{Z}$ itself is not. Another way of saying that A is bounded below in $\mathbb{Z}$ is to say that A is a subset of the ray $\{n \in \mathbb{Z} : b \le n\}$ for some integer b.

(1) Suppose b is a lower bound of $A \subset \mathbb{Z}$ in $\mathbb{Z}$. Show that if $b \in A$, then b is the least element of A. Is the converse true?

(2) Prove that $\mathbb{Z}$ has neither a lower bound in $\mathbb{Z}$ nor a least element.

(3) What are the lower bounds of the empty set $\varnothing$ in $\mathbb{Z}$?

(4) Prove: Each nonempty subset of $\mathbb{Z}$ that is bounded below in $\mathbb{Z}$ has a least element. (*Hint:* To reduce this to the Well-Ordering Principle, replace the set A in question with a suitable subset of $\mathbb{N}$.) [This is really a generalization of the Well-Ordering Principle, since a subset of $\mathbb{N}$ is certainly bounded below in $\mathbb{Z}$.]

(5) Define an appropriate notion of an *upper bound* of a subset of $\mathbb{Z}$ in $\mathbb{Z}$; define what it means for a subset of $\mathbb{Z}$ to be *bounded above* in $\mathbb{Z}$. Then state and prove an analog of (4) for upper bounds and greatest elements. (*Hint:* There is an easy way to prove this that

involves turning upper bounds and greatest elements of one set into lower bounds and least elements of another set.)

Exercise 1.4.9. The Well-Ordering Principle was deduced as a consequence of the Principle of Mathematical Induction. Show that, conversely, the Well-Ordering Principle implies the Principle of Mathematical Induction. (Thus, the Well-Ordering Principle is logically equivalent to the Principle of Mathematical Induction.)

Suppose you tried to use ordinary induction to prove that each integer greater than 1 is a prime or a product of primes. [A proof using well-ordering was requested in Exercise 1.4.7 (7).] For the inductive step, you would assume that a positive integer n greater than 1 is either a prime or a product of primes; from that you would want to deduce that $n + 1$ is a prime or a product of primes. But no relationship is apparent between the assumed property of n and the desired property of $n + 1$ (except when $n = 2$). For example, knowing that $n = 24$ is the product $n = 2 \cdot 2 \cdot 2 \cdot 3$ of primes provides no clue as to why $n + 1 = 25$ is the product $n + 1 = 5 \cdot 5$ of primes. A stronger assumption seems to be needed in the inductive step.

Exercise 1.4.10. Let n be a positive integer with $n > 1$. Assume that *every* integer k with $1 < k \leq n$ is a prime or a product of primes. Deduce that $n + 1$, too, must be a prime or a product of primes.

The inductive step in the original Principle of Mathematical Induction for a set $I \subset \mathbb{N}^*$ has the form

$$n \in I \implies n + 1 \in I,$$

which is a step from a single positive integer to the next one. As we just saw, in some situations such a step is not enough. In the following form of induction, the step is taken from *all* the positive integers up to and including a given n together to the next integer $n + 1$. Because of this difference, the original Principle of Mathematical Induction (Axiom 1.1.4) is often called the Principle of *Ordinary* Induction.

Theorem 1.4.11 (Principle of Strong Induction). *Let $I \subset \mathbb{N}^*$ such that:*

(a) $1 \in I$

(b) *For each positive integer n, if $\{ k \in \mathbb{N}^* : k \leq n \} \subset I$, then $n + 1 \in I$, too.*

Then $I = \mathbb{N}^*$.

Proof. Apply well-ordering to the set $A = \{ n \in \mathbb{N}^* : n \notin I \}$. □

Observe that the hypothesis about each n in (b) is: If $1 \in I$, $2 \in I$, $\ldots$, $n \in I$, then $n + 1 \in I$. Moreover, an alternate version of condition (b) is:

(c) For each integer $n > 1$, if $\{\, k \in \mathbb{N}^* : k < n \,\} \subset I$, then $n \in I$, too.

Exercise 1.4.12. (Continuation of Exercise 1.4.10) Use strong induction to prove that every integer greater than 1 is a prime or a product of primes. Give a formal proof where you explicitly identify and use the set I of the Principle of Strong Induction. Give also an informal proof where the set I is involved only implicitly.

Exercise 1.4.13. Suppose that condition (b) were replaced instead by:

(d) For each integer $n \in \mathbb{N}^*$, if $\{\, k \in \mathbb{N}^* : k < n \,\} \subset I$, then $n \in I$, too.

Explain why condition (a) would then become redundant.

Naturally, strong induction can be formulated in terms of subsets of $\mathbb{N}$ rather than of $\mathbb{N}^*$: replace 1 in (a) by 0, and $\mathbb{N}^*$ throughout by $\mathbb{N}$. More generally, we can start at any convenient base instead of 0 or 1.

Corollary 1.4.14 (Strong Induction—Generalized Form). *Let* s *be a natural number and let*

$$\mathbb{N}_s = \{\, n \in \mathbb{N} : n \geq s \,\}.$$

Suppose I is a subset of $\mathbb{N}_s$ such that:

(a) $s \in I$.

(b) *For each integer $n \geq s$, if $\{\, k \in \mathbb{N} : s \leq k \leq n \,\} \subset I$, then $n + 1 \in I$, too.*

Then $I = \mathbb{N}_s$.

Example 1.4.15. Each integer $n \geq 2$ can be expressed as $n = 2s + 3t$ for some *nonnegative* integers s and t.

Proof. Let

$$I = \{\, n \in \mathbb{N} : n \geq 2 \text{ and } n = 2s + 3t \text{ for some } s, t \in \mathbb{N} \,\}.$$

We use strong induction (generalized form) to show that $I = \{\, n \in \mathbb{N} : n \geq 2 \,\}$.

   Base step: Since $2 = 2 \cdot 1 + 3 \cdot 0$, then $2 \in I$.

   Inductive step: Let $n \geq 2$ and suppose $k \in I$ for each integer k in the range $2 \leq k \leq n$. We must deduce that $n + 1 \in I$, too. By the inductive assumption, $n - 1 \in I$, that is,

$$n - 1 = 2s + 3t$$

for some nonnegative integers s and t. Then

$$n + 1 = 2 + (n - 1) = 2 + (2s + 3t) = 2(1 + s) + 3t.$$

This means $n + 1 \in I$.   □

The result in the preceding example has an amusing interpretation: If you have an unlimited supply of 2-cent and 3-cent stamps only, then you can put any needed whole number of cents postage (except 1) on a letter or package—provided, of course, you don't run out of space on which to stick them.

Example 1.4.15 typifies one situation in which strong induction may be more appropriate than ordinary induction: when the desired property of n you want to show depends on the corresponding property of a smaller nonnegative integer—or perhaps two or several smaller nonnegative integers—but not necessarily on the immediate predecessor $n - 1$ of n. Another example of the same situation is our proof of the Fundamental Theorem of Arithmetic (each integer greater than 1 is a product of one or more primes)—see Theorem 2.3.9.

Exercises 1.4.16.   (1) There is a gap in the proof in Example 1.4.15: In the inductive step, if $n = 2$, then $n - 1 = 1$ is certainly less than n, but $n - 1 \not\geq 2$, and so the inductive assumption does not tell us that $n - 1 \in I$. (Recall that I consists only of integers that are at least 2.) Patch the gap.

(2) Give a proof of Example 1.4.15 that does not (at least directly) use induction. (*Hint:* A positive integer is either even, that is, divisible by 2, or else odd, in which case it is one more than an even integer.)

(3) Reformulate the Principle of Strong Induction, and its generalized form, in terms of predicates.

(4) Rewrite the (patched) proof of Example 1.4.15 in terms of predicates, without explicitly forming the set I.

(5) Which whole numbers of cents postage can be obtained using only 3-cent and 4-cent stamps? Formulate your answer as a mathematical proposition and prove it.

(6) In this exercise you will prove, in two different ways, that, for every nonnegative integer n,

$$(-1)^n = \begin{cases} 1 & \text{if } n \text{ is even,} \\ -1 & \text{if } n \text{ is odd.} \end{cases}$$

(Recall that an integer n is *even* if $n = 2k$ for some integer k, and n is *odd* otherwise.)

(a) Prove the above fact by means of induction. Do not use any laws of exponents except the recursive definition of powers (see Example 1.2.13); clearly identify any other arithmetic properties you use.

(b) Now prove the above fact again, but without induction. This time you may use any relevant laws of exponents, provided you make explicit which ones you use.

(7) Suppose $(s_n)_{n=1,2,...}$ is a sequence, where s_1 and s_2 are given positive integers, which satisfies the recurrence relation $s_{n+1} = s_n + s_{n-1}$ for all $n \geq 2$. Prove that s_n is a positive integer for every $n = 1, 2, \ldots$.

The Principle of Strong Induction and the Principle of Ordinary Induction involve quite different inductive assumptions; certainly some proofs by induction are more amenable to one of the principles than the other. Still, the two principles are equivalent to one another. In fact, we already know that the Principle of Ordinary Induction implies the Principle of Strong Induction: we proved the latter by deducing it from the former (through the intermediary of the Well-Ordering Principle). The converse implication is also true.

Exercise 1.4.17. Prove that the Principle of Strong Induction implies the Principle of Ordinary Induction. Be careful not to use in your proof either the Principle of Ordinary Induction or the Well-Ordering Principle!

The next exercises present more of the fascinating properties of the Fibonacci numbers F_n, which were defined by the second-order recursion $F_1 = F_2 = 1$ and $F_{n+1} = F_n + F_{n-1}$. One of these exercises provides a closed-form formula—a very surprising one involving, of all things, the irrational number $\sqrt{5}$—for the nth Fibonacci number.

Exercises 1.4.18. Using strong induction, or otherwise, establish the following properties of the Fibonacci numbers:

(1) $F_n = 2F_{n-4} + 3F_{n-3}$ for all $n \geq 5$.

(2) F_{5n} is divisible by 5 for every $n \geq 1$.

(3) $F_n \leq (13/8)^{n-1}$ for all $n \geq 1$, and strict inequality holds for all $n > 1$.

(4) $F_{n+1}^2 - F_n F_{n+2} = (-1)^n$ for all $n \geq 1$.

(5) For all nonnegative integers n,

$$F_n = \frac{1}{\sqrt{5}} \left[\left(\frac{1 + \sqrt{5}}{2} \right)^n - \left(\frac{1 - \sqrt{5}}{2} \right)^n \right].$$

(6) Let $(G_n)_{n=1,2,...}$ be the sequence defined recursively by $G_1 = G_2 = 1$ and $G_{n+1} = G_n + G_{n-1} + G_n G_{n-1}$ for $n \geq 2$. Then

$$G_n = 2^{F_n} - 1$$

for all $n = 1, 2, \ldots$.

Exercise 1.4.19. Discover the flaw in the following "proof" that $2^n = 1$ for every nonnegative integer n.

Proof. We use strong induction on n. First, $2^0 = 1$.

Now let $n \in \mathbb{N}$ and assume that $2^k = 1$ for all nonnegative integers $k \leq n$. We want to deduce that $2^{n+1} = 1$, too. Applying the inductive assumption to both $k = n$ and $k = n - 1$, we have

$$2^{n+1} = \frac{2^{2n}}{2^{n-1}} = \frac{2^n \cdot 2^n}{2^{n-1}} = \frac{1 \cdot 1}{1} = 1. \quad \square$$

One of the nicest, and most famous, applications of strong induction is the proof (requested in Exercise 1.4.12) that each integer greater than 1 is a product of primes. A stronger result appears in Chapter 2. An additional application is described in the next section.

1.5 Binomial Coefficients

Everybody knows that

$$(a + b)^1 = a + b,$$
$$(a + b)^2 = a^2 + 2ab + b^2.$$

You may even know that

$$(a + b)^3 = a^3 + 3a^2b + 3ab^2 + b^3,$$
$$(a + b)^4 = a^4 + 4a^3b + 6a^2b^2 + 4ab^3 + b^4.$$

In fact, the recursive definition $x^0 = 1$, $x^{n+1} = x \cdot x^n$ of powers given earlier yields these formulas and others. One of the aims of this section is to prove a generalization of all these special cases—the Binomial Theorem.

You may check that, in the expansion of $(a + b)^4$ above, the three distinct coefficients (including the implicit coefficient 1 of the leading term) appearing there can be expressed as

$$1 = \frac{4!}{0! \cdot 4!},$$
$$4 = \frac{4!}{1! \cdot 3!},$$
$$6 = \frac{4!}{2! \cdot 2!}.$$

(Where did these formulas come from? At the moment, this is not a question being considered!)

Exercise 1.5.1.   (a) Similarly express the coefficients 1 and 3 in the expansion of $(a + b)^3$ in terms of factorials.

(b) Repeat (a) for the expansions of $(a + b)^2$ and $(a + b)^1$.

(c) Does the pattern here work for $(a + b)^0$ as well?

(d) Calculate an expansion of $(a + b)^5$ and express the coefficients in the expansion in terms of factorials.

There seems to be a pattern here: The coefficient of $a^{n-j}b^j$ in the expansion of $(a + b)^n$ is

$$\frac{n!}{j!(n - j)!}.$$

The Binomial Theorem asserts that this is, in fact, true in general. Before proving it, we want to show that these coefficients are actually integers.

Example 1.5.2. If j, n are nonnegative integers with $j \le n$, then

$$n! \text{ is divisible by } j!(n - j)!.$$

(Recall that an integer b is said to be *divisible by* an integer a when $b = aq$ for some integer q.)

Proof. We use strong induction on n to prove, for each $n \in \mathbb{N}$, that $n!$ is divisible by $j!(n - j)!$ for all nonnegative integers $j \le n$.

First, take $n = 0$, so that $n! = 1$. Then $j \in \mathbb{N}$ with $j \le n$ means $j = 0$, and $j!(n - j)! = 1$ also, and so $n!$ is divisible by $j!(n - j)!$.

Now let $n \in \mathbb{N}$ and assume

$$(\forall k \in \mathbb{N}) \ (k \le n \implies n! \text{ is divisible by } k!(n - k)!).$$

Let $j \in \mathbb{N}$ with $j \le n + 1$. It remains to deduce that $(n + 1)!$ is divisible by $j!(n + 1 - j)!$. This is certainly true if $j = 0$ or if $j = n + 1$, and so suppose

$$0 < j < n + 1.$$

Then since $j \leq n$, the inductive assumption yields

$$n! \text{ is divisible by } j!(n-j)!.$$

Multiply by $(n + 1 - j)$ to obtain

$$(n!)(n+1-j) \text{ is divisible by } j!(n+1-j)!. \tag{*}$$

Next, since also $j - 1 \leq n$, the inductive assumption also yields

$$n! \text{ is divisible by } (j-1)![n-(j-1)]!.$$

Multiply by j to obtain

$$(n!)j \text{ is divisible by } j!(n+1-j)!. \tag{**}$$

But

$$(n!)(n+1-j) + (n!)j = n!(n+1) = (n+1)!.$$

From (*) and (**) it now follows that, indeed, $(n + 1)!$ is divisible by the product $j!(n + 1 - j)!$. $\square$

For $0 \leq j \leq n$, the quotient $n!/[j!(n-j)!]$ is thus a nonnegative *integer*, to which we give a name.

Definition 1.5.3. If n is a nonnegative integer and j is an integer with $0 \leq j \leq n$, the **binomial coefficient** $\binom{n}{j}$ is defined by

$$\binom{n}{j} = \frac{n!}{j!(n-j)!}.$$

The symbol $\binom{n}{j}$, which will often be squashed down in paragraphs of text to $\binom{n}{j}$, is read "n above j."

The reason for the name "binomial coefficient" is that $\binom{n}{j}$ is, as we shall see shortly, the coefficient of $a^{n-j}b^j$ in the expansion of the binomial $(a + b)^n$. [As we shall see later, $\binom{n}{j}$ is also the number of j-element subsets of an n-element set, and in that connection it is also denoted by C_j^n and called the **number of combinations of n things taken j at a time**.]

Exercises 1.5.4.   (1)   (a) Compute $\binom{2}{j}$ for $j = 0, 1, 2$.

(b) Compute $\binom{3}{j}$ for $j = 0, 1, 2, 3$.

(c) Compute $\binom{4}{j}$ for $j = 0, 1, 2, 3, 4$.

(d) Compute $\binom{5}{j}$ for $j = 0, 1, 2, 3, 4, 5$.

(e) What do $\binom{n}{0}$ and $\binom{n}{n}$ seem to be? Prove your answer.

(f) What is $\binom{n}{1}$ when $n \geq 1$? Prove your answer.

(g) Prove the *symmetry formula*

$$\binom{n}{n-j} = \binom{n}{j}$$

for $0 \leq j \leq n$.

(h) From the calculations you have already made, discover another pattern or relationship involving binomial coefficients.

(2) The final step in the proof in Example 1.5.2 had the form: If a is divisible by b and if c is divisible by b, then $a + c$ is divisible by b. Prove this in general.

(3) This exercise develops an especially efficacious way to compute $\binom{n}{j}$. Here n and j denote integers with $0 \leq j \leq n$.

(a) Show that

$$n! = [n(n-1)\cdots(n-j+1)](n-j)!.$$

(b) Deduce from (a) that

$$\binom{n}{j} = \frac{n(n-1)\cdots(n-j+1)}{j!},$$

that is,

$$\binom{n}{j} = \frac{n(n-1)\cdots(n-j+1)}{j(j-1)\cdots 1}$$

(This formula is easy to remember: there are the same number of factors in the numerator as in the denominator.)

(c) Use the preceding formula to calculate $\binom{6}{j}$ for $j = 0, 1, \ldots, 6$.

(d) Use the same formula to calculate $\binom{52}{5}$. (*Note:* This is the number of different poker hands one can be dealt where the order in which the cards are dealt does not matter.)

(4) Prove the *absorption formula*

$$j \binom{n}{j} = n \binom{n-1}{j-1}$$

for binomial coefficients.

Look at the following table of binomial coefficients $\binom{n}{j}$ for $n = 0, 1, 2, 3, 4$ and for $j = 0, 1, \ldots, n$ for each n:

$$
\begin{array}{ccccccccc}
 & & & & 1 & & & & \\
 & & & 1 & & 1 & & & \\
 & & 1 & & 2 & & 1 & & \\
 & 1 & & 3 & & 3 & & 1 & \\
1 & & 4 & & 6 & & 4 & & 1
\end{array}
$$

This table, which we can imagine extended indefinitely downward, is known as **Pascal's triangle** (after Blaise Pascal).

If we number the first row in the table as row 0 and the first non-blank column in each row as column 0, then the value of $\binom{n}{j}$ appears in row n and column j (not counting any blank columns). Except for the initial and final entry in each row, each entry in the table below the top row is the sum of the two entries straddling it in the row above it. If we imagine a zero appended at the end of each row and another zero prepended at the start of each row—

$$
\begin{array}{ccccccccccc}
 & & & & 0 & & 1 & & 0 & & \\
 & & & 0 & & 1 & & 1 & & 0 & \\
 & & 0 & & 1 & & 2 & & 1 & & 0 \\
 & 0 & & 1 & & 3 & & 3 & & 1 & & 0 \\
0 & & 1 & & 4 & & 6 & & 4 & & 1 & & 0
\end{array}
$$

—then each nonzero entry below the top row is the sum of the two entries straddling it in the row above it, without any exceptions for the initial or final 1 in each row. For this reason, it is convenient to extend the definition of $\binom{n}{j}$ slightly.

Definition 1.5.5. Let n be a nonnegative integer. If j is an integer with $j < 0$ or $j > n$, we define

$$\binom{n}{j} = 0.$$

Then what we have observed with the triangular table above is the following recurrence relation.

Proposition 1.5.6 (Addition Formula). *For all integers n and j with* $0 \le j \le n$,

$$\binom{n+1}{j} = \binom{n}{j-1} + \binom{n}{j}.$$

Proof. Exercise. □

Exercise 1.5.7. (a) Write a computer program to calculate binomial coefficients in terms of our original definition that used factorials (see Definition 1.5.3).[11]

(b) Write a computer program for calculating binomial coefficients without directly using factorials but instead based upon the formula in Exercise 1.5.4 (3)(b). (If you use that formula suitably, you can avoid forming intermediate results of large size.)

(c) Write a computer program for calculating binomial coefficients based upon the principle behind Pascal's triangle.

Exercise 1.5.8. (a) Establish the formula

$$\binom{m+n+1}{n} = \sum_{j=0}^{n} \binom{m+j}{j}.$$

(b) Use this formula to show that

$$\binom{5}{2} = \binom{4}{2} + \binom{3}{1} + \binom{2}{0}.$$

Exercise 1.5.9. Connect the Fibonacci numbers to the binomial coefficients by proving the identity

$$F_{n+1} = \sum_{j=0}^{k} \binom{n-j}{j},$$

where $k = \lfloor n/2 \rfloor$, the greatest integer not exceeding $n/2$.

The Addition Formula is the crux of a formula for the nth power of a binomial $(a + b)$.

---

[11]If you are using a "typed" language such as Pascal, then (as with any program involving factorials), be sure to declare the variables for factorials to be of type `Real` rather than `Integer`. This is to avoid "overflow" from factorials that become too large.

Theorem 1.5.10 (Binomial Theorem). *Let a and b be arbitrary numbers (real or complex). Then for each nonnegative integer n,*

$$(a + b)^n = \sum_{j=0}^{n} \binom{n}{j} a^{n-j} b^j.$$

Proof. Exercise. Use induction. You can simplify the algebra a bit by first proving the special case where $a = 1$ and $b = x$, in other words, by proving the formula for the expansion of $(1 + x)^n$. Obtain the general case for $a \neq 0$ by using $(a + b) = a(1 + b/a)$ and treat the case where $a = 0$ separately. □

Exercises 1.5.11.   (1) Use the Addition Formula to construct the next two rows (for $n = 5, 6$) of Pascal's triangle.

(2) Express 2^n in terms of binomial coefficients.

(3) Use the Binomial Theorem to calculate 11^5 quickly.

(4) Write the Binomial Theorem in terms of ascending powers a^j, for $j = 0, 1, \ldots, n$, of a rather than descending powers a^{n-j}.

(5) Fill in the details of the following proof of the special case $(1 + x)^n = \sum_{j=0}^{n} \binom{n}{j} x^j$ of the Binomial Theorem. Obviously (prove it!) $(1 + x)^n$ is a polynomial in x of degree n, and so we may write

$$(1 + x)^n = \sum_{j=0}^{n} a_j x^j \tag{*}$$

for *some* coefficients $a_0, a_1, \ldots, a_n$. We need only show that $a_j = \binom{n}{j}$ for each j.

To begin, take $x = 0$ to get $a_0 = 1$; then $a_0 = \binom{n}{0}$. Next, differentiate both sides of (*) and take $x = 0$ in the result to find a_1. Differentiate again and take $x = 0$ to find a_2. Continue in this way.

(6) Use the Binomial Theorem along with the definition of the derivative of a function as a limit of difference quotients to prove that the derivative of x^n is nx^{n-1}.

(7) Investigate an analog of the Binomial Theorem for powers $(a + b + c)^n$ of a trinomial $a + b + c$.

Chapter 2

Number Theory

Mathematics is the queen of the sciences, and number theory is the queen of mathematics.

— Carl Friedrich Gauss

Number theory deals broadly with properties of the set of all integers, especially the set of all positive integers. It includes, for example, the study of factoring positive integers into primes—numbers that cannot be further factored.

The English number theorist G. H. Hardy (1877–1947) celebrated how "useless" the subject was. He would be astonished at how applicable it has become now that, for example, questions of primality are so important to cryptography.

Part of number theory's great appeal is the ease with which the statements of so many of its problems can be understood. Yet number theory is famous—some would say infamous—for the difficulty of solving many of these very same problems. For example:

Are there integers a, b, and c, not all 0, that satisfy the equation

$$a^n + b^n = c^n$$

for a given integer $n > 2$?

There are certainly such integers when $n = 2$—such *pythagorean triples* as $a = 3, b = 4, c = 5$ and $a = 5, b = 12, c = 13$.[1] But, claiming

---

[1]Recall that, according to the Pythagorean Theorem, the lengths a and b of the two legs and length c of the hypotenuse of a right triangle satisfy the equation $a^2 + b^2 = c^2$. Conversely, if positive numbers a, b, c satisfy this equation, then a triangle having sides with lengths a, b, and c must be a right triangle whose hypotenuse has length c (why?).

that the answer is no for every $n > 2$, in 1637 Pierre de Fermat wrote in the margin of a book, "I have discovered a truly remarkable proof which this margin is too narrow to contain." Although Fermat never published a general proof, he did prove the result for $n = 4$, and over a century later Leonhard Euler proved it for $n = 3$. That the answer is *always* no—that there are *no* integers a, b, c, not all 0, for which $a^n + b^n = c^n$, no matter what $n > 2$ is—became known as *Fermat's Conjecture* or *Fermat's Last Theorem*.

In the more than 350 years since Fermat penned his famous marginal note, many fallacious proofs have been offered. But a correct general proof had never been given—until now! In 1993, Andrew Wiles announced a proof of Fermat's Last Theorem that relied upon very deep results by a number of other contemporary mathematicians. Experts at first believed the proof to be correct, but a gap was found in it. Finally, in 1994, Wiles and his former student Richard Taylor completed the proof.

Here is another simply stated problem. Observe that $8 = 2^3$ and $9 = 3^2$ are consecutive powers. Then:

> Are 8 and 9 the only two consecutive positive integers each of which is a power—with exponent an integer greater than 1—of a positive integer?

In 1842 Eugène Catalan claimed that this is true, and so the problem is known as *Catalan's Conjecture*. However, as of 1995 the problem remained open—the answer still unknown. Also unknown is whether there are any three consecutive positive integers whatsoever each of which is a power of a positive integer.

Although we shall prove several important theorems and establish some surprising facts about integers here, we shall, of course, merely scratch the surface of the subject.

Many of the interesting properties of integers will be consequences of more basic properties that may seem quite "obvious" to you—so obvious that they, in turn, seem to require no further proof. For example, we shall need to know that an integer that is not twice any integer is one more than twice some integer; we shall want to know that a prime number that divides the product of two integers (leaving remainder 0) must divide at least one of the two. Such properties are often obvious only in the sense that they are too familiar (from your concrete experience with specific integers) for you to doubt them. Please do not be too annoyed at being asked to justify such things you already "know"; it is often just those things you *think* you know that are the hardest to justify convincingly.

So—which things about integers may you safely assume and which must you justify? In short, where should you start? Look at Appendix D. There you will find a list of truly fundamental properties of integers that you should feel free to use as required—without any further justification. Please do not try, though, to memorize the list. Use it as a reference whenever you are in doubt about what to assume known. It should make your life among the integers a little easier.

One of the fundamental properties of integers listed in Appendix D is the Principle of Mathematical Induction, which was introduced in Chapter 1. At a number of key places, mathematical induction will be an essential tool of proof.

2.1 The Division Theorem

Everybody knows that if you try to divide one integer by another, the divisor may not go into the dividend an integral number of times, in other words, that you may need a remainder along with a quotient to express the result. For example,

$$\frac{23}{5} = 4 + \frac{3}{5}.$$

In elementary school, you doubtless spent considerable time learning the mechanics of carrying out such divisions, especially for multidigit numbers.

When dealing with properties of integers, it is desirable whenever possible to say whatever we want to say strictly in terms of integers, not in terms of other rational numbers such as 23/6 and 5/6. So, to keep in the right "spirit" of things, we prefer to write the above-displayed division result in the equivalent form

$$23 = 4 \cdot 5 + 3.$$

Our first theorem asserts the "obvious" fact that such a division is always possible and, moreover, that the quotient and remainder are uniquely determined by the divisor and dividend. If it is really obvious to you, it is probably because it is so familiar from all the examples you have seen. If you are certain that it is obvious, try to prove it yourself before peeking at the proof presented below.

Theorem 2.1.1 (Division Theorem). *Let n be any positive integer and m be any integer. Then there are unique integers q and r such that*

$$m = qn + r, \qquad 0 \le r < n.$$

Proof. The conclusion asserts both *existence*—there is *at least one* such pair q and r—and *uniqueness*—there is *at most one* such pair q and r. Accordingly, the proof is separated into two parts.

Existence. The idea of the proof, as illustrated for $m > 0$ in Figure 2.1, is to find the multiple kn of n that is still to the left of m yet closest to m, that is, for which the distance $|m - kn| = m - kn$ is least.

At least one of all the differences $m - kn$ is nonnegative. In fact, if, on the one hand, $m \geq 0$, then $m - (-1)n \geq 0$; if, on the other hand, $m < 0$, then $m - mn = -m(-1 + n) \geq 0$. Thus, the set A defined by

$$A = \{\, m - kn : k \in \mathbb{Z}, \, m - kn \geq 0 \,\}$$

is nonempty.

Figure 2.1: Idea behind the proof of the Division Theorem.

By the Well-Ordering Principle (Theorem 1.4.4), A has a least element r. Certainly $r \geq 0$. By definition of A, there is some $q \in \mathbb{Z}$ for which $r = m - qn$, and so

$$m = qn + r.$$

To show that $r < n$, we show, equivalently, that the difference $r - n$ is negative. (*Stop!* Try it before reading further.) We have

$$r - n = m - qn - n = m - (q + 1)n.$$

If $m - (q + 1)n \geq 0$, then this number belongs to A and yet is smaller than the least element $m - qn$ of A; that is impossible. Hence $m - (q + 1)n < 0$, that is, $r - n < 0$.

Uniqueness. Suppose also there are integers q' and r' with

$$m = q'n + r', \qquad 0 \leq r' < n.$$

We want to show that $q' = q$ and $r' = r$.

We cannot have $q' < q$ because if $q' < q$, then

$$r' = m - q'n \geq m - (q - 1)n = r + n \geq n,$$

which is contrary to one of the assumptions about r'. Similarly (give the details), we cannot have $q < q'$. Hence $q' = q$.

From $q' = q$, it follows at once (why?) that also $r' = r$. □

Exercise 2.1.2. Rewrite the proof of the existence part of the Division Theorem so as to use the least element of the set $\{\, k \in \mathbb{Z} : kn > m \,\}$.

The Division Theorem justifies the following definition, which you may have met in the guise of some standard functions in Pascal or other programming languages.

Definition 2.1.3. Given a positive integer n and any integer m, the unique integers q and r with $m = qn + r$ and $0 \le r < n$ are called the **quotient** and **remainder** for dividing m by n and are denoted by m div n and m mod n, respectively.

For example, 23 div 5 = 4 and 23 mod 5 = 3.

Exercises 2.1.4. (1) Calculate 14 div 4 and 14 mod 4. Calculate, also, (-14) div 4 and (-14) mod 4.

(2) For real x, let $\lfloor x \rfloor$ denote the *floor* of x, that is, the greatest integer less than or equal to x.

   (a) Determine $\lfloor \sqrt{2} \rfloor$, $\lfloor -\sqrt{2} \rfloor$, $\lfloor 1/2 \rfloor$, and $\lfloor -1/2 \rfloor$.

   (b) Prove that such a greatest integer less than or equal to x actually exists for arbitrary *rational* x.

   (c) Try to prove that such a greatest integer less than or equal to x actually exists for arbitrary real x. Indicate what properties of real numbers are required to complete the proof.

   (d) For positive integers n and m, show that m div $n = \lfloor m/n \rfloor$.

(3) This exercise involves long division.

   (a) Use long division (with paper and pencil) to find the quotient and remainder for dividing 791 by 3. Explain carefully how you got each digit in the quotient, as well as why the steps you did actually lead to the correct quotient and remainder.

   (b) Repeat (a), but for long division of 5864 by 23.

   (c) Now try to formulate long division formally in the form of an algorithm. (See also Exercise 2.1.12.)

You may recall that long division may be applied to polynomials as well as to integers. For polynomials, too, there is a division theorem—usually called the **Remainder Theorem:**

Let $p(x)$ be any polynomial and let $d(x)$ be a nonzero polynomial of degree n. Then there are unique polynomials $q(x)$ and $r(x)$ such that $p(x) = q(x)d(x) + r(x)$ and $r(x)$ either is the zero polynomial or else has degree $< n$.

(The polynomials involved may have rational, real, or even complex numbers as their coefficients.)

Exercises 2.1.5.   (1) Prove the Remainder Theorem. [*Suggestion:* Let
$d(x) = \sum_{j=0}^{n} b_j x^j$ have degree n, so that the coefficient $b_n \neq 0$.
Use strong induction or well-ordering on the degree m of $p(x) =$
$\sum_{j=0}^{m} a_j x^j$. The result is easy if $m < n$. If $m \geq n$, then the
polynomial

$$p(x) - \frac{a_m}{b_n} x^{m-n} d(x)$$

is zero or has degree less than m.]

(2) Does the Remainder Theorem remain true if we insist that all the
coefficients involved be integers?

(3) Deduce from the Remainder Theorem the following **Factor Theorem:**

Let $p(x)$ be a polynomial and let a be a number. Then a
is a root of $p(x)$ if and only if $x - a$ is a factor of $p(x)$.

[To say that a is a root of $p(x)$ means, of course, that $p(a) = 0$. To
say that $x - a$ is a factor of $p(x)$ means that $p(x) = (x - a)q(x)$
for some polynomial $q(x)$.]

(*Note:* A direct proof of the Remainder Theorem—in the case of integer coefficients and integer a—was indicated in Exercise 1.3.7.)

We are accustomed to writing integers in their **decimal** (base 10)
representation. For example,

$$20543 = 3 + 4 \cdot 10 + 5 \cdot 10^2 + 0 \cdot 10^3 + 2 \cdot 10^4$$
$$= 3 \cdot 10^0 + 4 \cdot 10^1 + 5 \cdot 10^2 + 0 \cdot 10^3 + 2 \cdot 10^4.$$

In general, for any digits $a_j \in \{0, 1, 2, 3, 4, 5, 6, 7, 8, 9\}$, where $j = 0, 1, \ldots, n$,

$$a_n \cdots a_2 a_1 a_0 = a_0 + a_1 \cdot 10^1 + a_2 \cdot 10^2 + \cdots + a_n \cdot 10^n$$
$$= \sum_{j=0}^{n} a_j 10^j.$$

From working with computers or otherwise, you may also be familiar with the **binary** (base 2) **representation** of integers. A subscript 2
is used. For example, the integer 23 (decimal) has the binary representation

$$(10111)_2 = 1 + 1 \cdot 2 + 1 \cdot 2^2 + 0 \cdot 2^3 + 1 \cdot 2^4$$
$$= 1 \cdot 2^0 + 1 \cdot 2^1 + 1 \cdot 2^2 + 0 \cdot 2^3 + 1 \cdot 2^4.$$

In general, for any **bits** (binary digits) $b_j \in \{0, 1\}$, $j = 0, 1, \ldots, n$,

$$(b_n \cdots b_2 b_1 b_0)_2 = b_0 + b_1 \cdot 2^1 + b_2 \cdot 2^2 + \cdots + b_n \cdot 2^n$$
$$= \sum_{j=0}^{n} b_j 2^j.$$

Both the binary and the decimal representations[2] are examples of **positional number representation**, where the value of a digit does not remain constant but rather depends upon its position. For example, the rightmost 5 in 555 stands for 5, the middle 5 for 50, and the leftmost 5 for 500.

Such positional representation contrasts with an *additive* representation, such as the Roman numeral system, where, say, X stands for 10 no matter where it appears in a number: XXX = X+X+X = 10+10+10 = 30.

Other positional representations, such as **octal** (base 8) and **hexadecimal** (base 16) are also common in computer work. For an octal representation, the permitted digits are $0, 1, \ldots, 7$. For a hexadecimal representation, one must invent symbols (typically A, B, C, D, E, F—having decimal values 10, 11, 12, 13, 14, 15, respectively) for the permitted digits beyond $0, 1, \ldots, 9$.

Why can we represent any given positive integer, as above, in base 2? For that matter, why can we represent it in base 10? Once again, don't let mere familiarity excuse the need for a proof! We shall handle an arbitrary base b, and not just 2 or 10. Such a proof—and an algorithm for obtaining the representation, can be motivated by an example: Find the binary representation of 23 (pretend you did not see the representation above). Here's what we do.

The highest power of 2 not exceeding 23 is $16 = 2^4$, and $23 - 16 = 7$, and so

$$23 = 2^4 + 7.$$

---

[2] Our decimal system is often mistakenly called the "Hindu-Arabic number system." In fact, the ideas of both zero and decimal positional representation were invented in China before 1030 B.C. By 876 A.D., Hindus in India were using a decimal positional system with the precursors of our current notation for the 10 digits. By the tenth century, the system had been transmitted to Europe by the Arabs. [Decimal *fractions*, however, came later. They were in use in China by the thirteenth century. The arithmetic of the decimal fraction system, essentially as we know it today, was developed and described by the Belgian Simon Stevin (1548–1620).]

The highest power of 2 not exceeding 7 is $4 = 2^2$, and $7 - 4 = 3$, and so

$$7 = 2^2 + 3.$$

The highest power of 2 not exceeding 3 is $2 = 2^1$, and $3 - 2 = 1$, and so

$$3 = 2^1 + 1.$$

Finally, $1 = 2^0$. Hence

$$23 = 2^4 + 2^2 + 2^1 + 2^0 = (10111)_2.$$

Exercise 2.1.6. Using the same method just illustrated, find:

(a) The base 2 expansion of decimal 29.

(b) The base 3 (ternary) expansion of decimal 259.

(c) The base 3 expansion of decimal 502.

The key to the method for base 2 expansion just illustrated is that there *is* a highest power of 2 not exceeding any given integer. This is true, more generally, for any base; it is a consequence of the following lemma, which says that a sufficiently high power of a given base $b > 1$ can be found to exceed any given number. (In terms of limits, $\lim_{m \to \infty} b^m = \infty$.)

Lemma 2.1.7. *Let b be an integer with $b > 1$. Then for each natural number a, there is some natural number m with $a < b^m$.*

Proof. Use a proof by contradiction: Just suppose there is some natural number a such that . . . . Apply well-ordering to $\{\, a \in \mathbb{N} : \dots \,\}$. To obtain the desired contradiction, note that $b^{m+1} \le a$ implies $b^m \le a/b$. $\quad\square$

Obviously the preceding result is false for $b = 1$. Moreover, since $1^n = 1$ for every n, number representation using base 1 is simply a **tally** system where, for example, we write twenty-three 1s (or other strokes) to represent 23.

Corollary 2.1.8. *Let b be an integer with $b > 1$. Then for each positive integer a there is a unique natural number n with*

$$b^n \le a < b^{n+1}.$$

Proof. Exercise. $\quad\square$

Exercise 2.1.9.   (a) Given integers k and n with $k \ge 1$ and $n \ge 2k$, show that some power of 2 lies between k and n, inclusive.

(b) The only partial sum $S_n = \sum_{j=1}^{n}(1/j)$ of the divergent harmonic series $\sum_{j=1}^{\infty}(1/j)$ that equals 1 is evidently the first one $S_1 = 1/1$. Are there any segments of consecutive terms farther out in the harmonic series that sum to 1? In other words, are there any positive integers n and k with $n \geq k$ for which the equality

$$\sum_{j=k}^{n} \frac{1}{j} = 1$$

holds, other than $n = k = 1$?

The possibility of positional representation of any number for an arbitrary base is guaranteed by the following theorem.

Theorem 2.1.10 (Base Representation). *Let b be an integer with $b > 1$. Then for each positive integer a there exists a unique natural number n and a unique $(n+1)$-tuple $(d_0, d_1, \ldots, d_n)$ of integers in $\{0, 1, \ldots, b-1\}$ with $d_n > 0$ such that*

$$a = \sum_{j=0}^{n} d_j b^j.$$

Proof. Existence. We use strong induction (Theorem 1.4.11) on a. For the base case $a = 1$, take $n = 0$ and $d_0 = 1$.

Let $a > 1$ and assume the result is true for all k with $1 \leq k < a$. (What, precisely, is being assumed for each such k?) By the preceding corollary, there is a natural number n with

$$b^n \leq a < b^{n+1}.$$

By the Division Theorem (2.1.1), there are integers q and r with

$$a = qb^n + r, \qquad 0 \leq r < b^n.$$

Now $q > 0$ because $qb^n = a - r > b^n - b^n = 0$; further, $q < b$ because $qb^n \leq a < b^{n+1}$. Thus

$$0 < q < b.$$

Consider two cases.

Case 1: $r = 0$. In this case, the desired representation of a is

$$a = \sum_{j=0}^{n-1} 0b^j + qb^n.$$

Case 2: $r > 0$. In this case, $1 \leq r < a$. By the inductive assumption, r has a representation

$$r = \sum_{j=0}^{m} c_j b^j$$

for some nonnegative integer m and integers c_j with $0 \leq c_j < b$ ($j = 0, 1, \ldots, m$) and $c_m > 0$. Now $m < n$ (why?), and so from $a = r + qb^n$ we obtain the desired representation of a:

$$a = \sum_{j=0}^{m} c_j b^j + \sum_{j=m+1}^{n-1} 0 b^j + q b^n.$$

Uniqueness. What must be proved is that, given two such representations

$$a = \sum_{j=0}^{n} d_j b^j$$

and

$$a = \sum_{i=0}^{m} c_i b^i,$$

then necessarily $m = n$ and $d_j = c_j$ for all j. The proof of this is left as an exercise. □

Even in the case of base $b = 10$, the Base Representation Theorem concerns only integers. Decimal expansion more generally of arbitrary real numbers—including noninteger rational numbers and irrational numbers—requires additional principles beyond induction. This more general expansion is discussed in Section 3.5 (see Example 3.5.36).

The Base Representation Theorem has an amusing interpretation in terms of weighing on a two-pan balance. We state it in the case of base $b = 2$: Given standard weights $1, 2, 2^2, \ldots, 2^n$, then any object with an integral weight less than 2^{n+1} can be weighed by putting it on one pan of the balance and placing a suitable combination of standard weights on the other.

Exercise 2.1.11.   (a) Express in precise mathematical terms the preceding statement about weights and then prove it.

(b) Show that no other set of $n + 1$ standard weights can accomplish the same thing. (*Hint:* Arrange such weights in order of increasing size: $w_0 \leq w_1 \leq \cdots \leq w_n$. Assume there is some j for which $w_j \neq 2^j$. Let l be the least j. Find a contradiction.)

Exercise 2.1.12. Try again to formulate long division formally in the form of an algorithm. (See page 65.)

2.2 Divisibility

The two fundamental operations on the set of integers are addition and multiplication, and there are many fascinating questions that can be asked about each. Concerning addition, we might ask, for example, "In how many ways can an arbitrary positive integer be written as the sum of smaller positive integers?"

The number-theoretic questions we shall investigate deal primarily with the other fundamental operation on integers—multiplication. For example, in how many ways can a positive integer be written as the product of positive integers that are not themselves products of still smaller positive integers?

In the preceding section, we proved that a given positive integer can be expressed as an integral multiple of another given integer plus a remainder. We now look at the situation when the remainder is zero.

Definition 2.2.1. Let d and n be integers. We say that n is a **multiple** of d when there is some integer k with $n = kd$. When n is a multiple of d and $d > 0$, we say that d **divides** n, or that n is **divisible** by d, call d a **divisor** of n, and write

$$d \mid n$$

(which is read "d divides n"). When positive integer d does *not* divide n, we write $d \nmid n$.

For example, 12 is a multiple of 6, and so $6 \mid 12$, but 12 is not a multiple of 5, and so $5 \nmid 12$. Each integer is a multiple of -1 but, since by definition a divisor must be positive, -1 does not divide any integer, nor does any negative integer, for that matter.[3] The only multiple of 0 is 0 itself, and $0 \nmid n$ no matter what n is.

Example 2.2.2. The integer 2 does not divide 1, in other words: $2 \nmid 1$.

"How can you possibly prove that—it's so *obvious*," you are doubtless thinking. Well, it certainly is *familiar* in the sense that everybody knows that $1/2$ is not an integer. Nonetheless, we still really ought to be able to justify it—provided we start at a sufficiently basic level of assumptions about integers. Everything we assume here about the integers appears in Appendix D or has already been deduced from those assumptions. That includes the Principle of Mathematical Induction

---

[3]We insist in the definition that divisors be positive just so that later on we will not have to keep including this as an additional condition when we do, in fact, wish to consider only positive divisors. Thus, there would have been nothing wrong had we defined $d \mid n$ to mean n is a multiple of d even when $d < 0$ (and then our definition would be the standard one used for the more general algebraic systems known as "rings").

and the consequences of it we have already deduced, such as the Gap Lemma 1.4.1. So here goes.

Just suppose 2 divides 1, that is, $1 = k \cdot 2$ for some integer k. Certainly $k \neq 0$ because $0 \cdot 2 = 0 \neq 1$; also, $k \not< 0$ because otherwise $1 = k \cdot 2 < 0$ whereas $1 > 0$. Thus $k > 0$. By the Gap Lemma, $k \geq 1$. Then $1 = k \cdot 2 \geq 1 \cdot 2 = 2$ so that $1 \geq 2$, which is absurd because $1 < 2$.

Exercise 2.2.3.   (a) Justify the assertion blithely made above that $5 \nmid 12$.

   (b) Generalize Example 2.2.2 by showing that $n \nmid m$ if m and n are positive integers with $m < n$.

Example 2.2.4. Suppose that the positive integer d divides each of the integers a and b. Then d also divides their sum $a + b$. Let us prove this in detail.

Since $d \mid a$, there is an integer k with $a = kd$. Since also $d \mid b$, there is a (possibly different!) integer l with $b = ld$. Then $a + b = kd + ld = (k + l)d$. Hence $d \mid (a + b)$.

Observe that we did *not* write "$a + b = kd$" at the end of the proof, even though to show that the definition of divisibility is satisfied for d and $a + b$ we need to find such a relation for *some k*. But we already had used k to mean some particular integer, and so we could not reuse it in the same context to mean something else. (If you wish, you can insert before the final sentence of the proof something like "Thus $a + b = k'd$, where $k' = k + l$," or "Thus $a + b = k'd$ for some k'.")

The preceding example is generalized in one of the following basic properties of the divisibility relation.

Proposition 2.2.5. *Let d be a positive integer. Then:*

   1. $d \mid 0$.

   2. $d \mid d$.

   3. *Let n and l be integers with $n > 0$. If $d \mid n$ and $n \mid l$, then $d \mid l$.*

   4. *Let a and b be any integers. If $d \mid a$ and $d \mid b$, then $d \mid (as + bt)$ for all integers s and t.*

   5. *Let a and b be any integers. If $d \mid a$ or $d \mid b$, then $d \mid ab$.*

Proof. Exercise.   □

Exercises 2.2.6.   (1) Deduce from the preceding proposition that a positive integer that divides each of two integers also divides their difference.

(2) If d divides each of two integers, must it divide their product?

(3) Is the converse of part 5 of Proposition 2.2.5 true? (Either prove that it is or else give a counterexample showing that it is not.)

(4) Formulate a statement of the form "If $d \nmid ab$, then ... ." that is equivalent to part 5 of Proposition 2.2.5.

One of the most familiar cases of divisibility is that by 2.

Definition 2.2.7. An integer n is said to be **even** when $2 \mid n$ and **odd** otherwise.

For example, 24 is even whereas 15 is odd. According to the Division Theorem (Theorem 2.1.1), each integer n has exactly one of the two forms $n = 2k$ or $n = 2k + 1$. Hence the following proposition holds.

Proposition 2.2.8. *An integer n is odd if and only if there is an integer k with $n = 2k + 1$.*

In other terms,

$$n \text{ is even} \quad \Longleftrightarrow \quad n \bmod 2 = 0,$$
$$n \text{ is odd} \quad \Longleftrightarrow \quad n \bmod 2 = 1.$$

Exercise 2.2.9.   (a) Show that the sum or difference of two even numbers must be even.

(b) What can you say about the sum or difference of two odd numbers?

(c) What can you say about the sum or difference of an odd number with an even number?

(d) What about products of even numbers? Odd numbers? An even and an odd number?

Exercise 2.2.10.   (a) Prove the "if" part of Proposition 2.2.8 (if $n = 2k + 1$ for some k, then n is odd) without using the Division Theorem.

(b) Prove the "only if" part of Proposition 2.2.8 without using the Division Theorem (but only its method of proof). You may want to flesh out the following skeleton:

It is enough to show the result for positive odd integers n because .... Just suppose there were an odd positive integer n such that $n \neq 2k + 1$ for each integer k. By well-ordering ... .

The next result will be used shortly when we give another proof that $\sqrt{2}$ is irrational.

Lemma 2.2.11. *Let m be an integer. If m is odd, then m^2 is odd.*

Proof. Exercise.   □

Exercise 2.2.12. Complete the following statement so that it is equivalent to the preceding lemma: *If m^2 is even, then . . . .*

A given integer may have many divisors. For example, the divisors of 24 are 1, 2, 3, 4, 6, 8, 12, and 24; the divisors of 36 are 1, 2, 3, 4, 6, 9, 12, 18, and 36. A given pair of integers may have several divisors in common. For example, the common divisors of 24 and 36 are 1, 2, 3, 4, 6, and 12.

Definition 2.2.13. Let m and n be integers that are not both 0. A positive integer d is called a **common divisor** of m and n if $d \mid m$ and $d \mid n$. The greatest integer among all the divisors of m and n is called the **greatest common divisor** of m and n and is denoted $\gcd(m, n)$.

For example,

$$\gcd(24, 36) = 12.$$

Note that $\gcd(m, n)$ is always a divisor of m and n.

The next exercise justifies the (obvious?) existence of a greatest common divisor. Recall that, for a subset A of $\mathbb{N}$ (or of $\mathbb{Z}$ or $\mathbb{Q}$ or $\mathbb{R}$), we call g a **greatest element** of A when $g \in A$ and each $a \in A$ satisfies $a \leq g$. When such a greatest element of a set A exists, it is necessarily unique [see Exercise 1.4.3 (8)], and then we may speak of *the* greatest element of A. We also call such g the **maximum** of A and then denote it by $\max A$. Thus the preceding definition may be stated:

$$\gcd(m, n) = \max\{\, d \in \mathbb{N}^* : d \mid m, d \mid n \,\}.$$

The principle we need, guaranteeing that such sets really have greatest elements, is the following.

Lemma 2.2.14. *Let A be a nonempty subset of $\mathbb{N}$. Suppose there exists some upper bound b of A in $\mathbb{N}$, that is, $b \in \mathbb{N}$ with $a \leq b$ for every $a \in A$. Then A has a greatest element g, and $g \leq b$ for any such upper bound b.*

Proof. Exercise.   □

Exercise 2.2.15.   (a) If n is a nonzero integer and $d \mid n$, show that $d \leq |n|$.

(b) For $b \in \mathbb{N}$, show that b is the greatest element of $\{\, a \in \mathbb{N} : a \le b \,\}$.

(c) Let m and n be any integers that are not both 0. Prove that the set

$$\{\, d \in \mathbb{N}^* : d \mid m, \, d \mid n \,\}$$

of all common divisors of m and n has a greatest element. (This justifies the definition of gcd given above.)

(d) Why, in the definition of the greatest common divisor of m and n, did we impose the restriction that m and n are not both 0?

Exercises 2.2.16. (1) In the notation of the Division Theorem (Theorem 2.1.1) and its proof, suppose the integer m is positive. Show that qn is the greatest multiple kn of n not exceeding m.

(2) Give a new proof of the existence statement of the Division Theorem based upon the preceding observation.

(3) In the notation of the proof of Corollary 2.1.8, show that n is the greatest natural number for which $b^n \le a$.

The next exercises give a bit of practice with the definition of greatest common divisor.

Exercises 2.2.17. (1) Find $\gcd(40, 100)$ and $\gcd(7, 23)$.

(2) Above we found $\gcd(24, 36) = 12$. Now find $\gcd(-24, 36)$ as well as $\gcd(24, -36)$ and $\gcd(-24, -36)$. Then generalize.

(3) What is $\gcd(1, n)$? What is $\gcd(0, n)$?

(4) Show that $\gcd(m, n) = \gcd(n, m)$.

(5) If $n > 0$ and $d \mid n$, what is $\gcd(d, n)$?

(6) Let $m > 0$ and $n > 0$.

    (a) Show that $\gcd(m + n, n) = \gcd(m, n)$.

    (b) Must $\gcd(m - n, n) = \gcd(m, n)$?

    (c) Must $\gcd(m \cdot n, n) = \gcd(m, n)$?

One of the uses of greatest common divisors is reducing fractions to lowest terms. For example, $\gcd(24, 36) = 12$ and

$$\frac{24}{36} = \frac{24/12}{36/12} = \frac{2}{3}.$$

To say that the fraction $2/3$ is in lowest terms means, of course, that its numerator and denominator have no common divisor except 1.

Definition 2.2.18. Integers m and n are said to be **relatively prime** to one another when $\gcd(m, n) = 1$, in other words, when their only common divisor is 1.

In this terminology, 2 is relatively prime to 3, whereas 24 is not relatively prime to 36. The reduction of the fraction $24/36$ to $2/3$ works in general.

Proposition 2.2.19. *Let m and n be integers that are not both 0. Then $m/\gcd(m, n)$ is relatively prime to $n/\gcd(m, n)$.*

Proof. [Note that $m/\gcd(m, n)$ and $n/\gcd(m, n)$ are actually integers since $\gcd(m, n)$ divides both m and n. Thus it is meaningful to ask whether these two quotients are relatively prime integers.]

Let $g = \gcd(m, n)$. Let d be a common divisor of m/g and n/g. We must show $d = 1$. There are integers s and t with $m/g = ds$ and $n/g = dt$. Then

$$m = dgs, \qquad n = dgt,$$

so that dg is a common divisor of m and n. Since g is the greatest such common divisor, necessarily $dg \le g$. But $g \le dg$ because $d \ge 1$. Hence $dg = g$, and so $d = 1$. $\square$

We are now ready to present a simple proof of the irrationality of $\sqrt{2}$ based upon divisibility. (A different proof, using well-ordering, appeared in Example 1.4.6.) Recall that a real number x is said to be **rational** when $x = m/n$ for some integers m and n with $n \ne 0$, and **irrational** otherwise.

Example 2.2.20. The number $\sqrt{2}$ is irrational.

Proof. Just suppose, to the contrary, that $\sqrt{2}$ is rational, that is,

$$\sqrt{2} = \frac{m}{n}$$

for some integers m and n with $n \ne 0$. In view of Proposition 2.2.19—if necessary, replace m and n by their quotients by their greatest common divisor—we may assume without loss of generality that m is relatively prime to n. In particular, m and n are not both even.

Square the above expression for $\sqrt{2}$ to get $2 = m^2/n^2$ so that

$$m^2 = 2n^2.$$

The right-hand side here is even, and so the left-hand side m^2 is even, too. By Lemma 2.2.11, m itself must be even. Thus there is some integer k with

$$m = 2k.$$

Then

$$2n^2 = m^2 = (2k)^2 = 4k^2,$$

and so

$$n^2 = 2k^2.$$

But then n^2 is even, and so again by Lemma 2.2.11, n itself must be even. Thus m and n are both even, and this is a contradiction.   □

Exercises 2.2.21.   (1) Prove that the real number $\sqrt{3}$ is irrational.

(2) What goes wrong if you use the same method as in (1) to try to prove that 4 or 9 is irrational?

(3) Prove: The square root $\sqrt{n}$ of a positive integer n is irrational unless n is a perfect square, that is, $n = r^2$ for some positive integer r.

(4) Prove that $\sqrt{2} + \sqrt{3}$ is irrational.

(5) Prove that $\sqrt[3]{2}$ and $\sqrt[3]{3}$ are irrational.

(6) Prove that the common (base 10) logarithm $\log_{10} 2$ is irrational.

Exercise 2.2.22. According to the usual definition of rational powers of a real number, $a^{1/2} = \sqrt{a}$. We now know that $2^{1/2}$ is irrational. It is interesting to know, more generally, when a^b is rational and when it is irrational.

(a) If a and b are both rational, show by example that a^b can be either rational or irrational.

(b) If a and b are both irrational, show that a^b can nonetheless be rational. (*Hint:* Let $x = \sqrt{2}^{\sqrt{2}}$. Look at x and $x^{\sqrt{2}}$.)

(c) Investigate the rationality of a^b when one of a and b is rational and the other irrational.

Look again at the common divisors 1, 2, 3, 4, 6, and 12 of 24 and 36 and at the greatest of these: $12 = \gcd(24, 36)$. Observe that each of the common divisors divides this greatest one. This phenomenon is true in general, not just for 24 and 36. To prove this, however, we need a bit of machinery. For motivation, notice that the sum of any two multiples of 12 is again a multiple of 12, and any multiple of a multiple of 12 is again a multiple of 12. In other words, the set $\{12k : k \in \mathbb{Z}\}$ has the properties specified in the following definition.

Definition 2.2.23. A subset J of $\mathbb{Z}$ is called an **ideal** in $\mathbb{Z}$ when

$$J \neq \varnothing,$$
$$m \in J, n \in J \implies m + n \in J, \text{ and}$$
$$m \in \mathbb{Z}, n \in J \implies mn \in J.$$

The word "ideal" is used for historical reasons but otherwise is quite arbitrary. You need not think of it as suggesting "perfect" or "ultimate."

Examples 2.2.24.   (1) The single-element set $\{0\}$ is an ideal in $\mathbb{Z}$.

(2) The set of all even integers is an ideal in $\mathbb{Z}$.

(3) More generally, if m is a given integer, then the set $\{km : k \in \mathbb{Z}\}$ of all multiples of m is an ideal in $\mathbb{Z}$.

(4) The set $\{24s + 36t : s, t \in \mathbb{Z}\}$ is an ideal in $\mathbb{Z}$.

   More generally, if m and n are two given integers, then the set

$$\{ms + nt : s, t \in \mathbb{Z}\}$$

of all "linear combinations" of m and n is an ideal in $\mathbb{Z}$. (Be sure to write out the details verifying that this is so!) [Notice that (3) is a special case of this example since we can take, in particular, $n = 0$.]

Here are some simple additional properties of ideals.

Lemma 2.2.25. *Let J be an ideal in $\mathbb{Z}$. Then:*

1. $0 \in J$.

2. If $n \in J$, then $-n \in J$.

3. If $m \in J$ and $n \in J$, then $m - n \in J$.

Proof.   1. Choose some $n \in J$; such exists because J is necessarily nonempty. Then $0 = 0n \in J$.

2. Let $n \in J$. Then $-n = (-1)n \in J$.

3. Let $m, n \in J$. From (2), we have $-n \in J$. Then $m - n = m + (-n) \in J$, too.   □

Exercises 2.2.26.   (1) Which of the following subsets of $\mathbb{Z}$ is an ideal?

(a) The set of all odd integers.

(b) For given integers m, n, and p, the set

$$\{ ms + nt + pu : s, t, u \in \mathbb{Z} \}.$$

(2) Definition 2.2.23 required that the product mn of an element n of J by an arbitrary element m of $\mathbb{Z}$ again belong to J. Suppose this condition is relaxed to require only that $m, n \in J$ implies $mn \in J$. Must J still be an ideal in $\mathbb{Z}$?

(3) Suppose a subset J of $\mathbb{Z}$ has the three properties listed in the Lemma 2.2.25. Must it be an ideal in $\mathbb{Z}$?

(4) Suppose a nonempty subset J of $\mathbb{Z}$ has the property that $m, n \in J$ implies $m + n \in J$ and $m - n \in J$. Must it be an ideal in $\mathbb{Z}$?

It turns out that ideals in $\mathbb{Z}$ actually have a very simple structure.

Theorem 2.2.27 (Principal Ideal Theorem). *If J is an ideal in $\mathbb{Z}$, then either $J = \{0\}$ or else J has a least positive element g. In the latter case, J is the set $\{ kg : k \in \mathbb{Z} \}$ of all multiples of g.*

Proof. Suppose $J \neq \{0\}$. Then J has at least one nonzero element, and so it has at least one positive element (why?). By well-ordering, J must have a least positive element g.

To show that $J = \{ kg : k \in \mathbb{Z} \}$, introduce the set

$$P = \{ kg : k \in \mathbb{Z} \}.$$

We want to show $P = J$. By the definition of an ideal, $P \subset J$. It remains to show the reverse inclusion $J \subset P$.

Let $m \in J$. By the Division Theorem, there are integers q and r with

$$m = qg + r, \qquad 0 \le r < g.$$

Then $r = m - qg \in J$ (why?). Since g is the least positive element of J, this is impossible unless $r = 0$. Thus $0 = m - qg$, and so $m = qg \in P$, as needed. $\square$

The notation g above was used not because g is the greatest element of anything but rather because we might say that g "generates" the ideal J.

Observe that $\gcd(24, 36) = 12$, and $12 = (-1)24 + 1 \cdot 36$. Likewise, $\gcd(45, 75) = 15$, and $15 = 2 \cdot 45 + (-1)75$. Those examples generalize:

Theorem 2.2.28 (Euclid's Divisor Theorem). *Let m and n be any two integers that are not both 0. Then there are integers s and t with*

$$\gcd(m, n) = ms + nt.$$

Proof. Form the ideal J in $\mathbb{Z}$ defined by

$$J = \{\, ms + nt : s, t \in \mathbb{Z} \,\}.$$

By the Principal Ideal Theorem, there is a positive g such that

$$J = \{\, kg : k \in \mathbb{Z} \,\}.$$

Then $g = 1 \cdot g \in J$, and so certainly

$$g = ms + nt$$

for some integers s and t. We claim that g is, in fact, the greatest common divisor of m and n.

Since $m, n \in J$, it follows that g is a divisor of both m and n. To see that g is the greatest of all common divisors of m and n, let d be any such common divisor. There are integers a and b with

$$m = da, \qquad n = db.$$

Then

$$g = ms + nt = das + dbt = d(as + bt).$$

This means $d \mid g$. Hence $d \leq g$. □

Corollary 2.2.29. *Let m and n be two integers that are not both 0. Then every common divisor of m and n divides their greatest common divisor.*

Proof. Extract this from the proof of Euclid's Divisor Theorem just given. Or prove it using the statement of that theorem. □

In view of this corollary, we may now say that $\gcd(m, n)$ is *the unique common divisor of m and n having the property that every common divisor of m and n divides it.*

Exercise 2.2.30. (a) Write a definition for the greatest common divisor, $\gcd(a, b, c)$, of three positive integers a, b, c. Justify the existence of such a greatest common divisor.

(b) Prove that $\gcd(a, b, c) = \gcd(\gcd(a, b), c)$.

(c) Deduce that $\gcd(\gcd(a, b), c) = \gcd(a, \gcd(b, c))$.

(d) Prove that $\gcd(a, b, c)$ is the unique positive integer that is a divisor of a, b, and c and that has the property that every divisor of a, b, and c divides it.

(e) If a, b, and c are arbitrary positive integers, what can you say (and prove) about the quotients $a/\gcd(a,b,c)$, $b/\gcd(a,b,c)$, $c/\gcd(a,b,c)$?

Exercise 2.2.31. In the notation of Euclid's Divisor Theorem, are s and t necessarily unique?

Knowing *that* we can express $\gcd(m,n)$ in the form $ms + nt$ is one thing; finding which s and t work is another. In fact, finding $\gcd(m,n)$ is itself not so simple as soon as m and n are not small. Fortunately, there is a procedure—the **Euclidean Algorithm**—for finding out both things together.

The Euclidean Algorithm is based upon two principles: first, the simple result that

$$\gcd(m,0) = m$$

for any nonzero integer m; second, the following proposition, which tells how to reduce computing the greatest common divisor of two positive integers to computing the gcd of smaller integers. It implies, for example, that $\gcd(110,30) = \gcd(30,20) = 10$.

Proposition 2.2.32. *Let m and n be integers with $0 < n < m$. Then*

$$\gcd(m,n) = \gcd(n, m \bmod n).$$

Proof. Use the Division Theorem to show that the common divisors of m and n are exactly the common divisors of $m \bmod n$ and n. $\square$

Exercise 2.2.33. Use Proposition 2.2.32 together with induction to give a new proof of Euclid's Divisor Theorem (Theorem 2.2.28) that does not involve the machinery of ideals.

The Euclidean Algorithm consists of applying the preceding proposition repeatedly. We first illustrate this algorithm in an example before stating it in general.

Example 2.2.34. Find $\gcd(144375, 21175)$ and express it in the form $21175s + 14475t$ for suitable integers s and t.

Solution. We omit the actual divisions involved, giving just the results. We have successively

$$144375 = 6 \cdot 21175 + 17325,$$
$$21175 = 1 \cdot 17325 + 3850,$$
$$17325 = 4 \cdot 3850 + 1925,$$
$$3850 = 2 \cdot 1925 + 0.$$

Observe that the divisor in each step becomes the dividend in the next step; and the remainder in each step becomes the divisor in the next step.

According to Proposition 2.2.32,

$$\begin{aligned}
\gcd(144375, 21175) &= \gcd(21175, 17325) \\
&= \gcd(17325, 3850) \\
&= \gcd(3850, 1925) \\
&= \gcd(1925, 0).
\end{aligned}$$

But $\gcd(1925, 0) = 1925$. Hence

$$\gcd(144375, 21175) = 1925.$$

Moreover, by starting at the next-to-last equation in the chain of divisions above and working backward, we have:

$$\begin{aligned}
1925 &= 17325 - 4 \cdot 3850 \\
&= 17325 - 4(21175 - 1 \cdot 17325) \\
&= (-4)21175 + 5 \cdot 17325 \\
&= (-4)21175 + 5(144375 - 6 \cdot 21175) \\
&= 5 \cdot 144375 + (-34)21175.
\end{aligned}$$

Hence

$$\gcd(144375, 21175) = 5 \cdot 144375 + (-34)21175.$$

The procedure just illustrated works in general.

Algorithm 2.2.35 (Euclidean Algorithm). *Let m and n be given integers with $0 < n < m$. Successively define pairs $(q_1, r_1), (q_2, r_2), \ldots$ with*

$$\begin{aligned}
m &= nq_1 + r_1, & 0 &\leq r_1 < n, \\
n &= r_1 q_2 + r_2, & 0 &\leq r_2 < r_1, \\
r_1 &= r_2 q_3 + r_3, & 0 &\leq r_3 < r_2, \\
&\ \ \vdots & &\ \ \vdots \\
r_{s-2} &= r_{s-1} q_s + r_s, & 0 &\leq r_s < r_{s-1}, \\
r_{s-1} &= r_s q_{s+1} + r_{s+1}, & 0 &= r_{s+1}.
\end{aligned}$$

Then the greatest common divisor of m and n is the final nonzero remainder:

$$\gcd(m, n) = r_s.$$

The Euclidean Algorithm[4] really works: The existence of the successive pairs (q_j, r_j) of quotients and remainders is guaranteed by the Division Theorem. The procedure really terminates in a finite number $s + 1$ of steps with a zero remainder because the remainder at each step must be at least 1 less than the remainder from the preceding step:

$$0 \le r_{s+1} < r_s < \cdots < r_2 < r_1 < n.$$

Finally, the last nonzero remainder r_s is actually the greatest common divisor according to Proposition 2.2.32.

Exercise 2.2.36. Use the Euclidean Algorithm to find each gcd and to express each as a linear combination $ms + nt$ of the given integers m and n:

(a) $\gcd(156, 42)$

(b) $\gcd(77760, 69800)$

As stated, the Euclidean Algorithm covers only the case $0 < n < m$. It is easy to reduce all other cases to this case when m and n are not both 0:

- If $n < 0$, then let $n = |n|$; if $m < 0$, then let $m = |m|$.

- If $m < n$, then interchange m and n (now $0 \le n \le m$).

- If $m = n$ or $n = 0$, then $\gcd(m, n) = m$; otherwise, $0 < n < m$, and then apply the Euclidean Algorithm as above.

With these reductions, an iterative function for computing greatest common divisors might be coded in Pascal as follows.

```
function gcd (m, n :  Integer) :  Integer;
   var
      r :  Integer;
   begin
      if n < 0 then n := Abs (n);
      if m < 0 then m := Abs (m);
      if m < n then Swap (m, n);
      (* now 0 <= n <= m *)
      if (m = n) or (n = 0)
         then gcd := m
      else (* now 0 < n < m *)
         begin
```

---

[4]Sometimes what we called Euclid's Divisor Theorem is itself referred to as the Euclidean Algorithm.

```
        r := m mod n;
        while r > 0 do
            begin
                m := n; n := r;
                r := m mod n
            end; (* while *)
            gcd := n
        end; (* else clause of if *)
    end; (* gcd *)
```

The preceding Pascal function requires that you define a procedure Swap, with two var integer parameters, that interchanges its two parameters. It assumes the Pascal version in use has built-in div and mod operators along with a built-in Abs function for finding absolute values; if any of these are not built in, you will, of course, need to define them yourself. Moreover, in order for the program to be more than marginally useful, you will need to use a version of Pascal whose Integer data type—or some similar data type such as Longint—can handle integers with more than a few digits.

One way to avoid the integer size limitation in Pascal or other traditional programming languages is to use a computer algebra system such as DERIVE or MATHEMATICA that allows arbitrary precision numbers. For example, here is a MATHEMATICA rendering of the Euclidean Algorithm:

```
mygcd[m_Integer, n_Integer] :=
    Module[{a = m, b = n, r},
        If[b < 0, b = Abs[b]];
        If[a < 0, a = Abs[a]];
        If[a < b, {a, b} = {b, a}];
        If[(a == b) || (b == 0), (* then *) a,
            (* else *)
            r = Mod[a, b];
            While[r > 0,
                a = b; b = r; r = Mod[a, b]
            ];
            Return[b]
        ]
    ]
```

Along with the iterative form of the Euclidean Algorithm described above and realized in the displayed Pascal and MATHEMATICA functions, there is a recursive form. This form is based upon Proposition 2.2.32. Here it is for finding the greatest common divisor when $0 \le n < m$:

$$gcd(m, n) = \begin{cases} m & \text{if } n = 0, \\ gcd(n, m \bmod n) & \text{if } n > 0. \end{cases}$$

Exercises 2.2.37.   (1) Explain why the preceding recursive form of the Euclidean Algorithm works even if $m \leq n$.

(2) Finish writing a complete program incorporating the iterative Pascal function gcd presented above (or its translation into another programming language).

Test your program on the examples and exercises given earlier.

(3) Code into a programming language the *recursive* form of the function gcd just displayed. Write and test a complete program incorporating it. (Don't forget to handle the other cases, too.)

(4) Formulate an algorithm that, for given integers m and n not both 0, determines integers s and t with $gcd(m, n) = ms + nt$. See the Euclidean Algorithm 2.2.35 and Example 2.2.34.

Implement and test your algorithm in a programming language.

In Section 2.3, prime numbers are used to give a different approach to finding greatest common divisors.

Euclid's Divisor Theorem (Theorem 2.2.28) has a number of significant consequences about divisibility. For the first of these, note that $d \mid ab$ does *not* in general imply that $d \mid a$ or $d \mid b$ (example?).

Proposition 2.2.38. *Let a and b be integers that are not both 0 and let d be a positive integer. If a is relatively prime to d and if $d \mid ab$, then $d \mid b$. In symbols,*

$$gcd(a, d) = 1 \; \& \; d \mid ab \implies d \mid b.$$

Proof. Assume $gcd(a, d) = 1$ and $d \mid ab$. There are integers s and t with

$$1 = as + dt.$$

Multiply by b:

$$b = (ab)s + d(bt).$$

By assumption, $d \mid ab$, and so d divides the first term on the right; certainly d divides the second term. Hence d divides the sum b.   □

For example, if $10 \mid ab$ but neither 2 nor 5 divides a, then $10 \mid b$.

The general result in Proposition 2.2.38 may seem obvious, but its proof here is surprisingly deep, requiring the apparatus of greatest common divisors and ideals in $\mathbb{Z}$ along with the Principal Ideal Theorem and Euclid's Divisor Theorem.

In general, if two integers divide a given integer, their product need not divide that integer (example?). However:

Corollary 2.2.39. *Let a and b be relatively prime positive integers and let n be any integer. If $a \mid n$ and $b \mid n$, then $(ab) \mid n$.*

Proof. Exercise. □

For example, if $10 \mid n$ and $9 \mid n$, then $90 \mid n$.

Here are some additional problems involving greatest common divisors.

Exercises 2.2.40. (1) Let $k > 0$. Prove or disprove: $\gcd(km, kn) = k\gcd(m, n)$. (Of course, for this to be meaningful we must suppose also that m and n are not both 0. We shall tacitly make such a supposition whenever we refer to the greatest common divisor of two integers.)

(2) Suppose $\gcd(m, n) = 1$. Deduce that $\gcd(m - n, m + n) = 1$ or 2.

(3) Suppose $\gcd(n, k) = 1$. Show that $\gcd(m, nk) = \gcd(m, n) \cdot \gcd(m, k)$.

Exercise 2.2.41. For two positive integers m and n, their *least common multiple*, denoted $\operatorname{lcm}(m, n)$, is the least positive integer that is a multiple of both. For example, $\operatorname{lcm}(12, 18) = 36$. When the two integers appear as denominators of two fractions, their least common multiple is what is also known as the fractions' *lowest common denominator* and is used to add the fractions; for example,

$$\frac{7}{12} + \frac{5}{18} = \frac{3 \cdot 7}{36} + \frac{2 \cdot 5}{36} = \frac{31}{36}.$$

(a) Explain why such a least common multiple must exist.

(b) If $m \mid a$ and $n \mid a$, show that $\operatorname{lcm}(m, n) \mid a$.

(c) Prove that $\operatorname{lcm}(m, n)$ is the unique common multiple of m and n that divides each common multiple.

(d) Prove: $\operatorname{lcm}(m, n) \cdot \gcd(m, n) = mn$.

(e) Prove: $\gcd(m, n) = \gcd(m + n, \operatorname{lcm}(m, n))$.

The next two exercises build toward determining all *pythagorean triples*—positive integers a, b, c satisfying $a^2 + b^2 = c^2$ (see page 61). One of the tools used is the greatest common divisor $\gcd(a,b,c)$ of three positive integers as introduced in Exercise 2.2.30.

Exercise 2.2.42. Say that three positive integers a, b, c are *mutually prime* when their only common divisor is 1, that is, when $\gcd(a,b,c) = 1$. Say that they are *pairwise relatively prime* when each pair of them is relatively prime, that is, when $\gcd(a,b) = \gcd(b,c) = \gcd(a,c) = 1$.

(a) Show that if a, b, c are pairwise relatively prime, then they are mutually prime. Give an example of three positive integers that are mutually prime but not pairwise relatively prime.

(b) Show that the quotients

$$a/\gcd(a,b,c), \quad b/\gcd(a,b,c), \quad c/\gcd(a,b,c)$$

of three positive integers a, b, c by their greatest common divisor must be mutually prime. Need the quotients be pairwise relatively prime?

(c) Let (a,b,c) be a pythagorean triple. Show that

$$\gcd(a,b) = \gcd(a,c) = \gcd(b,c) = \gcd(a,b,c).$$

(d) Show that positive integers a, b, c forming a pythagorean triple are mutually prime if and only if they are pairwise relatively prime.

Call a pythagorean triple (a,b,c) *primitive* when a, b, c are mutually prime—hence pairwise relatively prime.

(e) Show that if (a,b,c) is a primitive pythagorean triple and if g is an arbitrary positive integer, then (ga,gb,gc) is also a pythagorean triple. Show, conversely, that every pythagorean triple has the form (ga,gb,gc), where (a,b,c) is a primitive pythagorean triple.

According to (e), the "general" pythagorean triple is (ga,gb,gc), where (a,b,c) is an arbitrary primitive pythagorean triple. Thus to determine all pythagorean triples we need only find all primitive pythagorean triples.

(f) If (a,b,c) is a primitive pythagorean triple, show that a and b cannot both be even and cannot both be odd.

The determination of pythagorean triples begun above is concluded later, in Exercise 2.3.16.

2.3   Prime Numbers

Prime numbers are, with respect to multiplication, the building blocks
of all the positive integers. So states the **Fundamental Theorem of
Arithmetic**, which is proved in this section.

A prime number is an integer larger than 1 that cannot be written as
the product of two smaller positive integers. For example, 3 and 29 are
prime, whereas $21 = 3 \cdot 7$ and $16 = 4 \cdot 4$ are not. The formal meaning
of primality is included in the following definition.

Definition 2.3.1. An integer p is said to be **prime** if $p > 1$ and if the
only divisors of p are 1 and p itself. An integer $n > 1$ is said to be
composite if it is not prime.

Is 1 prime? According to our definition, definitely not: we explicitly
excluded it. The exclusion was, to some extent, arbitrary, although we
shall see below how convenient the exclusion can be.

People sometimes argue over whether 1 is or is not a prime. The
grounds for their argument are slightly differing definitions of "prime."
Perhaps they define a positive integer p as being prime by stating that
the only divisors of p are 1 and p itself. According to that definition,
yes, 1 is a prime. According to our definition, no, it is not. Again, it is
just a convention dictated by convenience.

According to our definition, each positive integer either is 1 or else
is prime or else is composite. In other words, the set $\mathbb{N}^*$ of all positive
integers is partitioned into three mutually exclusive subsets:

$$\mathbb{N}^* = P \cup C \cup \{1\},$$
$$P \cap C = \varnothing, \quad P \cap \{1\} = \varnothing, \quad C \cap \{1\} = \varnothing,$$

where

$$P = \{\, p : p \text{ is prime} \,\},$$
$$C = \{\, n : n \text{ is composite} \,\}.$$

Exercises 2.3.2.   (1) Restate the definition for p to be prime in a way
that explicitly uses the "divides" relation ($|$) and explicitly in-
cludes the quantifiers (in words or in symbols) appearing implicitly in the preceding definition.

(2) A composite integer $n > 1$ has the form $n = ab$ for some integers
a and b with $1 < a < n$, $1 < b < n$. Must a and b be unique?

(3) Formulate a reasonable definition for a negative integer to be
prime. (Note, however, that we shall *never* refer to any prime
numbers except those that are positive.)

(4) Find the first 25 prime numbers and the first 25 composite numbers.

(5) Prove that, for a positive integer k, the number $k^3 + 1$ is prime if and only if $k = 1$.

(6) If p and q are primes with $p \neq q$, then what is $\gcd(p, q)$?

If a nonprime integer $n > 1$ is a product $n = ab$ with $1 < a < n$, $1 < b < n$, then necessarily $a \leq \sqrt{n}$ or $b \leq \sqrt{n}$ (why?). Hence:

To test whether a given integer $n > 1$ is prime, we need only test for divisors not exceeding $\sqrt{n}$.

For example, $\sqrt{61} = 7.81024\ldots$, and so the only possible divisors we need to test are $2, 3, 4, 5, 6, 7$; since none of these six integers divides 61, then 61 is prime.

An effective method for generating at least relatively small primes is known as the **sieve of Eratosthenes.**[5] Suppose we want to determine all primes not exceeding 100. Begin by listing all the integers from 1 to 100:

1	2	3	4	5	6	7	8	9	10
11	12	13	14	15	16	17	18	19	20
21	22	23	24	25	26	27	28	29	30
31	32	33	34	35	36	37	38	39	40
41	42	43	44	45	46	47	48	49	50
51	52	53	54	55	56	57	58	59	60
61	62	63	64	65	66	67	68	69	70
71	72	73	74	75	76	77	78	79	80
81	82	83	84	85	86	87	88	89	90
91	92	93	94	95	96	97	98	99	100

Remove 1 from the list because it is not prime. Then 2 is the first prime. Now remove from the remaining list all multiples of 2 other than 2 itself (since they cannot be primes):

	2	3	5	7	9
11		13	15	17	19
21		23	25	27	29
31		33	35	37	39
41		43	45	47	49
51		53	55	57	59
61		63	65	67	69
71		73	75	77	79
81		83	85	87	89
91		93	95	97	99

[5] Named after the third-century B.C. Alexandrian scholar Eratosthenes.

Then 3 is the second prime. Now remove all remaining multiples of 3 other than 3 itself (the even multiples of 3 have already been removed):

2	3	5	7	
11	13		17	19
	23	25		29
31		35	37	
41	43		47	49
	53	55		59
61		65	67	
71	73		77	79
	83	85		89
91		95	97	

Then 5, the next remaining number after the primes 2 and 3, is the third prime. (Notice how quickly the list is being thinned out.) We do not need to remove multiples of 4, since they are multiples of 2, which have already been removed. Now remove all remaining multiples of 5 other than 5 itself:

2	3	5	7	
11	13		17	19
	23			29
31			37	
41	43		47	49
	53			59
61			67	
71	73		77	79
	83			89
91			97	

Then 7 is the fourth prime. We do not need to remove multiples of 6 (why not?). Now remove all remaining multiples of 7 other than 7 itself; note that we need remove only those remaining multiples $7k$ for $k \geq 7$ (why?):

2	3	5	7	
11	13		17	19
	23			29
31			37	
41	43		47	
	53			59
61			67	
71	73			79
	83			89
			97	

By now the way to continue this process should be clear. But, we know, it need be continued only up to removing multiples of m for $m \le 10 = \sqrt{100}$. And since we have already removed multiples of 8, 9, and 10 while removing multiples of 2 and 3, we are done! The remaining numbers

$$2, 3, 5, 7, 11, 13, 17, 19, 23, 29, 31, 37, 41, 43, 47, 53, 59, 61,$$
$$67, 71, 73, 79, 83, 89, 97$$

are the first 25 primes—all those primes not exceeding 100.

When applying the sieve of Eratosthenes, no multiplication or other arithmetic operation is actually needed—just counting! Thus, to remove the nonprime multiples of 2, delete every second number after 2 from the *original* list; to remove the nonprime multiples of 3, delete every third number after 3 from the *original* list; and so on.

Exercises 2.3.3.   (1) Apply the sieve of Eratosthenes to find all primes not exceeding 200.

(2) Write a computer program that implements the sieve of Eratosthenes to find all primes not exceeding a user-specified n (be sure to take advantage of some of the shortcuts used in the example above).[6] Apply your program to find all the primes not exceeding 1000; primes not exceeding 10000; and primes not exceeding 100000.

If you really want to search for primes, a powerful computer algebra system such as MATHEMATICA allowing arbitrary precision integers is nearly indispensable. In fact, MATHEMATICA includes, among other goodies, built-in functions for testing whether a given integer is prime and for finding the nth prime. Use of these functions is illustrated in Figure 2.2.

We now aim toward proving the Fundamental Theorem of Arithmetic: Each integer greater than 1 can be factored (in an essentially unique way) into a product of primes. One way to start is to show that such an integer that is not already prime has at least one prime divisor.

Proposition 2.3.4. *Each integer $n > 1$ has a prime divisor.*

Proof. Exercise. Use *strong* induction. Take as the inductive assumption that every integer k with $1 < k < n$ has a prime divisor. (Why would ordinary induction be inappropriate here, whereas strong induction works?)   □

---

[6]For additional efficiency, do not actually remove the composite numbers at each step. Rather, create a list of n 1s; at each step, change the 1 to a 0 in the position corresponding to each newly found composite number. After the final step, 1s mark the primes, and 0s the composite numbers.

```
In[1]:=
    PrimeQ[1234567]
Out[1]=
    False
In[2]:=
    Divisors[1234567]
Out[2]=
    {1, 127, 9721, 1234567}
In[3]:=
    Prime[100000] (* the 100,000th prime *)
Out[3]=
    1299709
In[4]:=
    NextPrime[n_Integer] :=
        Module[{k = n},
            While[!PrimeQ[k], k++]; Return[k]
        ]
In[5]:=
    NextPrime[1234567]
Out[5]=
    1234577
```

Figure 2.2: Using MATHEMATICA to find some primes.

Exercises 2.3.5.   (1) Show that each composite integer $n > 1$ has, in fact, some prime divisor $p \le \sqrt{n}$.

(2) Prove Proposition 2.3.4 again but by using the Well-Ordering Principle. For given $n > 1$, consider the set of all divisors of n greater than 1.

Of course if n is prime, then n itself is its *only* prime divisor. Otherwise, if n is not prime, it has some prime divisor p_1. Then $n = p_1 m_1$ for some $m_1 > 1$. If m_1 is prime, then n is already a product of two (not necessarily distinct) primes. Otherwise, if m_1 is not prime, it in turn has some prime divisor p_2. Then $m_1 = p_2 m_2$ for some $m_2 > 1$, and so $n = p_1 p_2 m_2$. We can continue this process until we obtain n completely factored into a product of primes.

Exercises 2.3.6.   (1) Apply the process just described to factor each of the following seven numbers into a product of primes: 273, 1617, 11, 111, 1111, 11111, 1234567.

(2) Formulate the process carefully as an algorithm.

The factoring process just described makes plausible the existence part of the Fundamental Theorem of Arithmetic (Theorem 2.3.9, below): Each integer greater than 1 can be factored into a product of primes.

Look at this example of a factorization into primes:

$$936 = 2 \cdot 2 \cdot 2 \cdot 3 \cdot 3 \cdot 13.$$

By simply rearranging the factors, we can get a number of forms that are different yet are essentially the same factorization. For example,

$$936 = 2 \cdot 3 \cdot 2 \cdot 13 \cdot 3 \cdot 2$$
$$= 13 \cdot 3 \cdot 3 \cdot 2 \cdot 2 \cdot 2.$$

(How many actually different such forms are there in this example? Note that one 2 is the same as any other 2—one is not colored red and the other blue—whereas 2 is distinct from 3.)

The uniqueness part of the Fundamental Theorem of Arithmetic is that the factorization of a given integer $n > 1$ into primes is essentially unique in the following sense: The prime factors involved and the number of times each prime appears as a factor are unique. The order in which the primes are listed in the factorization definitely is not unique (except when n is already a prime number). We express this by saying that the prime factors are *unique up to order*.

How should we prove that the prime factors of a given number n are unique up to order? Consider a simple special case: $n = p_1 p_2$ and $n = q_1 q_2$ for primes p_1, p_2, q_1, q_2. Since $p_1 \mid n$, then $p_1 \mid q_1 q_2$; we wish to deduce that $p_1 = q_1$ or $p_1 = q_2$.

Recall that a positive integer d can divide the product $m \cdot n$ of two integers without dividing either of the two. However, as we have found (in Proposition 2.2.38), if $d \mid m \cdot n$ and d is relatively prime to one of the two numbers m and n, then d must divide the other. We are interested in the case where the divisor d is a prime.

Proposition 2.3.7 (Prime Divisor Property). *Let p be a prime and let m and n be any integers. If $p \mid m \cdot n$, then $p \mid m$ or $p \mid n$.*

Proof. Exercise: deduce this as a special case of Proposition 2.2.38. □

In particular, if a prime p divides the product of two primes, then it must be one of these two primes. More generally, if a prime divides a product [7] of primes, then it must be one of these primes.

Lemma 2.3.8. *Let p and $q_1, q_2, \ldots, q_n$ be primes. If $p \mid \prod_{j=1}^{n} q_j$, then $p = q_j$ for some j with $1 \le j \le n$.*

Proof. Exercise. □

At last we are ready for the big result.

---

[7] For the product notation $\prod$, see Exercise 1.3.5.

Theorem 2.3.9 (Fundamental Theorem of Arithmetic). *Each integer* $n > 1$ *is a product of one or more (not necessarily distinct) primes. For a given n, these primes are unique up to order.*

Proof. *Existence.* Use *strong* induction on n.

Uniqueness. We use *strong* induction on n. There is nothing to prove if $n = 2$. Let $n > 2$ and assume that, for each integer $1 < k < n$, the primes in the factorization of k are unique up to order.

Suppose n has two factorizations,

$$n = p_1 p_2 \cdots p_s, \qquad (*)$$
$$n = q_1 q_2 \cdots q_t,$$

for some positive integers s and t with the p_i and the q_j being primes. We may assume without loss of generality (why?) that

$$p_1 \le p_2 \le \cdots \le p_s, \qquad (**)$$
$$q_1 \le q_2 \le \cdots \le q_t.$$

We make this assumption because now we need only prove that $s = t$ and $p_i = q_i$ for each $i = 1, 2, \ldots, s$.

First, we show that $p_1 = q_1$. (No, we are *not* beginning another induction!) If not, then $p_1 < q_1$ or $q_1 < p_1$—say, $p_1 < q_1$. Then $p_1 < q_j$ for each $j = 1, 2, \ldots, t$. Now $p_1 \mid n$, and so $p_1 \mid \prod_{j=1}^{t} q_j$. From Lemma 2.3.8, $p_1 = q_k$ for some $k \ge 1$. Now $p_1 \ne q_1$ by assumption, and so $p_1 = q_k$ for some $k \ge 2$. But $p_1 < q_1$, so $q_k < q_1$. This contradicts our assumption in $(**)$ about the ordering of the q_j.

Divide n by its factor p_1; according to what we have just established, the quotient is the same as the quotient of n by q_1. From $(*)$, we are left with

$$\prod_{i=2}^{s} p_i = \prod_{j=2}^{t} q_j.$$

This product is less than n. From the inductive assumption, it follows that $s = t$ and $p_i = q_i$ for each $i = 2, \ldots, s$. And we have already established that $p_1 = q_1$. $\square$

Exercise 2.3.10. Deduce Proposition 2.3.4 (each integer $n > 1$ has a prime divisor) as a special case of the Fundamental Theorem of Arithmetic.

As the proof of uniqueness indicates, the factorization of a given $n > 1$ into primes is unique if we insist that the primes be arranged in nondecreasing order. In other words, for a given $n > 1$, there is a unique s and unique primes $p_1, p_2, \ldots, p_s$ such that

$$n = p_1 p_2 \cdots p_s,$$
$$p_1 \leq p_2 \leq \cdots \leq p_s.$$

This is usually what we mean when we speak of *the* **prime factorization** (PF) of an $n > 1$. For example, the prime factorization of 236 is

$$236 = 2 \cdot 2 \cdot 2 \cdot 3 \cdot 3 \cdot 13. \tag{PF}$$

Another way of writing the prime factorization is as a product of positive powers of primes:

$$236 = 2^3 \, 3^2 \, 13^1. \tag{PF1}$$

If we allow exponent 0, too, then we can include the primes that are missing from among those in the preceding factorization:

$$236 = 2^3 \, 3^2 \, 5^0 \, 7^0 \, 11^0 \, 13^1. \tag{PF2}$$

Either one of these two forms, (PF1) or (PF2), is referred to as **the prime power representation** of 236.

Exercises 2.3.11.   (1) Give the prime factorization of each of the numbers in Exercise 2.3.6 (1).

(2) Give the prime power representation—in both forms (PF1) and (PF2)—of each of these numbers.

(3) State the prime power representation—in both forms (PF1) and (PF2)—as a general, precisely worded existence and uniqueness theorem.

The next exercises are intended to help you appreciate the significance of the Fundamental Theorem of Arithmetic's uniqueness assertion. They contrast the role of ordinary primes among all positive integers with the roles of the analogous numbers in two other number systems.

Exercises 2.3.12.   (1) Let E be the set of *even* positive integers. (Then E is *closed* under ordinary multiplication, that is, the product of any two members of E is again a member of E. However, the identity 1 for multiplication does not belong to E.)

Say that a number $p \in E$ is "E-prime" when p has no divisors that belong to E and are smaller than p. Thus the first two E-primes are 2 and 6.

(a) Find the first 20 numbers that are E-prime.

(b) Find a number in E that can be written as a product of E-primes in two essentially different ways.

(c) Is the analog of the Prime Divisor Property (Proposition 2.3.7) true in the number system E?

(2) Let H be the set of all integers of the form $4n+1$ for $n = 0, 1, 2, \ldots$. [Then H is closed under multiplication (why?). Moreover, the identity 1 for multiplication does belong to H.]

Say that a number $p \in H$ is "H-prime" when $p > 1$ and p has no divisors that belong to H and are smaller than p. Thus the first two H-primes are 5 and 9.

Repeat (1)(a)–(c) for H instead of E.

As promised in Section 2.2, we indicate (in the following exercise) how primes can be used to calculate a greatest common divisor in yet another way.

Exercise 2.3.13.   (a) Let

$$m = \prod_p p^{e_p}, \qquad n = \prod_p p^{f_p}$$

be the prime power representations of integers $m > 1$, $n > 1$. Show that the prime power representation of $\gcd(m, n)$ is

$$\gcd(m, n) = \prod_p p^{\min\{e_p, f_p\}}.$$

(By eliding subscripts on the primes in these products, we are tacitly extending the product notation $\prod$ introduced in Exercise 1.3.5. If p_k is the largest prime divisor of m, then we abbreviate $m = \prod_{j=1}^{k} p_j^{e_j}$ by $m = \prod_p p^{e_p}$. Of course, the product does not really involve infinitely many prime power factors: we understand that $e_p = 0$, and so $p^{e_p} = 1$, whenever p does not divide m.)

(b) Use this representation to find $\gcd(11250, 12150)$.

Exercises 2.3.14.   (1) What can you say about the prime power representation of n when n is a square?

(2) Suppose positive integers m and n are relatively prime. If their product $m \cdot n$ is a square, prove that m and n themselves must be squares. Is this implication true if m is not relatively prime to n?

(3) If p is prime, show that $\sqrt{p}$ is irrational.

Exercise 2.3.15. For a positive integer n, let $\sigma(n)$ denote the sum of its *proper* divisors, that is, those divisors d of n for which $d < n$. For example, $\sigma(12) = 16$ because the proper divisors of 12 are 1, 2, 3, 4, and 6; likewise, $\sigma(52) = 46$ because the proper divisors of 52 are 1, 2, 4, 13, and 26.

A positive integer $n > 1$ is said to be a *perfect number* if $\sigma(n) = n$— the sum of the proper divisors of the number is the number itself. For example, 6 is perfect because the proper divisors of 6 are 1, 2, and 3, which sum to 6.

  (a) Show that 28 and 496 are also perfect numbers. Find the prime power representation of each.

  (b) Observe that each of the perfect numbers 6, 28, and 496 has the form $2^m p$, where m is a positive integer and p is a prime of the form $p = 2^m - 1$. Show that each integer of such a form is a perfect number.

  (c) Find another perfect number of the form $2^m p$ with p a prime of the form $p = 2^m - 1$.

  (d) (*A programming problem*) Find all perfect numbers not exceeding 10,000.

Each of the perfect numbers in the preceding exercise was even. Are there any odd perfect numbers? How many perfect numbers are there—finitely many or infinitely many? These are open questions of long standing. As of June, 1995, the answers were not known!

Exercise 2.3.16. This exercise concludes the determination of all pythagorean triples begun in Exercise 2.2.42. Recall that positive integers a, b, c form a pythagorean triple when $a^2 + b^2 = c^2$. Recall, also, that each pythagorean triple has the form (ga, gb, gc), where g is a positive integer and (a, b, c) is a pythagorean triple that is *primitive* in the sense that a, b, and c have no common divisor except 1.

  (a) Let (a, b, c) be a primitive pythagorean triple. Then a and b cannot both be even and cannot both be odd (see Exercise 2.2.42). Without loss of generality, assume a is odd and b is even. Deduce that c must be odd.

   Let $u = (c + a)/2$ and $v = (c - a)/2$, so that $u + v = c$ and $u - v = a$. Show that $\gcd(u, v) \mid c$ and $\gcd(u, v) \mid a$. Deduce that u and v must be relatively prime. Check that $u \cdot v = (b/2)^2$. By Exercise 2.3.14(2), $u = r^2$ and $v = s^2$ for positive integers r and s.

Conclude that the primitive pythagorean triple (a, b, c) has the form

$$a = r^2 - s^2, \quad b = 2rs, \quad c = r^2 + s^2, \tag{*}$$

where r and s are integers with $r > s > 0$, $\gcd(r, s) = 1$, and one of r and s is odd whereas the other is even.

(b) In the preceding notation, what are r and s for the primitive pythagorean triple $(3, 4, 5)$? For the primitive pythagorean triple $(5, 12, 13)$?

(c) If a, b, c are positive integers having the form (*) for integers r and s such that $r > s > 0$, $\gcd(r, s) = 1$, and one of r and s is odd whereas the other is even, prove that the triple (a, b, c) is pythagorean. Must such a pythagorean triple be primitive?

(d) Construct several primitive pythagorean triples besides $(3, 4, 5)$ and $(5, 12, 13)$.

The Fundamental Theorem of Arithmetic says that the prime numbers are, with respect to multiplication, the building blocks of all the positive integers. Are they also the building blocks of the positive integers with respect to addition—that is, is every positive integer $n > 1$ a sum of primes? In 1742, C. Goldbach conjectured that every integer $n > 1$ is a sum of at most three primes—in fact, every even integer $n \geq 6$ is the sum of exactly two odd primes and every odd integer $n \geq 9$ is the sum of exactly three odd primes. In 1930, L. Schnirelmann proved that every integer $n > 1$ is the sum of at most 800,000 primes; by 1937, the bound 800,000 had been successively lowered to 67. Also in 1937, I. M. Vinogradov proved that every sufficiently large odd integer is the sum of three odd primes. Recent computer experimentation has verified the part of the Goldbach Conjecture about even integers up to 20 billion, but still this conjecture remains an open question.

Exercise 2.3.17. (a) Explain the restrictions $n \geq 6$ and $n \geq 9$ in the statement of the Goldbach Conjecture above.

(b) Check the Goldbach Conjecture for integers up to 100.

(c) Show that if the part of the Goldbach Conjecture about even integers is true, then so is the part about odd integers.

How many primes are there? Since there are infinitely many positive integers, and since each positive integer greater than 1 has a prime factorization, it seems plausible there must be infinitely many primes.

(You might try to write a careful proof based directly upon the Fundamental Theorem of Arithmetic. It can be done, but it may give you some trouble.)

What, exactly, is meant by saying that the set of primes—or any set, for that matter—is infinite? This question is answered in detail in Chapter 4. For now, we shall simply call a set **infinite** if it is not finite. And we shall call a set S **finite** if it either is empty or else has the form $S = \{x_1, x_2, \ldots, x_n\}$ for some positive integer n and some distinct elements $x_1, x_2, \ldots, x_n \in S$.

The following proof of the infinitude of the primes appears (with different language) in Euclid's *Elements*. The underlying idea is a procedure for obtaining a new prime from any number n of primes you already have.

Theorem 2.3.18 (Euclid's Theorem). *There are infinitely many primes.*

Proof. Just suppose there were only finitely many primes. The set P of all primes is certainly not empty, and so P has the form

$$P = \{p_1, p_2, \ldots, p_n\}$$

for some positive integer n. In other words, $p_1, p_2, \ldots, p_n$ are, by supposition, *all* the prime numbers there are. (These *could* be arranged in increasing order so that $p_1 = 2$, $p_2 = 3$, etc., but that is not needed for the proof.) Form the number

$$M = p_1 p_2 \cdots p_n + 1.$$

This number M has some prime divisor p. Now $p_j \nmid M$ for each $j = 1, 2, \ldots, n$ (why?), and so $p \neq p_j$ for each such j. Then $p > p_j$ for every $j = 1, 2, \ldots, n$, and so $p \notin P$. This is a contradiction. □

Observe that, in the proof of Euclid's Theorem, we did *not* claim that the number M itself is a prime.

Exercises 2.3.19. (1) In the notation of the preceding proof, show that M need not be a prime. More precisely, if $p_1, p_2, \ldots, p_n$ are the first n primes, show that $p_1 p_2 \cdots p_n + 1$ need not be prime.

(2) Use the *method* of the proof of Euclid's Theorem—of obtaining a new prime from any n given primes—to generate 20 primes starting with just $p_1 = 2$.

(3) As suggested by the proof of Euclid's Theorem, define a sequence $(E_n)_{n=1,2,\ldots}$ recursively by

$$E_1 = 2,$$
$$E_{n+1} = E_1 \cdots E_n + 1 \qquad (n \geq 1).$$

    (a) Calculate $E_1, E_2, \ldots, E_7$.

    (b) Show that E_n need not be a prime.

    (c) Prove that gcd $(E_m, E_n) = 1$ whenever $m \neq n$.

    (d) For each n, let p_n be the smallest factor of E_n greater than 1. Prove that $p_1, p_2, \ldots$ is a sequence consisting of infinitely many distinct primes.

    (e) Show that the sequence $p_1, p_2, \ldots$ does not include every prime.

(4) Give a new proof of Euclid's Theorem by fleshing out the following skeleton. For $n \geq 1$, let $F_n = n! + 1$. For each n, the number F_n has some prime divisor; select some such prime divisor and denote it by q_n. Then $(q_n)_{n=1,2,\ldots}$ is a sequence of infinitely many distinct primes.

(5) Let $(p_n)_{n=1,2,\ldots}$ be the sequence of all primes arranged in increasing order. Prove:

    (a) For each k, the kth prime $p_k \leq \prod_{j=1}^{k-1} p_j + 1$.

    (b) For each k, the kth prime $p_k \leq \prod_{j=1}^{k-1} p_j - 1$.

As Exercise 2.3.19 (3)(d) indicates, it is fairly straightforward to create sequences consisting of infinitely many distinct primes, in other words, including distinct primes and only primes. Moreover, it is trivial to produce a sequence that includes all the primes—the sequence $1, 2, 3, 4, \ldots$ of all positive integers is one such—but, of course, this includes all composite numbers as well. Yet nobody knows an algebraic formula for generating a sequence consisting of all the primes and only primes!

Exercise 2.3.20. Prove that a nonconstant polynomial with integer coefficients cannot have only prime numbers as its values at integers. [*Hint:* Let c be the constant term of such a polynomial $p(x)$ having only prime values. Write $p(x)$ in the form $p(x) = xq(x) + c$ for a polynomial $q(x)$ of degree 1 less that the degree of $p(x)$. (Why does such a $q(x)$ exist?) Then $p(0)$ is prime. So is $p(m \cdot c)$ for $m = 1, 2, \ldots$.]

Exercise 2.3.21. Except for 2, every prime is odd. Then to form primes, it is natural to add or subtract 1 from an even number, and the simplest kind of even number is a power of 2. Thus numbers of the form $2^n - 1$ seem good candidates to be primes.

    (a) Show that $2^n - 1$ is prime only if n is prime. (*Hint:* If $n = k \cdot d$, write $2^n - 1$ as a product with $2^k - 1$ as one of the factors.)

(b) A **Mersenne number**[8] is an integer of the form $2^p - 1$, where p is a prime. For example, the first three Mersenne numbers are 3, 7, and 31 (for $p = 2, 3$, and 5, respectively). By examining the first six Mersenne numbers (for $p = 2, 3, 5, 7, 11, 13$), show that $2^p - 1$ *can* be composite—but *may* be prime—even when p is a prime.

(c) Express the Mersenne number $2^n - 1$ in binary (base 2) notation. (See pages 66-67.)

Mersenne numbers have been a prime source of new primes ever since Mersenne discovered the first one. As of early 1994, there were 33 known Mersenne primes. The two largest and most recently discovered (in 1992 and 1994, respectively) are

$$M_{756839} = 2^{756839} - 1$$

and

$$M_{859433} = 2^{859433} - 1$$

(the exponents 756839 and 859433 are readily checked to be prime). Both these Mersenne primes—consisting of 227,832 and 258,716 decimal digits, respectively—were discovered using supercomputers.[9] As of this writing, the latter number not only remains the largest known Mersenne prime but also holds the world's record as the largest known prime.

Exercise 2.3.22. Without actually computing M_{756839} or M_{859433}, confirm the statement in the preceding paragraph about the number of decimal digits in each by estimating these numbers of digits. (*Hint:* Approximate $2^n - 1$ by 2^n. What is $\log_2 2^n$?)

Testing very large integers for primality and factoring very large composite numbers into products of primes are subjects of intense interest today. One reason for the interest is the application to cryptography, especially to "public-key" ciphers such as the one devised by Ronald L. Rivest, Adi Shamir, and Leonard Adleman which in practice seems to require unachievably large amounts of computing to crack.[10]

---

[8] So named after the seventeenth-century monk Marin Mersenne, who studied which of these numbers are primes.

[9] Calculating such a large Mersenne number is a triviality on any computer with arbitrary precision arithmetic. (With MATHEMATICA on a high-end microcomputer, the entire calculation of M_{756839} took me under 20 seconds—although it took several minutes for all its digits to be displayed.) However, searching for this number and verifying that it is, in fact, prime is what requires massive computing power.

[10] In his August, 1977, "Mathematical Games" column in *Scientific American*, Martin

We know there are infinitely many primes (as well as infinitely many nonprimes), but how are they distributed among all the positive integers? Are they distributed very thickly or very thinly? Here are two ways in which the question can be posed more precisely:

- How long are the gaps between consecutive primes as compared with the gaps between consecutive composite numbers?

- How "dense" are the primes up to some x—that is, what is the ratio between the number of primes up to x and that x? (This is akin to asking for any ordinary density in the physical world, such as the number of parts per million of a pollutant in a lake.)

As to gaps, the following exercise establishes that there are arbitrarily long gaps between consecutive prime numbers. In other words, there are arbitrary long lists of consecutive composite numbers.[11] On the other hand, the shortest gap between two primes is 1, between 2 and 3. But the pair $(2, 3)$ is exceptional in that 2 is the only even prime. What about just odd primes? The shortest possible gap, namely, 2, is attained between the primes 5 and 7, between 101 and 103, and between members of other pairs. How many such *twin primes*—two consecutive odd numbers both of which are primes—are there? Many number theorists believe the *Twin Prime Conjecture:* There are infinitely many twin primes. But nobody knows a proof!

Exercise 2.3.23.   (a) For each positive integer n, prove that the n numbers

$$(n + 1)! + 2, (n + 1)! + 3, \ldots, (n + 1)! + (n + 1)$$

are all composite. Hence, for arbitrary n, there are at least n consecutive composite numbers.

(b) Apply (a) to calculate five consecutive composite numbers. Are these the smallest five consecutive composite numbers?

As to the density of primes, the answer is known—in the limit. Define the function $\pi \colon \mathbb{R} \to \mathbb{N}$ by

$$\pi(x) = \text{the number of primes } p \text{ with } p \leq x.$$

Gardner challenged his readers to crack a Rivest-Shamir-Adleman cipher by factoring its 129-digit public key, named RSA-129. The challenge was unmet for 17 years. Then, on April 26, 1994, a factorization of RSA-129 was announced. The work had taken 8 months, including contributions by over 600 volunteers and their computers worldwide and culminating in 45 hours calculation on a massively parallel computer. In actual practice, the Rivest-Shamir-Adleman public keys are 150 digits or more, and factoring such numbers seems to present a challenge orders of magnitude greater.

[11] For an even stronger result, see Example 2.4.27.

[Actually, we are really interested in $\pi(x)$ only when x is a positive integer, but for deriving results about π it is convenient to allow arbitrary real values of x.] For example, $\pi(1) = 0$ because 1 is not a prime; $\pi(2.34) = \pi(2) = 1$ because 2 is the only prime not exceeding 2; $\pi(3) = 2$; and $\pi(100) = 25$, as we found earlier with the sieve of Eratosthenes.

For a given $x \geq 1$, the ratio $\pi(x)/x$ is the relative number of primes not exceeding x, in other words, the density of primes up to that x. Table 2.1—calculated by MATHEMATICA (with all numbers in the third and fourth columns rounded to six significant digits)—shows, for increasing powers of 10, the corresponding values of $\pi(x)$, the function $x/\log x$ (where log denotes the *natural* logarithm, that is, to base e), and the ratio of these two functions. Evidently $\pi(x)$ gets closer and closer to $x/\log x$ as x increases. In fact, the last column—the ratio

$$\frac{\pi(x)/x}{1/\log x} = \frac{\pi(x)}{x/\log x}$$

of the density of primes up to x to the quotient $1/\log x$—seems to be approaching 1 as x grows.

Table 2.1: Density of primes.

x	$\pi(x)$	$\dfrac{x}{\log x}$	$\dfrac{\pi(x)}{x/\log x}$
10^2	25	21.7147	1.15129
10^3	168	144.765	1.16050
10^4	1229	1085.74	1.13195
10^5	9592	8685.89	1.10432
10^6	78498	72382.4	1.08449
10^7	664579	620421.	1.07117
10^8	5761455	$5.42868 \cdot 10^6$	1.06130
10^9	50847534	$4.82549 \cdot 10^7$	1.05373
10^{10}	455052511	$4.34295 \cdot 10^8$	1.04780

Such numerical evidence strongly suggests that the ratio has limit 1. This is the content of the celebrated **Prime Number Theorem**, which was conjectured by Gauss in 1793.

Theorem 2.3.24 (Prime Number Theorem).

$$\lim_{x \to \infty} \frac{\pi(x)}{x/\log x} = 1.$$

A proof of this theorem is completely beyond the scope of this book. Gauss himself had no proof. Chebyshev took what was regarded as a significant step by showing merely that the ratio $\pi(x)/(x/\log x)$ is bounded between two numbers enclosing 1, namely,

$$a \le \frac{\pi(x)}{x/\log x} \le b,$$

where

$$a = \frac{\log 2}{4} = 0.1732\ldots,$$
$$b = 32\log(2) = 22.18\ldots.$$

(The proof of Chebyshev's result is readily accessible in an undergraduate course in number theory.) Still, the difference between the upper bound b and the lower bound a is quite substantial!

So intractable did a proof of the Prime Number Theorem seem, that it was said that whoever first proved it would become immortal. The theorem was finally proved only in 1896, and independently, by C. J. de la Vallée-Poussin and Jacques Hadamard. And nearly immortal it seemed they might be: de la Vallée-Poussin and Hadamard lived until 1962 and 1963, reaching the ages of 96 and 98, respectively!

2.4 Congruence

Congruence is a generalization of clock arithmetic. If it is precisely 9 A.M. today, then exactly 2 days from now it will again be 9 A.M., and exactly half a day after that it will be 9 P.M. Count the numbers of hours a, b, c of these three times since the immediately preceding midnight. Then

$$a = 9,$$
$$b = 9 + 2 \cdot 24 = 57,$$
$$c = 9 + 2 \cdot 24 + 12 = 69.$$

Observe that $a - b = (-2)24$ whence $24 \mid (a-b)$; also, $a - b = (-4)12$, whence $12 \mid (a-b)$. Similarly, $a - c = (-2)24 - 12 = (-5)12$, whence $12 \mid (a-c)$; however, $24 \nmid (a-c)$. On a 24-hour clock, the hour counts a and b show the same time, whereas a and c do not; on a 12-hour clock (with no A.M./P.M. indicator), all three hour counts a, b, and c show the same time. With respect to 24-hour time intervals, a and b are alike, whereas a and c are not; with respect to 12-hour time intervals, a, b, and c are all alike. We shall say that a and b are "congruent modulo

24," whereas a and c are not congruent modulo 24. Similarly, we shall say that a and b are "congruent modulo 12," as are a and c. (The formal definition is coming in a moment.)

Congruence also arises naturally from divisibility as studied in Section 2.2. Let m be a given positive integer. (If you wish, take $m = 12$ or $m = 24$ for concreteness.) For an arbitrary integer a, the Division Theorem (Theorem 2.1.1) guarantees the existence of unique integers b and d such that $a = dm + b$ and $0 \le b < m$. The remainder b is related to a with respect to m by $a - b = dm$ with quotient d, in other words, $m \mid (a - b)$. Then with respect to the divisor m, we can consider a and b as related to one another. Again we shall say that "a is congruent to b modulo m."

Congruence, as described in these two situations, is just a certain way of classifying pairs of integers as being alike. This relation of congruence and the resulting notion of *congruence classes* (introduced below) illustrate the key ideas of *equivalence relation* and *equivalence classes* that permeate much of mathematics.[12] The same ideas arise in Chapter 3, where we examine the real numbers and various related number systems.

Here is the formal definition of congruence.

Definition 2.4.1. Let m be a positive integer. We say that integer a is **congruent to** integer b **modulo** m when $m \mid (a - b)$, and then we write

$$a \equiv b \pmod{m}.$$

The positive integer m is called the **modulus** of such a congruence.

If a is *not* congruent to b, then we write

$$a \not\equiv b \pmod{m}.$$

The word "modulo" is often abbreviated as "mod," as in the notation shown; then the statement $a \equiv b \pmod{m}$ may be read, "a is congruent to b mod m." Sometimes "mod" is omitted from the congruence notation, which is therefore shortened to

$$a \equiv b \pmod{m}.$$

At times, when we are dealing with a fixed m, even the "(mod m)" or "(m)" may be omitted, so that the congruence is further shortened to

$$a \equiv b.$$

From our clock example, $9 \equiv 57 \pmod{24}$ but $9 \not\equiv 69 \pmod{24}$. Some more examples are $15 \equiv -9 \pmod{12}$ and $30 \equiv 0 \pmod{5}$.

Modulus 1 is quite uninteresting (why?), and so we now stipulate the following.

---

[12]Equivalence relations and equivalence classes in general are treated in Appendix C.

In congruences, *the modulus m is always an integer greater than 1.*

Earlier, in Section 2.2, the mod notation was used in a different way, to denote an *operation* upon two integers a and m with $m > 0$, namely, a mod m is the remainder—an integer—when a is divided by m. By way of contrast, the notation mod in $a \equiv b$ (mod m) is a *relation* between the two integers a and b. In short, a mod b is a *number*, whereas $a \equiv b$ (mod m) is a *statement*. Please do not confuse these two uses of mod. Nonetheless, there is an intimate connection between the two.

Proposition 2.4.2. *Let a and b be integers. Then*

$$a \equiv b \quad (\text{mod } m) \iff a \text{ mod } m = b \text{ mod } m.$$

Proof. Exercise.   □

The preceding proposition reinforces an interpretation of congruence already mentioned: For a fixed modulus m, we may consider integers a and b in a certain sense alike when they are congruent modulo m—not necessarily exactly the same, but the same with respect to this modulus. Of course,

$$a = b \implies a \equiv b \quad (\text{mod } m)$$

(see the following proposition) but *not* conversely. Thus congruence modulo m is an analog—indeed, a generalization—of actual equality. For equality we have the familiar properties that $a = a$ (anything equals itself); if $a = b$, then $b = a$ (if one thing equals a second, then the second equals the first); and if $a = b$ and $b = c$, then $a = c$ (if one thing equals a second and the second equals a third, then the first equals the third). These three properties of equality—**reflexivity, symmetry**, and **transitivity**—also hold, not surprisingly, for congruence:

Proposition 2.4.3. *Let $m > 1$. For all integers a, b, and c:*

1. (reflexive property) $a \equiv a$ (mod m).

2. (symmetric property) *If $a \equiv b$ (mod m), then $b \equiv a$ (mod m).*

3. (transitive property) *If $a \equiv b$ (mod m) and $b \equiv c$ (mod m), then $a \equiv c$ (mod m).*

Proof (transitive property). Assume that $a \equiv b$ (mod m) and $b \equiv c$ (mod m). Then $m \mid (a - b)$ and $m \mid (b - c)$. We know that a divisor of each of two numbers also divides their sum, and so

$$m \mid [(a - b) + (b - c)] = (a - c).$$

This means that $a \equiv c$ (mod m).   □

The proof of the reflexive and symmetric properties of congruence are left for you to supply. We summarize the preceding proposition by saying that *congruence modulo m is an* **equivalence relation** *on the set* $\mathbb{Z}$ *of all integers.*[13]

Because of transitivity and symmetry, a string of several congruences such as

$$a \equiv b \equiv c \equiv d \pmod{m}$$

should now be as unambiguous as a string of several equalities such as $a = b = c = d$.

Exercises 2.4.4. (1) When are two integers a and b congruent modulo 2? (Try to state and prove an "if and only if" condition.)

(2) Given m, when is $a \equiv 0 \pmod{m}$?

(3) Show that $a \equiv b \pmod{m}$ if and only if $a - b \equiv 0 \pmod{m}$.

(4) Suppose $a \equiv b \pmod{m}$. Explain why, then, $m \mid a$ if and only if $m \mid b$.

(5) Let integer $m > 1$. For a given integer a, describe the set

$$\{\, b \in \mathbb{Z} : a \equiv b \pmod{m} \,\}$$

of *all* integers that are congruent to a modulo m. Go back to first principles in your description: do *not* use explicitly any of the terms "congruent," "divides," or "multiple."

(6) Suppose also $n > 1$. If $n \mid m$, what is the connection between $a \equiv b \pmod{m}$ and $a \equiv b \pmod{n}$?

(7) Solve the congruence $x^2 \equiv 0 \pmod{4}$, in other words, determine *all* integers x for which it is true. Similarly, solve the congruence $x^2 \equiv 1 \pmod{4}$.

Equals added to equals gives equals, in other words,

$$a = b \text{ and } c = d \implies a + c = b + d,$$

and similarly equals multiplied by equals gives equals. In short, addition and multiplication preserve equality. Since congruent numbers (for a fixed modulus), although not necessarily equal, are alike—equal in so far as division by m goes—we should not be surprised to find that addition and multiplication also *preserve congruence,* as indicated by the following proposition.

---

[13]This is a good time to take a look at some other examples of equivalence relations in Appendix C.

Proposition 2.4.5. *Let $m > 1$. Suppose a, b, c, and d are integers with*

$$a \equiv b \pmod{m} \quad \text{and} \quad c \equiv d \pmod{m}.$$

Then

$$a + c \equiv b + d \pmod{m},$$
$$a - c \equiv b - d \pmod{m},$$
$$a \cdot c \equiv b \cdot d \pmod{m}.$$

Proof. Exercise.   □

In particular, by taking $c = d$ above, we conclude

$$a \equiv b \pmod{m} \implies a \cdot c \equiv b \cdot c \pmod{m}.$$

In other words, if you multiply a congruence by a constant, you will still get a congruence.

Also, in particular, if $a \equiv b \pmod{m}$, then by taking $c = a$ and $d = b$ above, we conclude $a^2 \equiv b^2 \pmod{m}$. More generally (by induction) the following corollary holds.

Corollary 2.4.6. *If $a \equiv b \pmod{m}$, then $a^n \equiv b^n \pmod{m}$ for every $n = 0, 1, 2, \ldots$.*

A polynomial

$$p(x) = c_n x^n + c_{n-1} x^{n-1} + \cdots + c_2 x^2 + c_1 x + c_0$$
$$= \sum_{j=0}^{n} c_j x^j$$

of degree n with *integer* coefficients $c_n, c_{n-1}, \ldots, c_2, c_1, c_0$ arises by the repeated operations of raising the variable x to powers, multiplying by constants, and adding. We just saw that all three of these operations preserve congruence. Hence we have the following proposition.

Proposition 2.4.7. *Let $p(x)$ be a polynomial with integer coefficients and let $m > 1$. For integers a and b,*

$$a \equiv b \pmod{m} \implies p(a) \equiv p(b) \pmod{m}.$$

Proof. Exercise.   □

This general rule about polynomials preserving congruence has an interesting interpretation in a special case.

Example 2.4.8. Apply the preceding proposition with $a = 10$, $b = 1$, and $m = 9$. The hypothesis $a \equiv b \pmod{m}$ certainly holds. Then so does the conclusion that, for any polynomial $p(x) = \sum_{j=0}^{n} c_j x^j$ with integer coefficients, $p(a) \equiv p(b) \pmod{m}$, that is,

$$p(10) \equiv p(1) \pmod 9. \tag{*}$$

Assume now that each coefficient c_j is one of the decimal digits 0, 1, 2, 3, 4, 5, 6, 7, 8, 9. Then

$$p(10) = \sum_{j=0}^{n} c_j 10^j$$

is just the integer k whose decimal (base 10) representation is

$$k = (c_n c_{n-1} \ldots c_2 c_1 c_0)_{10} \tag{**}$$

(see page 66 and Theorem 2.1.10). And

$$p(1) = \sum_{j=0}^{n} c_j$$

is just the sum of the decimal digits of that integer k. From congruence (*), it follows that

$$(c_n c_{n-1} \ldots c_2 c_1 c_0)_{10} \equiv \sum_{j=0}^{n} c_j \pmod 9.$$

Hence

$$9 \mid (c_n c_{n-1} \ldots c_2 c_1 c_0)_{10} \iff 9 \mid \sum_{j=0}^{n} c_j. \tag{***}$$

Since any positive integer k has a decimal representation of the form (**), then from (***) we obtain the following **criterion for divisibility by 9**:

An integer k is divisible by 9 if and only if the sum of its digits is divisible by 9.

For example, the sum $5 + 6 + 7$ of the digits of 567 is 18, which is divisible by 9, and so 567 is divisible by 9. The sum $2 + 4 + 1 + 7$ of the digits of 2417 is 14, which is not divisible by 9, and so 2417 is not divisible by 9.

Exercises 2.4.9.   (1) Use the criterion above to test the following for divisibility by 9.

(a) the numbers 4172, 341721, 1234567, 12345678;

(b) the numbers

$$1234567890123456789012345 6789$$

and

$$122333444455555666666677777778888888 8999999999.$$

(2) Apply Proposition 2.4.7 to state and prove a criterion for divisibility of a decimal integer by 11.

(3) Show that $2^n \equiv (-1)^n$ (mod 3) and from that deduce that $2^n - 1$ is a multiple of 3 if and only if n is even.

Exercise 2.4.10. Here is a stronger result than that of Exercise 2.3.20: A nonconstant polynomial $f(x) = \sum_{j=0}^{n} c_j x^j$ with integer coefficients cannot assume only prime values at all integers $x \geq N$ no matter how large N is.

Fill in the details of the following proof: Suppose $f(k)$ is prime for all integers $k \geq N$. Let $p = f(N)$. For any integer k of the form $k = N + mp^2$, we have $f(k) \equiv p$ (mod p^2). There exists an integer m with $f(k) > p$, where $k = N + mp^2$. Then p divides $f(k)$ but $p \neq f(k)$. It follows that $f(k)$ is not prime.

How far can we stretch the analogy between actual equality of integers and their mere congruence modulo some $m > 1$? For equality, we have the familiar cancellation laws

$$a + c = b + c \implies a = b$$

and

$$a \cdot c = b \cdot c \implies a = b.$$

Because cancelling c in $a + c = b + c$ amounts to adding $-c$ to both sides, the analog

$$a + c \equiv b + c \pmod{m} \implies a \equiv b \pmod{m}$$

is true since addition (of $-c$) does preserve congruence. However, cancelling c in $a \cdot c = b \cdot c$ amounts to multiplying both sides by $1/c$. Now $1/c$ will not in general be an integer. Hence, even though multiplication (by integers) does preserve congruence, we should *not* expect that $a \cdot c \equiv b \cdot c$ (mod m) in general implies $a \equiv b$ (mod m).

Indeed, in general we *cannot* cancel factors in congruences. For example,

$$7 \cdot 10 \equiv 1 \cdot 10 \pmod{12} \quad \text{but} \quad 7 \not\equiv 1 \pmod{12}.$$

By contrast,

$$27 \cdot 35 \equiv 3 \cdot 35 \pmod{12} \quad \text{and} \quad 27 \equiv 3 \pmod{12}.$$

(Notice that the latter holds even though, if $c = 35$, then $1/c$ is definitely not an integer.) An evident difference between these two examples is that 10 is not relatively prime to the modulus 12 in the first, whereas 35 is relatively prime to the modulus 12 in the second. And this is precisely the difference that matters. The common factor c in $ac \equiv bc$ (mod m) *can* be cancelled provided that it is relatively prime to the modulus m.

Proposition 2.4.11 (Congruence Cancellation Law). *Let $a, b, c,$ and m be integers with $m > 1$. Assume*

$$\gcd(c, m) = 1.$$

Then

$$ac \equiv bc \pmod{m} \implies a \equiv b \pmod{m}.$$

Proof. Exercise. □

Exercise 2.4.12. What does the preceding proposition say in the case where $b = 0$? Have you ever seen this result before (in different language)?

Early in algebra, you learn that a linear equation

$$ax = b$$

always has a solution—in fact, a unique solution—provided that $a \neq 0$. (In this chapter we are interested only in integers and, of course, if a and b are integers, there is no guarantee of a solution x to $ax = b$ that is itself an integer.) What happens if we replace equality by congruence modulo a given m? That is, given $a \neq 0$ and b, must the **linear congruence**

$$ax \equiv b \pmod{m}$$

have a solution at all, and if so does it have only one solution?

The answer to each of our questions is no. For example, the linear congruence

$$3x \equiv 1 \pmod{6}$$

has no solution (why?), and the linear congruence

$$3x \equiv 1 \quad (\text{mod } 25)$$

has solution $x = 17$ among the infinitely many solutions

$$x = \ldots, 17 - 2 \cdot 25 = -33, 17 - 25 = -8, 17,$$
$$17 + 25 = 42, 17 + 2 \cdot 25 = 67, \ldots.$$

Do you have a general conjecture yet about the answers? (Try it before reading further!)

Theorem 2.4.13 (Linear Congruence Theorem). *Let a and b be integers. Suppose a is relatively prime to the modulus $m > 1$. Then the linear congruence*

$$ax \equiv b \quad (\text{mod } m)$$

has integral solutions. Moreover, solutions are "unique modulo m" in the sense that any two solutions are congruent to one another modulo m.

Proof. Existence. By Euclid's Divisor Theorem (Theorem 2.2.28) there are integers s and t with

$$1 = as + mt.$$

Multiply by b to get

$$b = abs + bmt.$$

Then

$$a(bs) \equiv b \quad (\text{mod } m),$$

and so $x = bs$ is a solution.

Uniqueness modulo m. Use the Congruence Cancellation Law (Proposition 2.4.11). □

Suppose you have some solution c of a linear congruence $ax \equiv b$ (mod m) whose coefficient a of x is relatively prime to the modulus m. Then you actually know all solutions: not only is each solution congruent to c (according to "uniqueness modulo m"), but *every* integer congruent to c modulo m is a solution (why?). In other words, the set

$$\{ k \in \mathbb{Z} : k \equiv c \quad (\text{mod } m) \}$$

is the set of all solutions.

Exercises 2.4.14. (1) Solve the linear congruence $3x \equiv 2 \pmod 5$ directly, by inspection. Then solve it using the method of proof of the Linear Congruence Theorem.

(2) Find the set of all solutions of the congruence in (1).

(3) If a linear congruence $ax \equiv b \pmod m$ has a solution, show that a need not necessarily be relatively prime to m.

(4) Generalize the Linear Congruence Theorem: The linear congruence $ax \equiv b \pmod m$ has a solution if and only if $\gcd(a, m) \mid b$.

The Linear Congruence Theorem takes a special form in the case where the modulus is prime.

Corollary 2.4.15. *Let p is a prime. If $a \not\equiv 0 \pmod p$, that is, if $p \nmid a$, then for any b the linear congruence*

$$ax \equiv b \pmod p$$

has a solution that is unique modulo p.

Exercise 2.4.16. Must p be a prime for the result in the preceding corollary to hold? If p is a prime, is the converse of the implication there also true?

Exercises 2.4.17. Prove or disprove each of the following logical equivalences. If the implication is true in only one direction, also prove it in that direction alone.

(1) Suppose $d \neq 0$. Then

$$ad \equiv bd \pmod{md} \iff a \equiv b \pmod m.$$

(2) Again suppose $d \neq 0$. Then

$$ad \equiv bd \pmod m \iff a \equiv b \pmod{m / \gcd(d, m)}.$$

We shall need the following result in a moment, when we try to solve simultaneously several linear congruences with different moduli.

Proposition 2.4.18 (Relatively Prime Moduli Law). *Assume m is relatively prime to n. Then*

$$a \equiv b \pmod{mn} \iff a \equiv b \pmod m \text{ and } a \equiv b \pmod n.$$

Proof. Exercise. □

Exercise 2.4.19. Extend the preceding Relatively Prime Moduli Law to 3 moduli. Generalize it to k moduli.

We have not and will not deal with the solving of quadratic congruences, but here is one little result.

Exercise 2.4.20. Let n be an integer. Prove that $n^2 \equiv 0, 1,$ or $4 \pmod 8$.

For a system

$$\begin{cases} a_1 x = b_1 \\ a_2 x = b_2 \end{cases}$$

of two linear equations in a single unknown x, the story is pretty simple: Provided $a_1 \neq 0$ and $a_2 \neq 0$, there is a simultaneous solution x if and only if $b_1/a_1 = b_2/a_2$, and then this common ratio is the unique solution (which will not be an integer unless $a_1 \mid b_1$).

What about systems of linear congruences in a single unknown? Suppose we are given two moduli m_1 and m_2 along with coefficients a_1, a_2, b_1, and b_2 and we wish to solve simultaneously the system

$$\begin{cases} a_1 x \equiv b_1 \pmod{m_1} \\ a_2 x \equiv b_2 \pmod{m_2} \end{cases}$$

of two linear congruences. Of course, this is possible only if each congruence individually has a solution, and so we want to assume that a_1 and a_2 are relatively prime to m_1 and m_2, respectively. In this case, by the Linear Congruence Theorem there are integers c_1 and c_2 with

$$a_1 c_1 \equiv b_1 \pmod{m_1},$$
$$a_2 c_2 \equiv b_2 \pmod{m_2}.$$

Then the original system of congruences has the same solutions as the system

$$\begin{cases} a_1 x \equiv a_1 c_1 \pmod{m_1} \\ a_2 x \equiv a_2 c_2 \pmod{m_2}. \end{cases} \tag{*}$$

Since a_1 is relatively prime to m_1 and a_2 is relatively prime to m_2, then according to the Congruence Cancellation Law (Proposition 2.4.11) the solutions x of system (*) are exactly the same as those of the simpler system

$$\begin{cases} x \equiv c_1 \pmod{m_1} \\ x \equiv c_2 \pmod{m_2}. \end{cases} \tag{**}$$

In view of this observation, we might as well deal with such a simpler system of linear congruences—where the coefficient of x in each is 1— from the start. And that is what we shall do.

A linear congruence of the form

$$x \equiv b \pmod{m} \tag{***}$$

such as appears in systems of the form (**) is particularly nice in that we can immediately write down the *general solution* to it:

$$x = b + tm,$$

where t can be any integer. That is, each integer of the form $b + tm$ for some t is a solution of (***), and each solution of (***) is of this form for some t.

Exercise 2.4.21. Write a system of linear congruences of the form (**) above having no solution at all.

Theorem 2.4.22 (Chinese Remainder Theorem).[14] *Let the moduli m_1 and m_2 be relatively prime to one another. Let b_1 and b_2 be any integers. Then the system*

$$\begin{cases} x \equiv b_1 \pmod{m_1} \\ x \equiv b_2 \pmod{m_2} \end{cases}$$

of linear congruences has a solution. Moreover, any two solutions are congruent modulo the product $m_1 m_2$.

Proof. Existence. The general solution of the first of the two congruences is

$$x = b_1 + tm_1,$$

where t is an arbitrary integer. It remains only to find such a number that is also a solution of the second of the two congruences.

Such an x is a solution of the second congruence if and only if

$$b_1 + tm_1 \equiv b_2 \pmod{m_2},$$

that is,

$$tm_1 \equiv b_2 - b_1 \pmod{m_2}.$$

Consider this a linear congruence with unknown t. Since by assumption m_1 is relatively prime to the modulus m_2 here, then by the Linear Congruence Theorem (2.4.13), a solution t does exist.

Uniqueness modulo $m_1 m_2$. Use the Relatively Prime Moduli Law (Proposition 2.4.18). □

---

[14]This theorem is so named because the method of solution was described by the fourth-century Chinese mathematician Sun Tsu.

Exercises 2.4.23.   (1) Solve the system

$$\begin{cases} x \equiv 1 & (\text{mod } 27) \\ x \equiv 2 & (\text{mod } 10). \end{cases}$$

(2) Formulate and prove an analog of the Chinese Remainder Theorem for a system of three linear congruences.

Please read no further until you have done (2)!

Corollary 2.4.24. *Let $m_1, m_2, \ldots, m_n$ be n moduli that are pairwise relatively prime. Let $b_1, b_2, \ldots, b_n$ be any integers. Then the system*

$$\begin{cases} x \equiv b_1 & (\text{mod } m_1) \\ x \equiv b_2 & (\text{mod } m_2) \\ \quad\vdots & \quad\vdots \\ x \equiv b_n & (\text{mod } m_n) \end{cases}$$

of n linear congruences has a solution, and any two solutions are unique modulo the product $m_1 m_2 \cdots m_n$.

(Saying that the moduli $m_1, m_2, \ldots, m_n$ are *pairwise relatively prime* means, of course, that m_i is relatively prime to m_j whenever $i \neq j$.)

Proof. Exercise.   □

Now we return to systems of linear congruences of the original general form, with each congruence of the system looking like $ax \equiv b$ $(\text{mod } m)$.

Corollary 2.4.25. *Let $m_1, m_2, \ldots, m_n$ be n moduli that are pairwise relatively prime. Let $a_1, a_2, \ldots, a_n$ be integers such that a_i is relatively prime to the modulus m_i for each i and let $b_1, b_2, \ldots, b_n$ be any integers. Then the system*

$$\begin{cases} a_1 x \equiv b_1 & (\text{mod } m_1) \\ a_2 x \equiv b_2 & (\text{mod } m_2) \\ \quad\vdots & \quad\vdots \\ a_n x \equiv b_n & (\text{mod } m_n) \end{cases}$$

of n linear congruences has a solution, and any two solutions are unique modulo the product $m_1 m_2 \cdots m_n$.

Proof. Exercise. (Look again at our discussion on page 114 about reducing a system of two linear congruences to a system where the unknown x in each has coefficient 1.)   □

Exercises 2.4.26.   (1) Solve the system

$$\begin{cases} x \equiv 1 \pmod 3 \\ x \equiv 2 \pmod 4 \\ x \equiv 3 \pmod 5. \end{cases}$$

(2) Solve the system

$$\begin{cases} 2x \equiv 1 \pmod 3 \\ 3x \equiv 2 \pmod 4 \\ 4x \equiv 3 \pmod 5. \end{cases}$$

(3) Say that numbers $m_1, m_2, \ldots, m_n$ are *n-wise mutually prime* when their only common divisor is 1. Of course, n pairwise relatively prime numbers are n-wise mutually prime. Give an example of $n \geq 3$ numbers that are n-wise mutually prime but not pairwise relatively prime.

(4) (The egg problem.) A farmer has collected a number of eggs that she can divide into 2 equal piles and have 1 egg left over; or divide into 3 equal piles with 1 egg left over; or divide into 4 equal piles with 1 left over; or 5 equal piles with 1 left over; or 6 equal piles with 1 left over; or 7 equal piles with *none* left over. What is the least number of eggs she can have?

Recall from Section 2.3 that there are arbitrarily large gaps between successive prime numbers. In fact, according to Exercise 2.3.23, for a positive integer n, the n consecutive integers

$$(n + 1)! + 2, (n + 1)! + 3, \ldots, (n + 1)! + (n + 1)$$

are all composite.

How far from being primes are the numbers in such a list of non-primes? Some of the numbers might be simply the product of two different primes or the square of a prime. (Look at the list for, say, $n = 2$ or $n = 5$.) For a given k, is it possible to construct such a list of n consecutive composite numbers each of which is divisible by k or more different primes? The answer is yes, and the argument in the following example, due to P. Schumer, uses the Chinese Remainder Theorem.

Example 2.4.27. Let n and k be given positive integers. Then there are n consecutive composite numbers each of which is divisible by k or more distinct primes.

Proof. Let $\left(p_j\right)_{j=1,2,\ldots}$ be the sequence of all primes in ascending order (so that $p_1 = 2$, $p_2 = 3$, etc.). Let m_1 be the product of the first k primes $p_1, p_2, \ldots, p_k$, let m_2 be the product of the next k primes $p_{k+1}, p_{k+2}, \ldots, p_{2k}$, and, in general, let

$$m_i = p_{(i-1)k+1} \cdot p_{(i-1)k+2} \cdots p_{ik} \qquad (i = 1, 2, \ldots).$$

Consider the system of linear congruences

$$\begin{cases} x & \equiv & -1 & (\text{mod } m_1) \\ x & \equiv & -2 & (\text{mod } m_2) \\ & \vdots & & \vdots \\ x & \equiv & -n & (\text{mod } m_n). \end{cases}$$

The moduli here are pairwise relatively prime (why?), and so the Chinese Remainder Theorem provides a solution x of this system. Then (provide the details) $x + 1$, $x + 2$, $\ldots$, $x + n$ is a list of n consecutive composite numbers each of which is divisible by at least k different primes. $\square$

Exercises 2.4.28.   (1) Carry out the solution of the system of congruences in the preceding example for the case where $n = 5$ and $k = 1$ to obtain a list of 5 consecutive composite numbers divisible by the first 5 primes 2, 3, 5, 7, and 11, respectively.

(2) Modify the proof in Example 2.4.27 to show that, for given n and k, there is a list of n composite numbers each of which is divisible by k or more different primes and whose successive differences are the first n Fibonacci numbers.

Our final few results here concerning congruence deal with prime powers. Observe that

$$a^2 \equiv a \quad (\text{mod } 2)$$

for every a (why?), whereas $a^6 \not\equiv a$ (mod 6) for, say, $a = 2$. The difference is that the modulus 2 in the first congruence is prime, whereas the modulus 6 in the second is not. We are going to prove that $a^p \equiv a$ (mod p) for all a provided that the modulus p is a prime. Our proof will use binomial coefficients (Definition 1.5.3) and the Binomial Theorem (Theorem 1.5.10).

Lemma 2.4.29. *Let p be a prime. Then*

$$p \mid \binom{p}{j} \qquad (j = 1, 2, \ldots, p - 1).$$

Proof. Let $1 \le j \le p - 1$. By definition,

$$\binom{p}{j} = \frac{p!}{j!\,(p-j)!}.$$

This binomial coefficient is an integer, yet no factor in its denominator divides p except 1. It follows that

$$\binom{p}{j} = p\frac{(p-1)!}{j!\,(p-j)!}$$

with the fraction on the right being an integer. $\square$

Theorem 2.4.30 (Fermat's Little Theorem).[15] *Let p be a prime number. Then for every integer a,*

$$a^p \equiv a \pmod{p}.$$

Proof. The case $p = 2$ was treated earlier. Assume now that p is an odd prime. Then it suffices (why?) to establish the result just for integers $a \ge 0$. For $a = 0$, the claimed congruence is, in fact, an equality. We now establish the result for $a = 1, 2, \ldots$ by induction on a.

Certainly $1^p \equiv 1 \pmod{p}$, and so the base case is true.

Let $a \ge 1$ and make the inductive assumption

$$a^p \equiv a \pmod{p}.$$

We want to deduce that $(a + 1)^p \equiv (a + 1) \pmod{p}$. Expand $(a + 1)^p$ by the Binomial Theorem and separate out the first and last terms a^p and $a^0 = 1$:

$$(a + 1)^p = a^p + \sum_{j=1}^{p-1} \binom{p}{j} a^j + 1.$$

By Lemma 2.4.29, for each term $\binom{p}{j} a^j$ in the sum, p divides the coefficient $\binom{p}{j}$ so that the term is congruent to 0 modulo p. Hence

$$(a + 1)^p \equiv a^p + 0 + 1 \pmod{p}.$$

Now finish the proof of the inductive step yourself. $\square$

In Fermat's Little Theorem, by the Congruence Cancellation Law (Proposition 2.4.11) we may cancel a from both sides of the congruence provided that a is relatively prime to p. But p is already a prime. Hence we obtain the following equivalent form of Fermat's Little Theorem.

---

[15]This is called Fermat's *Little* Theorem to distinguish it from his "big" theorem—Fermat's Last Theorem (see page 62).

Corollary 2.4.31. *Let p be a prime number. Then for every integer a that is not divisible by p,*

$$a^{p-1} \equiv 1 \pmod{p}.$$

Another way of stating the hypothesis about a is, of course, that $a \not\equiv 0 \pmod{p}$.

Exercise 2.4.32. What is the last (units place) digit in the decimal representation of 3^{4000}? (*Hint:* Apply Fermat's Little Theorem with $p = 5$ and $a = 3$. And to what is 3^2 congruent modulo 2?)

One of the banes of math students' existence is that the equality

$$(a + b)^2 = a^2 + b^2$$

and, more generally, the equality

$$(a + b)^n = a^n + b^n \tag{*}$$

are *not true* in general. (As a serious student of math—perhaps a math major—you of course know better than to expect them to be true: things are just not meant to be that simple.) You may therefore be delighted to learn that a congruence corresponding to (*) *is* true—provided just that the exponent n is a prime:

$$p \text{ prime} \implies (a + b)^p \equiv a^p + b^p \pmod{p} \tag{**}$$

for all integers a and b. In short, when you compute modulo a prime p, the Binomial Theorem with exponent p collapses to a triviality.

Exercise 2.4.33. Prove (**). (Wow—this one is really easy!)

Exercise 2.4.34. Let n be an integer. Suppose prime p divides n^p. Deduce that p^2 divides n^p, too.

2.5 Congruence Classes

Suppose a standard 12-hour clock, with no A.M./P.M. indicator, shows the time as precisely 9 A.M. today. What time will it show 8 hours later? 20 hours later (after 9 A.M.)? This is easier to visualize with an analog clock—see Figure 2.3.[16] The answers are as depicted in the figure because

$$9 + 8 \equiv 5 \pmod{12},$$
$$9 + 20 \equiv 5 \pmod{12}. \tag{*}$$

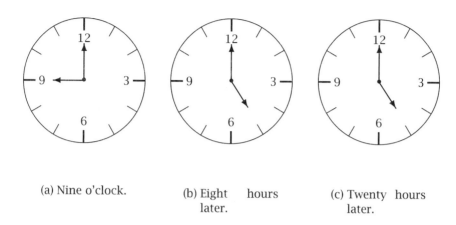

(a) Nine o'clock.   (b) Eight   hours   (c) Twenty hours
                              later.   later.

Figure 2.3: Clock addition.

In the context of a clock, the whole apparatus of the "≡" and "(mod 12)" notation seems superfluous and perhaps even awkward. We would really like to write just

$$9 + 8 = 5,$$
$$9 + 20 = 5. \tag{**}$$

You know and I know what these two equalities mean in this context, but unfortunately they are just not literally true: $9 + 8 = 17$, not 5; $9 + 20 = 29$, not 5. What we are going to do is enclose the numbers in brackets like this:

$$[9] + [8] = [5],$$
$$[9] + [20] = [5]. \tag{***}$$

There is no longer any problem of these being manifestly false, like (**); we still understand what the meaning is—that addition "modulo 12" is at issue; and the original congruence notation in (*) is still gone.

Now all we have to do is define carefully what equalities like (***) really mean. Of course, we know what equality means, and so the real issue is what terms like [9] and [20] and [5] should mean.

Let us tentatively define [9] to stand for the totality of *all* the numbers of hours (since the immediately preceding midnight) that show on

---

[16]When teaching a class such "modular arithmetic," the author can no more assume an analog clock will be at hand than, when teaching logarithms, he can assume a slide rule will be at hand!

the clock the same way that 9 A.M. does—with the big hand on 12 and the little hand on 9. In other words, define

$$[9] = \{\ldots, 9 - 2 \cdot 12 = -15, 9 - 12 = -3, 9,$$
$$9 + 12 = 21, 9 + 2 \cdot 12 = 33, \ldots\}$$

(where -3 hours is 9 P.M. and -15 is 9 A.M. the previous day, etc.). Then $[9]$ is just the set

$$[9] = \{k \in \mathbb{Z} : k \equiv 9 \pmod{12}\}$$

of all the infinitely many integers that are congruent to 9 modulo 12. Similarly, define $[5]$ and $[20]$ to be the sets of all integers congruent modulo 12 to 5 and 20, respectively.

With these tentative definitions, it is apparent that

$$[9] + [20] = [5]$$

is a very convenient shorthand for the correct assertion:

> The sum of any integer congruent to 9 and any integer congruent to 20 is an integer congruent to 5.

(Why is this assertion true?) Of course, we still have to keep in mind that the modulus is 12.

The way to define $[a]$ in general, for any integer a—and for any modulus, not just 12—is now evident.

Definition 2.5.1. Let $m > 1$ be a given modulus. For an integer a, its **congruence class modulo** m, denoted by $[a]$, is the set of all integers congruent to a modulo m, that is,

$$[a] = \{k \in \mathbb{Z} : a \equiv k \pmod{m}\}.$$

Thus, for each integer b,

$$\boxed{b \in [a] \iff a \equiv b \pmod{m}.}$$

In particular, always

$$a \in [a].$$

But there are many other—infinitely many other—elements of $[a]$ than a because

$$[a] = \{a + jm : j \in \mathbb{Z}\}.$$

For denoting the congruence class of an $a \in \mathbb{Z}$, some people prefer to put a *vinculum*—a bar—over a, as in $\overline{a}$, rather than to surround a with brackets as we have done, as in $[a]$. In fact, you may find the overbar notation easier to write; it takes one stroke of the pen instead of two. Unfortunately, it can become awkward in print, as in $\overline{a + b}$ to mean what we denote by $[a + b]$.

Ordinarily when we use the $[a]$ notation for a congruence class, we shall be working with a single, fixed modulus m. If several different moduli happen to be involved, then the notation must be made more cumbersome by writing, for example,

$$[a]_{12} \quad \text{and} \quad [a]_{24}.$$

to mean the congruence classes of a modulo 12 and modulo 24, respectively.

For a given modulus m, what $[a]$ does is lump together many different individual numbers into a new, single object. If you have studied vector algebra, then you have seen this process before. In the plane, all the directed line segments having, say, length 3 and direction 45° clockwise from the positive x-axis are lumped together into the single vector $\vec{v} = \overrightarrow{(3/\sqrt{2}, -3/\sqrt{2})}$. This process is a general one: for any equivalence relation, the set of all elements considered to be like a given one for that equivalence relation constitute an **equivalence class**—see Appendix C.

The congruence class notation $[a]$ conceals one serious trap.

Caution: $[a] = [b]$ **does *not* mean the same thing as** $a = b$.

For example, if $m = 12$, then

$$[8] = [20] \quad \text{but} \quad 8 \neq 20.$$

The correct statement is the following.

Lemma 2.5.2. *Let* $m > 1$. *For integers* a *and* b,

$$\boxed{[a] = [b] \iff a \equiv b \pmod{m}.}$$

Proof. First assume $[a] = [b]$. Since $b \in [b]$, then also $b \in [a]$. But $b \in [a]$ means $a \equiv b \pmod{m}$.

Conversely, assume $a \equiv b \pmod{m}$. We shall prove that $[a] \subset [b]$; the proof of the reverse inclusion $[b] \subset [a]$ is similar. (Exactly why is it similar? Be careful!) Let $x \in [a]$ be arbitrary. (We want to show $x \in [b]$.) Then

$$a \equiv x \pmod{m}.$$

But we know, by assumption, that

$$a \equiv b \pmod{m}.$$

It follows that

$$b \equiv x \pmod{m}.$$

(Precisely why does this follow—what properties of congruence are involved?) But this means $x \in [b]$.   □

An immediate consequence of the last two boxed results is the following.

Lemma 2.5.3. *Let $m > 1$. For integers a and b,*

$$\boxed{b \in [a] \iff [a] = [b].}$$

Exercises 2.5.4.   (1) Let $m = 12$.

    (a) Is $41 \in [5]$? Is $5 \in [41]$?

    (b) Repeat (a) for 36 and 5.

    (c) Generalize what you saw in (a) and (b).

(2) Did you notice the sneaky change from the particular definition

$$[9] = \{ k \in \mathbb{Z} : k \equiv 9 \pmod{12} \}$$

to the general definition

$$[a] = \{ k \in \mathbb{Z} : a \equiv k \pmod{m} \}$$

in Definition 2.5.1? (For $[9]$, the k is to the left of $\equiv$ and the 9 to the right; for $[a]$, the k is to the right and the a to the left.) Explain why this is alright—why our original definition of $[9]$ is really a special case of the general definition of $[a]$.

(3) How many *different* congruence classes are there modulo m if:

    (a) $m = 12$? (Keep in mind the caution on page 123: $[a] = [b]$ does not necessarily mean that $a = b$.)

    (b) $m = 2$?

    (c) m is an arbitrary integer with $m > 1$?

(4) Suppose we allowed $m = 1$ as the modulus. What would the congruence class $[9]$ be? How many congruence classes in all would there be?

(5) What is the relationship between $[a]_{12}$ and $[a]_{24}$ for an arbitrary integer a? Generalize your answer.

(6) Write a critique of the following alleged proof of the transitive law for congruence modulo m:

Assume $a \equiv b$ (mod m) and $b \equiv c$ (mod m). Then $[a] = [b]$ and $[b] = [c]$. By transitivity of equality (of sets), $[a] = [c]$. Hence $a \equiv c$ (mod m).

In Figure 2.3, the clock hands are in the same position both 8 hours after 9 A.M. and 20 hours after 9 A.M., namely, at 5 P.M. and 5 A.M., respectively. In other words, $17 \in [5]$ and $29 \in [5]$ as well as $5 \in [5]$. In accordance with the following definition, we may say that each of the numbers 17, 29, and 5 *represents* the congruence class $[5]$—as do infinitely many other integers.

Definition 2.5.5. Let $m > 1$. Given an integer a, an integer b is said to **represent** the congruence class $[a]$ and is called a **representative** of that congruence class when $b \in [a]$, in other words, when $a \equiv b$ (mod m).

In view of the caution on page 123 ($[a] = [b]$ does *not* mean the same thing as $a = b$), it is sometimes a good idea to use a name for a congruence class that does not explicitly refer to any particular representative of that congruence class. Thus, if $a = 9$ and $b = 20$, we might let $A = [a] = [9]$ and $B = [b] = [20]$. (This is in keeping with our usual practice of denoting individual numbers and similar elements with lowercase letters but denoting sets of such elements with uppercase letters.)

What is the relationship between two arbitrary congruence classes modulo m? According to the next proposition, two such congruence classes that are distinct are necessarily *disjoint*, that is, have no elements in common whatsoever.

Proposition 2.5.6. *Let $m > 1$ be a fixed modulus. If A and B are congruence classes modulo m, then either $A = B$ or else $A \cap B = \varnothing$.*

Proof. Let A and B be congruence classes modulo m. We shall show that if $A \cap B \neq \varnothing$, then $A = B$. (Why will this prove the "either ... or else ... " assertion of the proposition?)

Assume $A \cap B \neq \varnothing$. Then there exists some $x \in A \cap B$. Since $x \in A$, then $[x] = A$; similarly, $[x] = B$. It follows that $A = [x] = B$. □

Once again, recall our caution that $[a] = [b]$ does not necessarily mean $a = b$. Thus it would be *wrong* to rephrase the preceding proposition as "$a \neq b$ if and only if $[a] \cap [b] = \varnothing$."

For a given modulus $m > 1$, what are all its possible congruence classes?

Example 2.5.7. Take the modulus $m = 3$ (for a change). Then $A = [0]$, $B = [1]$, and $C = [2]$ are distinct congruence classes (why?). Moreover, they are the *only* congruence classes modulo 3 because, by the Division Theorem (Theorem 2.1.1), each $k \in \mathbb{Z}$ is congruent to one of 0, 1, or 2 modulo 3. Thus the set

$$\{ E : E \text{ is a congruence class modulo 3 } \}$$

of *all* the congruence classes—which is the same thing as

$$\{ [k] : k \in \mathbb{Z} \} \qquad (*)$$

—really boils down just to the three-element set

$$\{A, B, C\} = \{[0], [1], [2]\}.$$

Once again, you can see the importance of distinguishing between distinct integers and their corresponding congruence classes, which need not be distinct. The set $(*)$ looks like it might have infinitely many different members $[k]$, but it really does not; it has only three different members. Moreover, we could just as well have said that this set $(*)$ of all congruence classes is

$$\{[-27], [25], [8]\},$$

because $[-27] = [0]$, $[25] = [1]$, and $[8] = [2]$ modulo 3. In fact, clearly we could take *any* representatives a, b, c of A, B, C, respectively, and describe the set $(*)$ of all congruence classes as $\{[a], [b], [c]\}$.

Here is one more observation about this example. Each integer belongs to at least one of the three congruence classes A, B, C (why?), and no integer belongs to more than one of these congruence classes—because two congruence classes having an element in common are necessarily identical (why?). Thus the three congruence classes split up the set $\mathbb{Z}$ of all integers into three mutually exclusive sets:

$$\mathbb{Z} = A \cup B \cup C, \quad A \cap B = A \cap C = B \cap C = \emptyset.$$

In simpler terms, all the integers appear in the following three rows, and each integer appears in exactly one of these three rows:

$$\ldots, -15, -12, \ -9, \ -6, \ -3, 0, \ 3, \ 6, \ 9, 12, \ldots,$$
$$\ldots, -14, -11, \ -8, \ -5, \ -2, 1, \ 4, \ 7, 10, 13, \ldots,$$
$$\ldots, -13, -10, \ -7, \ -4, \ -1, 2, \ 5, \ 8, 11, 14, \ldots.$$

These three rows are just the three congruence classes modulo 3.

Exercise 2.5.8. Carry out a similar analysis as in the preceding example for:

(a) modulus $m = 12$.

(b) modulus $m = 2$.

(c) modulus $m = 1$ (here we relax our usual insistence that $m > 1$).

There was nothing special about the modulus 3 in Example 2.5.7, and so it should generalize. First, we present a definition.

Definition 2.5.9. Let m be an integer with $m > 1$. The set of all congruence classes modulo m is called the **integers modulo** m and is denoted $\mathbb{Z}_m$.

Thus, in general,

$$\mathbb{Z}_m = \{ [a] : a \in \mathbb{Z} \}.$$

For example,

$$\mathbb{Z}_3 = \{[0], [1], [2]\}.$$

Notice what we are up to here. We are no longer dealing just with individual numbers. And we are not even dealing just with individual congruence classes—sets of such numbers. Rather, we are now dealing with a set of congruence classes—a set of sets of numbers. (To get the hierarchy straight, you might wish to call a set such as $\mathbb{Z}_3$—whose elements are themselves sets—a *collection* of sets.) Going up a three-level hierarchy like this should cause no great discomfort because it is something you are accustomed to doing all the time. For example, you deal with individual words on a page, pages in a book (which is a collection of such words), and a shelf of books (which is a collection of such books).

From previous results, we know the following about the integers modulo m.

Theorem 2.5.10. *Let m be an integer with $m > 1$. Then $\mathbb{Z}_m$ is a collection of subsets of $\mathbb{Z}$—the congruence classes modulo m—with the following three properties:*

1. *Each set belonging to $\mathbb{Z}_m$ is nonempty.*

2. *Each element of $\mathbb{Z}$ belongs to at least one member of $\mathbb{Z}_m$.*

3. *No element of $\mathbb{Z}$ belongs to more than one member of $\mathbb{Z}_m$, that is, each two distinct members of $\mathbb{Z}_m$ are disjoint.*

Proof. Exercise. (Just cite whatever previous results are relevant to proving each property.)   □

Property 2 in this theorem means that the set of all integers $\mathbb{Z}$ is the *union* of all the members of $\mathbb{Z}_m$. Property 3 means that the collection $\mathbb{Z}_m$ is *pairwise disjoint*. (See Appendix A for the general notion of the union of a collection of sets and the general notion of a pairwise disjoint collection.)

Theorem 2.5.10 may be summarized by saying that *the set $\mathbb{Z}_m$ of all congruence classes is a **partition** of the set $\mathbb{Z}$ of integers*. Moreover, the set $\mathbb{Z}_m$ is also referred to as the **quotient** of the set $\mathbb{Z}$ under the equivalence relation of congruence. What we have here are just special cases of the more general notions of partition of a set and quotient of a set under an equivalence relation—see Appendix C.

Exercises 2.5.11.   (1) What is $\mathbb{Z}_m$ if we allow modulus $m = 1$?

(2) List the members of the set $\mathbb{Z}_2$. How many are there? How many elements are there in each of these members?

(3) Repeat (2) for $\mathbb{Z}_{12}$.

(4) In general, for any integer $m > 1$, how many members does $\mathbb{Z}_m$ have?

(5) Show that each member of $\mathbb{Z}_6$ is a subset of exactly one member of $\mathbb{Z}_3$, but not conversely. Is $\mathbb{Z}_6$ a subset of $\mathbb{Z}_3$?

Finally, let us return to our original question of giving meaning to $[9] + [20]$, where we are again working modulo 12. The language of representatives helps. We tentatively defined $[9] + [20] = [5]$ because the representatives 9, 20, and 5 of the congruence classes $[9], [20]$, and $[5]$, respectively, satisfy $9 + 20 \equiv 5 \pmod{12}$, that is, $9 + 20 \in [5]$. But what if we had chosen different representatives of $[9]$ and $[20]$, say 45 and -4? Now $45 + (-4) = 41 \in [5]$, too. Thus the answer, $[5]$, would still have been the same had we used these different representatives.

What we have just observed is true in general and will allow us to define not only addition but also, similarly, multiplication, of congruence classes. The general observation is as follows.

Lemma 2.5.12. *Let $m > 1$ be a fixed modulus. Let $A, B \in \mathbb{Z}_m$ be congruence classes modulo m. For any representatives $a, a' \in A$ and any representatives $b, b' \in B$,*

$$[a + b] = [a' + b'],$$
$$[a \cdot b] = [a' \cdot b'].$$

Proof. We supply the proof only for addition; you should carry out the proof for multiplication.

Let $a, a' \in A$ and $b, b' \in B$. Then

$$a \equiv a' \pmod{m} \quad \text{and} \quad b \equiv b' \pmod{m}.$$

Since addition preserves congruence (see Proposition 2.4.5),

$$a + b \equiv a' + b' \pmod{m}.$$

But this congruence is the same thing as saying $[a + b] = [a' + b']$.   □

With this lemma, we are now justified in making the general definition we wanted all along.

Definition 2.5.13. Let $m > 1$ be a fixed modulus. For congruence classes $A, B \in \mathbb{Z}_m$, define their **sum** $A + B$ and **product** $A \cdot B$, respectively, as follows: Choose any representatives $a \in A$ and $b \in B$ and let

$$A + B = [a + b],$$
$$A \cdot B = [a \cdot b].$$

The point of Lemma 2.5.12 is that the sum $A + B$ is **well-defined** in the sense that *this congruence class does not depend upon the choice of the representatives a of A and b of B*; similarly for the product $A \cdot B$. Until we knew this lemma, it would have been dangerous to try to write the definitions simply as $[a] + [b] = [a + b]$ and $[a] \cdot [b] = [a \cdot b]$. After the sum and product are actually defined, these equalities are true, but only because of Lemma 2.5.12. Thus

$$
\begin{aligned}
[a] + [b] &= [a + b], \\
[a] \cdot [b] &= [a \cdot b].
\end{aligned}
$$

Congruence classes modulo m are elements of the set $\mathbb{Z}_m$. When we add or multiply a pair of such congruence classes, we obtain another such congruence class, that is,

$$
\begin{aligned}
A, B \in \mathbb{Z}_m &\implies A + B \in \mathbb{Z}_m, \\
A, B \in \mathbb{Z}_m &\implies A \cdot B \in \mathbb{Z}_m.
\end{aligned}
$$

Thus the set $\mathbb{Z}_m$ is **closed** under addition and multiplication as defined above, just as the set $\mathbb{Z}$ is closed under ordinary addition and multiplication of integers. In short, addition and multiplication between elements of $\mathbb{Z}_m$ are **operations** on $\mathbb{Z}_m$, just as ordinary addition and multiplication are operations on $\mathbb{Z}$. (For the general concept of operation, see Definition B.1.6 in Appendix B.)

With these two operations of addition and multiplication, $\mathbb{Z}_m$ becomes an algebraic object of study. We can ask whether multiplication is associative, or whether multiplication is distributive over addition, and so forth. Such algebraic questions are taken up in Chapter 3, where we compare the algebraic objects $\mathbb{Z}_m$, for various values of m, with the algebraic system of all real numbers.

Here is one such example of an algebraic property of addition in $\mathbb{Z}_m$.

Example 2.5.14. Addition in $\mathbb{Z}_m$ is associative; that is, if $A, B, C \in \mathbb{Z}_m$, then $(A + B) + C = A + (B + C)$.

To prove this, let $A, B, C \in \mathbb{Z}_m$. Choose representatives a, b, c of A, B, C, respectively. Then

$$A + B = [a] + [b] = [a + b],$$
$$(A + B) + C = [a + b] + [c],$$
$$B + C = [b] + [c] = [b + c],$$
$$A + (B + C) = [a] + [b + c].$$

Then

$$(A + B) + C = [a + b] + [c] = [(a + b) + c],$$
$$A + (B + C) = [a] + [b + c] = [a + (b + c)].$$

But, ordinary addition of integers is associative, and so $(a + b) + c = a + (b + c)$. Since these two integers are equal, then their congruence classes $[(a + b) + c]$ and $[a + (b + c)]$ are equal. Thus $(A + B) + C = A + (B + C)$.

According to our earlier definition, we can add only two elements of $\mathbb{Z}_m$ at a time (addition is a *binary* operation). Now that we know addition in $\mathbb{Z}_m$ is associative, then notation such as $A + B + C$ for adding three elements becomes unambiguous: whichever way you parenthesize $A + B + C$, you are only adding a pair of elements at each step—whether as $(A + B) + C$ or as $A + (B + C)$—and so you get exactly the same answer. Similarly for adding four or more elements of $\mathbb{Z}_m$. For example, in $\mathbb{Z}_{12}$, the sum $[5] + [2] + [1] + [10]$ is defined, and equals $[6]$.

The following exercises give a taste of studying additional algebraic properties of $\mathbb{Z}_m$.

Exercises 2.5.15. Let $m > 1$ be a given modulus. Prove each of the following properties of addition and/or multiplication in $\mathbb{Z}_m$:

(1) Addition is commutative; that is, if $A, B \in \mathbb{Z}_m$, then $A + B = B + A$.

(2) There is a unique identity element for addition; that is, there is a unique $N \in \mathbb{Z}_m$ such that $A + N = N + A = A$ for every $A \in \mathbb{Z}_m$. Which congruence class *is* this element N? (Thus, N plays the same role for addition in $\mathbb{Z}_m$ as 0 does in $\mathbb{Z}$.)

(3) Multiplication is associative; that is, if $A, B, C \in \mathbb{Z}_m$, then $(A \cdot B) \cdot C = A \cdot (B \cdot C)$.

(4) Multiplication is distributive over addition; that is, if $A, B, C \in \mathbb{Z}_m$, then $A \cdot (B + C) = (A \cdot B) + (A \cdot C)$.

(5) There is a unique identity element for multiplication; that is, there is a unique $U \in \mathbb{Z}_m$ such that $A \cdot U = U \cdot A = A$ for every $A \in \mathbb{Z}_m$. Which congruence class *is* this element U?

(6) Another property, similar to one for ordinary addition and/or multiplication in $\mathbb{Z}$.

Do not be misled into believing that the algebraic properties of $\mathbb{Z}_m$ under addition and multiplication are all exactly like those of $\mathbb{Z}$, or vice versa.

Exercise 2.5.16. In $\mathbb{Z}_m$, let N be the unique identity element for addition (see the preceding exercise).

(a) In $\mathbb{Z}_2$, find an element $A \neq N$ for which $A + A = N$. (Compare: In $\mathbb{Z}$, if $a + a = 0$, then $a = 0$.) How many such elements A can you find?

(b) In $\mathbb{Z}_3$, find an element $A \neq N$ for which $A + A + A = N$. (Compare: In $\mathbb{Z}$, if $a + a + a = 0$, then $a = 0$.) How many such elements A can you find?

(c) Generalize (a) and (b).

(d) In $\mathbb{Z}_{12}$, are there elements $A \neq N$ and $B \neq N$ for which $A \cdot B = N$? (Compare: In $\mathbb{Z}$, if $ab = 0$, then $a = 0$ or $b = 0$.) How many such A and B pairs can you find?

(e) Repeat (d) for $\mathbb{Z}_5$.

(f) Generalize your answers to (d) and (e).

(g) In $\mathbb{Z}_{12}$, are there elements A, B, C with $C \neq N$ and $A \cdot C = B \cdot C$ yet $A \neq B$? (Compare: In $\mathbb{Z}$, if $c \neq 0$ and $ac = bc$, then $a = b$.) Are there any elements C for which always $A \cdot C = B \cdot C$ implies $A = B$?

(h) Repeat (g) for $\mathbb{Z}_5$.

(i) Generalize your answers to (g) and (h).

Chapter 3

The Real Numbers

Was sind und was sollen die Zahlen?

— Richard Dedekind

What is a real number? There are two quite different ways of answering this question. The first way—the *constructive* approach—builds the set of all real numbers in several steps out of simpler systems. One starts with the set $\mathbb{N}$ of natural numbers and their fundamental properties as expressed, say, by the Peano Postulates[1] (see page 27). Then one constructs the set $\mathbb{Z}$ of all integers; next, from that, the set $\mathbb{Q}$ of all rational numbers; and finally, the set $\mathbb{R}$ of all real numbers—by some sort of completion process to "fill in the gaps" between rational numbers.

(For the constructive approach, the step to get from natural numbers to integers is indicated in Appendix D. The step from integers to rationals is suggested in Example 3.4.10. Two alternative steps to get from rationals to reals are suggested in Exercises 3.3.18 and 3.5.46.)

The constructive approach certainly does tell what the real numbers really are. However, carrying out all the details of all the steps in this approach is somewhat long and tedious.

A second way of answering our question is taken in this chapter—the *axiomatic* approach. This approach is suggested by rephrasing our original question to read: What are the basic properties enjoyed by the set of all real numbers? To answer this question, we shall specify real numbers as objects whose nature is completely described by a small number of fundamental properties—the **axioms** (or **postulates**)—that we assume them to have: nothing is known about these objects except

---

[1] Or, taking things back even one step further, one starts with the theory of sets and constructs the natural numbers within it.

133

what is explicitly stated in these axioms or is logically derivable from them.

This axiomatic approach bypasses the question of whether a set of objects satisfying the assumed properties actually exists. To establish existence, it is indeed possible to follow the constructive approach, and then each property we list as an axiom for the real numbers in this approach is derivable as a theorem in the constructive approach. And this is perhaps the only good reason to carry out the constructive approach. Indeed, in practice we normally do not need to know what real numbers actually are but only how they behave with respect to addition, ordering, etc.

Throughout the development that follows, we must continually remind ourselves which of two complementary attitudes we are taking toward the real numbers.

- When, on the one hand, we are motivating definitions or statements of theorems, or citing the set of real numbers as an example, we must and do rely upon our prior familiarity with the real numbers.

- When, on the other hand, we are deriving consequences of the axioms, we must pretend we do not know anything else about the real numbers except what is expressed in the axioms. That is, we must be sure that what we prove follows strictly from the axioms (and previous consequences of them).

There is one qualification to the latter. Even when following the axiomatic approach, at several crucial points we shall need to use prior knowledge of the set $\mathbb{N}$ of natural numbers, including such things as the Principle of Mathematical Induction. This will be so even though we can identify, within any set satisfying all our axioms, a subset essentially the same as $\mathbb{N}$.

3.1 Fields

The fundamental properties we shall postulate about the real numbers say together that $\mathbb{R}$ is an *archimedean ordered field having the Nested Interval Property*. We now proceed to explain the meaning of these terms. We investigate first, in this section, the notion of a *field*, and then, in the next section, the notion of an *ordered field* and, finally, the notions of an ordered field that is archimedean and has the Nested Interval Property.

A field is, roughly speaking, a set of numbers in which:

- You can add, subtract, multiply, and divide numbers to get numbers in the same set—provided, of course, you do not try to divide by zero:

  > Said Nero unto Julius Caesar, "This problem really should be easier."
  > Said Julius Caesar unto Nero, "You *never* can divide by zero!"

- Addition, subtraction, multiplication, and division obey the usual rules of algebra.

Here is a more precise definition of a field. (Only addition and multiplication are explicitly mentioned in the definition; negatives and reciprocals, also mentioned, lead later to subtraction and division.)

Definition 3.1.1. Let F be a set on which there are given two binary operations: an operation called **addition**, which assigns to each pair (x, y) of elements of F an element of F denoted by $x + y$ and called the **sum** of x and y; and an operation called **multiplication**, which assigns to each pair (x, y) of elements of F an element of F denoted by $x \cdot y$ or simply xy and called the **product** of x and y.[2]

The system consisting of the set F and these two operations is called a **field** provided the following axioms are satisfied:

(A1) (associative law for addition) *For all $x, y, z \in F$,*

$$(x + y) + z = x + (y + z).$$

(A2) (existence of zero element) *There exists an element $0 \in F$ such that, for every $x \in F$,*

$$0 + x = x.$$

(A3) (existence of negatives) *For each $x \in F$ there exists an element $-x \in F$ such that*

$$x + (-x) = 0$$

[here 0 is an element given by Axiom (A2)].

(A4) (commutative law for addition) *For all $x, y \in F$,*

$$x + y = y + x.$$

---

[2]Strictly speaking, the operations of addition and multiplication are *functions* $+: F \times F \to F$ and $\cdot: F \times F \to F$. See Definition B.1.6 in Appendix B.

(M1) (associative law for multiplication) *For all $x, y, z \in F$,*

$$(x \cdot y)z = x(y \cdot z).$$

(M2) (existence of unity) *There exists an element $1 \in F$ such that $1 \neq 0$ [where 0 is an element given by Axiom (A2)] and, for every $x \in F$,*

$$1 \cdot x = x.$$

(M3) (existence of reciprocals) *For each $x \in F$ with $x \neq 0$ [where 0 is an element given by Axiom (A2)], there exists an element $x^{-1} \in F$ such that*

$$x(x^{-1}) = 1$$

[where 1 is an element given by Axiom (M2)].

(M4) (commutative law for multiplication) *For all $x, y \in F$,*

$$x \cdot y = y \cdot x.$$

(D) (distributive law) *For all $x, y, z \in F$,*

$$x(y + z) = (x \cdot y) + (x \cdot z).$$

Examples 3.1.2. (1) The set $\mathbb{R}$ of all real numbers, with its usual addition and multiplication operations, is a field. (This is not something we can prove, since at this point we have no definition of the reals. Rather, these axioms have been set up to express some of the basic properties of the real number system with which we are already familiar.)

(2) The set $\mathbb{Q}$ of all rational numbers, with its usual addition and multiplication operations, is a field.

(3) The set $\mathbb{C}$ of all complex numbers, with its usual addition and multiplication operations, is a field. We shall examine—in fact, formally define—this field in Section 3.6.

(4) The set $\mathbb{Z}$ of all integers, with its usual addition and multiplication operations, is *not* a field. In fact, every one of the nine axioms for a field holds except Axiom (M3); although $0 \neq 2 \in \mathbb{Z}$, there is no element $r \in \mathbb{Z}$ such that $2 \cdot r = 1$.

(5) Let α and ω be any two distinct elements. (What α and ω are is quite irrelevant; the only thing that matters is that $\alpha \neq \omega$.) Then

the two-element set $F = \{\omega, \alpha\}$ becomes a field when addition and multiplication are defined by the rules

$$
\begin{aligned}
\omega + \omega &= \omega, & \omega \cdot \omega &= \omega, \\
\omega + \alpha &= \alpha, & \omega \cdot \alpha &= \omega, \\
\alpha + \omega &= \alpha, & \alpha \cdot \omega &= \omega, \\
\alpha + \alpha &= \omega, & \alpha \cdot \alpha &= \alpha.
\end{aligned}
$$

These lists of sums and products in F can be conveniently summarized in the following **addition table** and **multiplication table**:

+	ω	α
ω	ω	α
α	α	ω

$\cdot$	ω	α
ω	ω	ω
α	ω	α

That all the axioms are satisfied is just a straightforward, if tedious, checking of all relevant cases for each axiom. (Do it!) For example, the commutative law $x + y = y + x$ for addition holds because

$$\alpha + \omega = \alpha = \omega + \alpha$$

(the only other cases are those where $x = y$, when necessarily $x + y = y + x$). Which of α and ω is the zero $0 \in F$? The unity $1 \in F$?

(6) Consider the quotient set $\mathbb{Z}_2 = \{[0], [1]\}$ consisting of the congruence classes of the integers modulo 2 (see Section 2.5). The set $\mathbb{Z}_2$ becomes a field when addition and multiplication are defined in the natural way—by the rules

$$
\begin{aligned}
[a] + [b] &= [a + b], \\
[a] \cdot [b] &= [a \cdot b].
\end{aligned}
$$

These operations of addition and multiplication are well-defined, that is, do not depend on the choice of representatives of the congruence classes. (See Section 2.5.)

The addition and multiplication tables for $\mathbb{Z}_2$ are:

+	[0]	[1]
[0]	[0]	[1]
[1]	[1]	[0]

$\cdot$	[0]	[1]
[0]	[0]	[0]
[1]	[0]	[1]

For most of the field axioms, the stated property holds for $\mathbb{Z}_2$ because the corresponding property holds for $\mathbb{Z}$. For example, Axiom (A2) holds because $[0] \in \mathbb{Z}_2$ satisfies

$$[0] + [b] = [0 + b] = [b]$$

for every $[b]$ (or, look at the addition table to see that always $[0] + [b] = [b]$).

The commutative law $a + b = b + a$ for addition in $\mathbb{Z}$ implies

$$[a] + [b] = [a + b] = [b + a] = [b] + [a],$$

which verifies Axiom (A4). Similarly, the property $1 \cdot a = a$ of multiplication in $\mathbb{Z}$ implies

$$[1] \cdot [a] = [1 \cdot a] = [a],$$

which, together with the fact that $[1] \neq [0]$, verifies Axiom (M2).

Field Axiom (M3) for $\mathbb{Z}_2$ holds even though the corresponding property does *not* hold for $\mathbb{Z}$ [see (4)]. Just look at the multiplication table for $\mathbb{Z}_2$.

Exercise 3.1.3. Is each of the following a field? Why or why not?

(a) The set $\mathbb{R}^+$ consisting of the nonnegative real numbers with their usual addition and multiplication.

(b) The set $\mathbb{R}^2$ of all pairs (x, y) of real numbers with the usual (vector) addition defined by $(x, y) + (u, v) = (x + u, y + v)$ and with multiplication likewise defined "coordinatewise" by $(x, y) \cdot (u, v) = (xu, yv)$.

(c) The set F consisting of all integers together with the reciprocals of all nonzero integers and with elements of F added and multiplied in the usual way.

(d) The set $\mathbb{Q}(\sqrt{2}) = \{q + r\sqrt{2} : q, r \in \mathbb{Q}\}$ with its elements added and multiplied in the usual way as real numbers.

(e) The set $\mathcal{P}$ of all polynomial functions $p: \mathbb{R} \to \mathbb{R}$. [Recall that such a polynomial has the form $p(x) = a_n x^n + a_{n-1} x^{n-1} + \cdots + a_2 x^2 + a_1 x + a_0$ for some nonnegative integer n and some coefficients $a_0, a_1, \ldots, a_n \in \mathbb{R}$.] Elements of $\mathcal{P}$ are added and multiplied "pointwise" as functions; that is, for $p, q \in \mathcal{P}$, the elements $p + q$ and $p \cdot q$ are defined by

$$(p + q)(x) = p(x) + q(x) \qquad (x \in \mathbb{R}),$$
$$(p \cdot q)(x) = p(x) \cdot q(x) \qquad (x \in \mathbb{R}).$$

As formulated, the axioms for a field are in a "weak" form, with no obvious redundancies. This is the most useful form to use when verifying that a particular example satisfies them: we don't have to do any extra work. When using the axioms to derive further properties of a field, on the other hand, we may want to make use of the following "stronger" versions of the field axioms.

Axioms 3.1.4 (Field Axioms—Strong Form).

(A1′) (associative law for addition) *For all* $x, y, z \in F$,

$$(x + y) + z = x + (y + z).$$

(A2′) (existence of zero element) *There exists a* unique *element* $0 \in F$ *such that, for every* $x \in F$,

$$0 + x = x = x + 0.$$

(A3′) (existence of negatives) *For each* $x \in F$ *there exists a* unique *element* $-x \in F$ *such that*

$$x + (-x) = 0 = (-x) + x.$$

(A4′) (commutative law for addition) *For all* $x, y \in F$,

$$x + y = y + x.$$

(M1′) (associative law for multiplication) *For all* $x, y, z \in F$,

$$(x \cdot y)z = x(y \cdot z).$$

(M2′) (existence of unity) *There exists a* unique *element* $1 \in F$ *such that* $1 \neq 0$ *and, for every* $x \in F$,

$$1 \cdot x = x = x \cdot 1.$$

(M3′) (existence of reciprocals) *For each* $x \in F$ *with* $x \neq 0$ *there exists a* unique *element* $x^{-1} \in F$ *such that*

$$x(x^{-1}) = 1 = (x^{-1})x.$$

(M4′) (commutative law for multiplication) *For all* $x, y \in F$,

$$x \cdot y = y \cdot x.$$

(D′) (distributive laws) *For all* $x, y, z \in F$,

$$x(y + z) = (x \cdot y) + (x \cdot z),$$
$$(x + y)z = (x \cdot z) + (y \cdot z).$$

Together these stronger versions of the field axioms are equivalent to the original ones in Definition 3.1.1. In fact, Axioms (A1′)–(D′) obviously imply the original Axioms (A1)–(D). The additional identities in Axioms (A2′), (A3′), (M2′), (M3′), and (D′) follow from the corresponding identities in the original, weaker form together with the commutative law (A4) for addition or (M4) for multiplication. Finally, that the original axioms imply the uniqueness assertions in Axioms (A2′), (A3′), (M2′), and (M3′) is the content of the following two lemmas.

Lemma 3.1.5 (Uniqueness of Zero and Unity). *Let F be a field. Then:*

1. *There is exactly one element $0 \in F$ satisfying Axiom* (A2).

2. *There is exactly one element $1 \in F$ satisfying Axiom* (M2).

Proof.   1. Suppose $0 \in F$ satisfies

$$(\forall x \in F)\,(0 + x = x)$$

and $z \in F$ satisfies

$$(\forall x \in F)\,(z + x = x).$$

We must show that $z = 0$. In the first identity take $x = z$ to get

$$0 + z = z,$$

and in the second take $x = 0$ to get

$$z + 0 = 0.$$

By the commutative law (A4) for addition, also

$$0 + z = 0.$$

Then $z = 0 + z = 0$.

2. Exercise.   □

Lemma 3.1.6 (Uniqueness of Negatives and Reciprocals). *Let F be a field. Then:*

1. *For each $x \in F$ there is exactly one element $-x \in F$ satisfying Axiom* (A3).

2. *For each $x \in F$ with $x \neq 0$ there is exactly one element $x^{-1} \in F$ satisfying Axiom* (M3).

Proof. Exercise.   □

Definition 3.1.7. In a field F, the unique element $0 \in F$ satisfying Axiom (A2 ′) is called the **zero** element of the field; all other elements of F are, of course, said to be **nonzero**. The unique element $1 \in F$ satisfying Axiom (M2 ′) is called the **unity** element of the field.

For $x \in F$, the unique element $-x \in F$ satisfying Axiom (A3) is called the **negative** of x; when $x \neq 0$, the unique element $x^{-1} \in F$ satisfying Axiom (M3) is called the **reciprocal** of x.

For example, in the field $\mathbb{Z}_2$, the zero is $[0]$ and the unity is $[1]$. Of course, in the field $\mathbb{Q}$, the zero and unity are what we usually denote by 0 and 1.

Exercise 3.1.8.   (a) Prove that $-0 = 0$ in every field.

  (b) Is it possible to have $-1 = 1$ in a field?

  (c) Let x belong to a field F. Prove that $x = 0$ if and only if $-x = 0$.

  (d) Prove a similarly interesting "if and only if" statement about $x = 1$. (Yes, $x = 1$ if and only if $-x = -1$, but that would not be very interesting. Why not?)

Exercise 3.1.9. In $\mathbb{Z}_2$, what is

  (a) the negative $-[1]$ of the only nonzero element $[1]$?

  (b) the reciprocal $[1]^{-1}$ of $[1]$?

  (c) the sum $1 + 1$, where 1 is the unity of this field?

Exercise 3.1.10. In the quotient set $\mathbb{Z}_m$ consisting of the congruence classes of the integers modulo m, define operations of addition and multiplication in the natural way—by the rules

$$[a] + [b] = [a + b],$$
$$[a] \cdot [b] = [a \cdot b].$$

Just as we saw earlier in the special case $m = 2$, it is immediate that all the field axioms hold for $\mathbb{Z}_m$ with the possible exception of (M3), which asserts the existence of reciprocals.

  (a) Show that reciprocals exist in $\mathbb{Z}_3$ so that $\mathbb{Z}_3$ is a field.

  (b) Show that reciprocals do not exist for all elements of $\mathbb{Z}_4$ so that $\mathbb{Z}_4$ is not a field.

  (c) Is $\mathbb{Z}_6$ a field?

  (d) For which positive integers m is $\mathbb{Z}_m$ a field?

The notational abbreviations in the following definitions are convenient.

Definition 3.1.11. Let x, y be elements of a field F. The **difference** of x and y is the element $x - y \in F$ defined by

$$x - y = x + (-y).$$

If $y \neq 0$, then the **quotient** of x and y is the element $x/y \in F$ (also denoted $\dfrac{x}{y}$) defined by

$$x/y = x(y^{-1}).$$

In particular, the negative of an $x \in F$ is a difference:

$$-x = 0 - x.$$

And the reciprocal of a nonzero $y \in F$ is a quotient:

$$y^{-1} = 1/y.$$

Subtraction—forming differences—is an operation $(x, y) \mapsto x - y$ on F (that is, a function $F \times F \to F$ whose domain $F \times F$ consists of all pairs of elements of F). However, division—forming quotients—is not quite an operation on F since the domain of $(x, y) \mapsto x/y$ must exclude all pairs of the form $(x, 0)$.

The following proposition lists a number of familiar properties of addition and multiplication. You should supply proofs for at least some of these, being careful in each case to use only the nine axioms for a field (in either their original weak form or their derived strong form) together with previously established properties.

Proposition 3.1.12. *Let F be a field. Let $x, y, z, w \in F$. Then:*

1. *(cancellation law for addition) If $x + z = y + z$, then $x = y$.*

2. $-(-x) = x.$

3. $x \cdot 0 = 0 = 0 \cdot x.$

4. $(-1)x = -x.$

5. $x(-y) = -(xy) = (-x)y.$

6. $(-x)(-y) = xy.$

7. $x(y - z) = xy - xz.$

8. *(cancellation law for multiplication) If $xz = yz$ and if $z \neq 0$, then $x = y$.*

9. *If $xy = 0$, then $x = 0$ or $y = 0$.[3] Hence if $x \neq 0$ and $y \neq 0$, then $xy \neq 0$.*

---

[3] As usual, the "or" here is inclusive: it allows for the possibility that both $x = 0$ and $y = 0$.

10. *If $x \neq 0$, then $x^{-1} = 1/x$.*

11. *If $x \neq 0$ and $y \neq 0$, then $(xy)^{-1} = x^{-1}y^{-1}$.*

12. *If $x \neq 0$, then $x^{-1} \neq 0$ and $(x^{-1})^{-1} = x$.*

13. *If $x \neq 0$, then $-x \neq 0$ and $(-x)^{-1} = -(x^{-1})$.*

14. *If $y \neq 0$ and $z \neq 0$, then $(zx)/(zy) = x/y$.*

15. (cross-multiplication rule) *If $y \neq 0$ and $w \neq 0$, then $x/y = z/w$ if and only if $xw = yz$.*

16. *If $y \neq 0$ and $w \neq 0$, then*

$$\frac{x}{y} + \frac{z}{w} = \frac{xw + yz}{yw}.$$

17. *If $y \neq 0$ and $w \neq 0$, then*

$$\frac{x}{y} \cdot \frac{z}{w} = \frac{xz}{yw}.$$

Proof. Proofs or hints of proofs appear here for only a few of the parts. Be sure to prove the others yourself.

1. Assume $x + z = y + z$. Add $-z$ (on the right) to both sides:

$$(x + z) + (-z) = (y + z) + (-z).$$

By associativity of addition,

$$x + (z + -z) = y + (z + -z).$$

But $z + -z = 0$, and so

$$x + 0 = y + 0.$$

Finally, since 0 is the identity for addition,

$$x = y.$$

[*Note:* Once you get the knack of doing all these steps separately, you may wish to combine them into a form such as follows. Although this concise form does not leave space to cite each axiom used, still the way it is written makes clear which ones are involved. Here it is.

Assume $x + z = y + z$. Then,

$$x = x + 0 = x + (z + -z) = (x + z) + (-z)$$
$$= (y + z) + (-z) = y + (z + -z) = y + 0 = y.$$

(You might still wish to include before the "Then" in such a write-up a statement of the key step involved, such as, "Add $-z$ to both sides.")]

3. Multiply x by $0 + 0$.

4. Be careful. Does your proof really makes sense? See the next exercise for a common fallacious proof.

5. Again be careful. And again see the next exercise.

6. Use part 5.

9. Start something like this: Assume $xy = 0$ but $x \neq 0$. Then deduce that $y = 0$. (Why would this prove what is asserted?) □

Exercise 3.1.13.   (a) Criticize the following "proof" of part 4 of Proposition 3.1.12:

$$(-1)x = (-1)(x) = -(1)(x) = -(1x) = -(x) = -x.$$

(b) Criticize the following "proof" of the right equality in part 5 of Proposition 3.1.12:

$$(-x)y = (-x)(y) = -(x)(y) = -(xy).$$

There are many other elementary properties of addition and multiplication easily derivable from the field axioms and Proposition 3.1.12. One example is $(x - y) - z = x - (y + z)$; some others are indicated in the following exercises. In the future we shall not hesitate to use such properties without explicitly mentioning them.

Exercise 3.1.14. In a field F, a *fraction* is an element of the form a/b for $b \neq 0$. Establish the following properties concerning fractions.

(a) If $y \neq 0$, then $-(x/y) = (-x)/y = x/(-y)$. If, moreover, $x \neq 0$, then $x/y \neq 0$ and $(x/y)^{-1} = y/x$.

(b) If $y \neq 0$ and $w \neq 0$, then $x/y - z/w = (xw - yz)/yw$. If, moreover, $z \neq 0$, then $(x/y)/(z/w) = (xw)/(zy)$.

(c) If $y \neq 0$ and $z \neq 0$, then $(x/y)/(z/y) = x/z$.

Exercise 3.1.15. For an arbitrary element x of a field F, its *square* x^2 is defined to be $x \cdot x$ and its *double* $2 \times x$ to be $x + x$. (Please temporarily resist the temptation to denote $2 \times x$ simply by $2x$. Observe that $2 \times x = 0$ is possible in a field even when $x \neq 0$—for example, in $\mathbb{Z}_2$—whereas we are accustomed to thinking that $2x \neq 0$ whenever $x \neq 0$.)

For $x, y \in F$, prove,

$$(x + y)^2 = x^2 + 2 \times (xy) + y^2,$$
$$x^2 - y^2 = (x + y)(x - y).$$

We want to generalize the notion of double by defining $n \times x$ for every $n \in \mathbb{N}$. We do so recursively. Fix $x \in F$. By ordinary recursion (Theorem 1.2.9), there is a unique sequence $(s_n)_{n \in \mathbb{N}}$ such that

$$
\begin{cases}
s_0 &= 0, \\
s_{n+1} = s_n + x & (n \in \mathbb{N}).
\end{cases}
\tag{*}
$$

Definition 3.1.16. For an element x of a field F and a natural number n, the element $n \times x$ of F is defined to be the element s_n given by (*), above. In particular, when $x = 1$, then $n \times x$ is denoted by n_F or simply by n.

By definition, for $x \in F$,

$$0 \times x = 0,$$
$$1 \times x = x,$$
$$2 \times x = x + x,$$
$$3 \times x = 2x + x,$$
$$\vdots$$
$$(n + 1) \times x = (n \times x) + x \qquad (n \in \mathbb{N}).$$

In particular, $0 = $ the zero element of F, $1 = $ the unity element of F, $2 = 1 + 1$, $3 = 2 + 1$, etc. In view of associativity of addition, we may write

$$2 \times x = x + x,$$
$$3 \times x = x + x + x,$$
$$4 \times x = x + x + x + x, \text{ etc.}$$

In particular, $2 = 1 + 1$, $3 = 1 + 1 + 1$, $4 = 1 + 1 + 1 + 1$, etc.

From the start, we used the same notation, 1, for both the first positive natural number and the unity element of any field. Although the context usually indicates which of the two meanings is intended, still the ambiguity can at times cause confusion. For example, $1 + 1 \neq 0$ in $\mathbb{N}$, whereas $1 + 1 = 0$ in $\mathbb{Z}_2$. Likewise, we now are using the same notation n for a natural number and the associated element $n \times 1$ of any field. This ambiguity, too, can cause confusion. For example, $3 \neq 0$ in $\mathbb{N}$ as well as in $\mathbb{Z}_2$, whereas $3 = 0$ in $\mathbb{Z}_3$. When the context is not sufficient to remove ambiguity and prevent confusion, we revert to the more precise notation n_F for the element of a field F associated with a natural number n.

Definition 3.1.16 produced a way to multiply an element x of a field F by a natural number n, and the "product" was denoted $n \times x$. It is tempting to denote $n \times x$ merely by nx, but that leads to yet another possible ambiguity. After all, nx (that is, $n \cdot x$) also denotes the product of the elements n_F and x of F. Fortunately, there is no actual ambiguity. For example,

$$2 \times x = x + x = 2_F \cdot x$$

(why?). More generally,

$$\boxed{n \times x = n_F \cdot x}$$

for every $n \in \mathbb{N}$, as you are asked to prove in the first part of the next exercise. Accordingly, there is no harm in writing $n \cdot x$, or even just nx, instead of $n \times x$.

Exercise 3.1.17. Let x be an element of a field F.

(a) Prove that, for every nonnegative integer n, the element $n \times x$, as defined recursively above, is the same as the product $n_F \cdot x$ of the elements n_F and x of F.

(b) Show that $(m + n)_F = m_F + n_F$ and $(mn)_F = m_F n_F$ for all $m, n = 0, 1, 2, \ldots$.

(c) Give a definition of the multiples nx for $n = -1, -2, \ldots$. (Thus nx is defined for every integer n.) Then show that $(m + n)x = mx + nx$ for all integers m and n.

Like the notion of double, so the notion of square of a field element, introduced in Exercise 3.1.15, also generalizes.

Exercise 3.1.18. Let F be a field.

(a) For $x \in F$, give a definition of powers x^n for $n = 0, 1, 2, \ldots$. If $x \neq 0$, also give a definition of powers x^n for $n = -1, -2, \ldots$. (Your definitions should, of course, generalize the usual ones for powers of real numbers.)

(b) From your definitions, prove that, for $x \neq 0$,

$$x^{-n} = 1/x^n \qquad (n \in \mathbb{Z}).$$

(c) From your definitions, prove that, for $x \in F$,

$$x^{m+n} = x^m x^n \qquad (x \in F; m, n \in \mathbb{Z}).$$

Remember that m or n can be negative, too; you may wish to separate the proof into several cases.

(d) State and prove some other laws of exponents.

(e) Does the law $(a + b)^3 = a^3 + 3a^2b + 3ab^2 + b^3$ always hold for elements a, b of a field F?

(f) More generally, does the Binomial Theorem (Theorem 1.5.10) hold in an arbitrary field F?

Earlier, we noted that $n_F = 0$ is possible in a field F even when the natural number $n \neq 0$. (For example, $2_F = 0$ in $F = \mathbb{Z}_2$.) Now such an n cannot be 1 because $1_F = 1$ and, by a field axiom, the unity and zero elements of a field are distinct. Hence such an n, if it exists at all, must be at least 2. A definition is in order.

Definition 3.1.19. If $n_F = 0$ in a field F for some positive integer n, then the least such positive integer is called the **characteristic** of the field. If no such positive integer n exists, then we say that F has **characteristic 0**.

For example, $\mathbb{Z}_2$ has characteristic 2 and $\mathbb{Z}_3$ has characteristic 3. After we state more axioms for the field of real numbers, it will of course turn out that $\mathbb{R}$ has characteristic 0. For now, we look at some properties of the characteristic of a field and, in particular, at the distinction between characteristic > 0 and characteristic 0.

Proposition 3.1.20. *Let F be a field.*

1. *If F has characteristic $n > 0$, then $nx = 0$ for every $x \in F$.*

2. *If F has characteristic 0, then $nx \neq 0$ for all positive integers n and all $x \in F$ with $x \neq 0$.*

Proof. 1. Exercise

2. Let the field F have characteristic 0. Just suppose there exists $0 \neq x \in F$ and a positive integer n such that $nx = 0$, that is, $n_F \cdot x = 0$. Multiply by x^{-1} to get $n_F = 0 \cdot x^{-1} = 0$. $\square$

Exercise 3.1.21. Generalize Proposition 3.1.20, part 2, by proving that, if x is a nonzero element of a field F of characteristic 0, then $mx \neq nx$ for all natural numbers m, n with $m \neq n$.

The fields $\mathbb{Z}_2$ and $\mathbb{Z}_3$ are finite.[4] The field $\mathbb{Z}_2$ of two elements has characteristic 2, and the field $\mathbb{Z}_3$ of three elements has characteristic 3. More generally, the following holds.

Proposition 3.1.22. *A finite field has characteristic $n > 0$.*

Proof. Exercise. □

Put differently, Proposition 3.1.22 asserts that a field of characteristic 0 must be infinite. In particular, then, the fields $\mathbb{Q}$ and $\mathbb{R}$ are infinite.

Does the converse of Proposition 3.1.22 hold; that is, must a field of characteristic $n > 0$ be finite? The answer is no, and you can construct a counterexample—an infinite field of characteristic $n > 0$—by suitably altering a later example in this chapter. Be on the lookout for it.

Exercises 3.1.23.   (1) Prove that if a field F has characteristic $n > 0$, then n is necessarily a prime number.

(2) This exercise demonstrates that a field can be finite—hence have prime characteristic—yet not have a prime number of elements.

Let 0, 1, α, and β be any four distinct elements. (What α and β are is quite irrelevant.) Form the four-element set $F = \{0, 1, \alpha, \beta\}$.

(a) Verify that F becomes a field when addition and multiplication are defined by the following tables.

+	0	1	α	β
0	0	1	α	β
1	1	0	β	α
α	α	β	0	1
β	β	α	1	0

·	0	1	α	β
0	0	0	0	0
1	0	1	α	β
α	0	α	β	1
β	0	β	1	α

(b) Determine the characteristic of F.

(3) If F is a finite field, prove that the characteristic p of F must divide the number of elements of F.

---

[4]A set is **finite** if it either is empty or else has the form $\{x_1, x_2, \ldots, x_n\}$ for some positive integer n and some distinct elements $x_1, x_2, \ldots, x_n$. And a set is **infinite** if it is not finite. (Finite and infinite sets are discussed in detail in Chapter 4.)

3.2 Ordered Fields

Besides their operations of addition and multiplication, the fields $\mathbb{Q}$ of rational numbers and $\mathbb{R}$ of real numbers have *orderings* which allow us to say which of two elements is the greater and, in particular, whether a given element is positive, that is, greater than zero. We wish to abstract this notion of ordering.

Definition 3.2.1. Let F be a field. Suppose F has a subset P with the following properties:

(O1) (trichotomy law) *If $x \in F$, then one and only one of the following is true:*

$$x = 0, \qquad x \in P, \qquad -x \in P.$$

(O2) *If $x, y \in P$, then $x + y \in P$.*

(O3) *If $x, y \in P$, then $xy \in P$.*

Then we call F an **ordered** field and say that P **orders** F. If $x \in P$, then x is said to be **positive**. If $x \in F$ with $x \neq 0$ and $x \notin P$ (which, according to the trichotomy law, means that $-x \in P$), then x is said to be **negative**. If x is not negative, that is, if either $x = 0$ or x is positive, then x is said to be **nonnegative**.

Exercises 3.2.2. (1) Let F be an ordered field.

  (a) Prove: If x is a nonzero element of F, then its square $x^2 = x \cdot x$ must be positive.

  (b) Show that the unity 1 of F is positive.

  (c) When is the product xy of two elements of F positive?

  (d) When is the inverse x^{-1} of a nonzero element x of F positive?

  (e) Prove: If $x, y \in F$ with $y \neq 0$, then x/y is positive if and only if xy is positive.

(2) Let F be the field $\mathbb{Q}(\sqrt{2})$ consisting of all real numbers of the form $q + r\sqrt{2}$ with q and r rational [see Exercise 3.1.3 (d)]. Let P be the set of those elements of F that are positive as real numbers. Show that P orders F.

(3) (a) Can the field $\mathbb{Z}_2$ [Example 3.1.2 (6)] be ordered? That is, is there a subset P of this field satisfying the order axioms (O1)–(O3)?

  (b) Repeat (a) for the field $\mathbb{Z}_3$.

(c) Repeat (a) for the field $\mathbb{Z}_m$—for those m for which it is a field. (See Exercise 3.1.10.)

(d) Repeat (a) for the field $\mathbb{C}$ of all complex numbers. (If you are not familiar with complex numbers, then you may want to wait to do this part until you have read Section 3.6.)

(4) Can a field F be ordered in more than one way? That is, can there be two different subsets P_1 and P_2 each satisfying the axioms (O1)–(O3) for an ordered field?

From its set P of positive elements in an ordered field F we construct the corresponding **order relation**.

Definition 3.2.3. Let F be an ordered field whose set of positive elements is P. Let $x, y \in F$. Then each of the expressions

$$x > y, \qquad y < x$$

is defined to mean that $x - y \in P$. Each of the expressions

$$x \geq y, \qquad y \leq x$$

is defined to mean that either $x > y$ or else $x = y$. As usual, $x \not> y$ is defined to mean the negation of $x > y$, etc.

(What we have really done here is to define $<$ and $\leq$ as **relations** in the set F—see Section A.3. And $>$ and $\geq$ are just the relations **opposite** $<$ and $\leq$, respectively.)

In particular, in an ordered field F, the expressions $x > 0$ and $0 < x$ both mean that x is positive. In other words,

$$P = \{ x \in F : x > 0 \}.$$

Also, $x \leq x$ for each $x \in F$.

Proposition 3.2.4. *Let F be an ordered field. Let $x, y, z \in F$. Then:*

1. (trichotomy law) *Exactly one of the following is true:*

$$x = y, \qquad x > y, \qquad x < y.$$

2. *If $x < y$, then $x + z < y + z$.*

3. *If $x < y$ and $z > 0$, then $xz < yz$; if $x < y$ and $z < 0$, then $xz > yz$.*

4. *If $x < y$, then $-y < -x$.*

5. *If $x < y$ with $x \neq 0$ and $y \neq 0$, then $y^{-1} < x^{-1}$.*

6. (transitive law) *If $x < y$ and $y < z$, then $x < z$.*

7. (antisymmetric law) *If $x \le y$ and $y \le x$, then $x = y$.*

8. *If $x \ne 0$, then $x \cdot x > 0$.*

9. *$1 > 0$.*

Proof. Exercise. □

We shall often use without explicit mention the assertions of the preceding proposition.

According to the trichotomy law, above, for distinct elements $x, y \in F$, either $x < y$ or $y < x$. Then, just as we are accustomed to doing when $F = \mathbb{R}$, we may represent elements of F along a number line and mark the point representing x to the left of the point representing y when $x \le y$.

In the language of Section A.3, the properties

$$
\begin{aligned}
x &\le x & &\text{(reflexivity),} \\
x &\le y \,\&\, y \le x \;\Longrightarrow\; x = y & &\text{(antisymmetry),} \\
x &\le y \,\&\, y \le z \;\Longrightarrow\; x \le z & &\text{(transitivity),} \\
x &\le y \text{ or } y \le x & &\text{(comparability)}
\end{aligned}
$$

together say that the relation $\le$ in an ordered field F is a **total ordering** of the set F.

Because of transitivity, writing $x \le y \le z$ (and similarly with $<$ replacing $\le$) to mean $x \le y \,\&\, y \le z$ is both acceptable and useful. However, writing something like $3 < 5 > 4$ is confusing and unacceptable.

For an ordered field F—as for any set with a total ordering—the notion of **least element**, or **minimum**, of a subset A of F has an obvious meaning (compare page 43); the minimum of A, if it exists, is denoted $\min A$. For example, 0 is the least element of the set of nonnegative elements of F. Similarly, the notion of **greatest element**, or **maximum**, of $A \subset F$ has an obvious meaning (state it!); the maximum of A, if it exists, is denoted $\max A$.

The comparability property, above, implies that, for $x, y \in F$, either $x \le y$ or $y \le x$ (or both—in which case $x = y$ by antisymmetry). In other words, the two-element (or one-element) set $\{x, y\}$ has both a minimum and a maximum. If $x \le y$, then $\min\{x, y\} = x$ and $\max\{x, y\} = y$; otherwise, $\min\{x, y\} = y$ and $\max\{x, y\} = x$.

The properties that, for all $x, y, z \in F$,

$$
\begin{aligned}
x \le y \;&\Longrightarrow\; x + z \le y + z, \\
x \le y \,\&\, z \ge 0 \;&\Longrightarrow\; x \cdot z \le y \cdot z,
\end{aligned}
$$

which follow immediately from parts 2 and 3 of Proposition 3.2.4, say that addition and multiplication (by a nonnegative element) both **preserve** the total ordering. Together these two properties say that the total ordering $\leq$ in an ordered field F is **compatible with** the field operations.

Exercise 3.2.5. Let x, y, z be nonzero elements in an ordered field F. If $y < z$, how do the quotients x/y and x/z compare? The quotients y/x and z/x?

Just as for real numbers, so for elements of any ordered field there is a notion of absolute value.

Definition 3.2.6. Let F be an ordered field. For $x \in F$, its **absolute value** $|x|$ is defined by

$$|x| = \begin{cases} x & \text{if } x \geq 0, \\ -x & \text{if } x < 0. \end{cases}$$

This definition provides an absolute-value function

$$\begin{array}{ccc} F & \to & F \\ x & \mapsto & |x| \end{array}$$

on F whose values $|x|$ are always nonnegative. (At times the absolute-value function is considered a function from F to $P \cup \{0\}$, where P is the set of positive elements of F.)

The algebraic properties of absolute value in any ordered field are familiar ones:

Proposition 3.2.7. *Let F be an ordered field. Let $x, y \in F$. Then:*

1. $|x| \geq 0$; further, $|x| = 0$ if and only if $x = 0$.

2. $|-x| = |x|$.

3. $|x + y| \leq |x| + |y|$.

4. $|x \cdot y| = |x| \cdot |y|$.

5. $-|x| \leq x \leq |x|$.

6. If $y \geq 0$, then $|x| \leq y$ if and only if $-y \leq x \leq y$.

Proof. It is simplest to prove the parts of the proposition in an order different from that in which they are listed. (For later reference, we want parts 1–3 listed together.)

1. This is obvious from the definition.

2. Consider separately the cases $x \geq 0$ and $x < 0$. (Another way is to deduce part 2 as a special case of 4.)

5. Consider separately the cases $x \geq 0$ and $x < 0$.

6. Exercise.

3. Use parts 5 and 6: in part 6, replace x with $x + y$ and y with $|x| + |y|$.

4. Consider four different cases: $x \geq 0$ and $y \geq 0$; $x \geq 0$ and $y < 0$; $x < 0$ and $y \geq 0$; and $x < 0$ and $y < 0$. (The four cases can be boiled down to only three essentially different cases. How?) □

Exercises 3.2.8. Let x and y be elements in an ordered field.

(1) When is $|x + y| = |x| + |y|$? (Compare part 3 of the preceding proposition.)

(2) Prove that $\bigl| |x| - |y| \bigr| \leq |x - y|$. Draw figures illustrating this inequality. When is the inequality an equality?

(3) Show that $\max\{x, y\}$ can be expressed as $(x + y + |x - y|)/2$. Similarly express $\min\{x, y\}$.

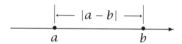

Figure 3.1: Absolute value as distance.

For $z \in F$, an ordered field, we may think of $|z|$ as being the magnitude, or size, of z irrespective of the sign of z. Then for $x, y \in F$, we may think of the magnitude $|x - y|$ of the difference between x and y as the *distance* between x and y. See Figure 3.1. Accordingly, for $x, a \in F$ and $0 < \varepsilon \in F$, we may think of the set

$$\{x \in F : |x - a| < \varepsilon\}$$

as consisting of all elements of distance less than ε from a.

Exercise 3.2.9. Let x, a, ε be elements of an ordered field F with $\varepsilon > 0$. Prove that the set $\{x \in F : |x - a| < \varepsilon\}$ is the *open interval* $(a - \varepsilon, a + \varepsilon)$, that is, the set $\{x \in F : a - \varepsilon < x < a + \varepsilon\}$ consisting of all elements of F that are strictly between $a - \varepsilon$ and $a + \varepsilon$.

In Figure 3.1 and others we shall draw, it appears that the ordered field has infinitely many elements. This appearance is not deceptive, as the next proposition and its corollary indicate. Recall that a field F has characteristic > 0 when there is some positive integer n such that $n \cdot 1 = 0$, and F has characteristic 0 otherwise. Recall also that the elements $n_F \in F$ for $n \in \mathbb{N}$ were defined recursively by $0_F = 0 =$ the zero element of F and $(n + 1)_F = n_F + 1$ for all $n \geq 0$.

Lemma 3.2.10. *If F is an ordered field, then $n_F > 0$ for each positive integer n.*

Proof. The definition of n_F was by recursion, so you should, of course, try to use induction to prove this result. $\square$

The following is an immediate consequence of Lemma 3.2.10.

Proposition 3.2.11. *An ordered field has characteristic 0.*

Corollary 3.2.12. *An ordered field has infinitely many elements.*

Proof. By Proposition 3.1.22, a finite field has characteristic > 0. $\square$

In particular, finite fields of the form $\mathbb{Z}_m$ cannot be ordered.

In our program for characterizing the reals axiomatically, our next task is to identify the natural numbers, integers, and rational numbers within an ordered field F. We begin with the natural numbers.

Definition 3.2.13. For an ordered field F, its **nub** $N(F)$ is the subset of F defined by

$$N(F) = \{ n_F : n \in \mathbb{N} \}.$$

This definition generalizes from the cases $F = \mathbb{Q}$ and $F = \mathbb{R}$, in both of which $N(F) = \mathbb{N}$, the set of natural numbers.

This is a good place to recall what was said, in the chapter's introduction, about the two complementary attitudes being taken toward the real numbers. When we motivate the definition of the nub $N(F)$ or look at examples, we may freely use our familiarity with $\mathbb{R}$ (and $\mathbb{Q}$). When, however, we prove properties of $N(F)$, we must pretend to not yet know about $\mathbb{R}$—although we certainly do need to know about its subset $\mathbb{N}$—and, instead, must rely exclusively upon the definitions, axioms, and previously proved properties.

An ordered field F has characteristic 0, and then

$$m \neq n \implies m_F \neq n_F \qquad (m, n \in F)$$

(see Exercise 3.1.21). Thus the elements $0, 1, 2 = 1 + 1, 3 = 2 + 1 = 1 + 1 + 1, 4 = 3 + 1 = 1 + 1 + 1 + 1$, etc., constituting $N(F)$ are all distinct. In other words, the map $\mathbb{N} \to N(F)$ given by $n \mapsto n_F$ is a one-to-one correspondence (that is, a bijection) between $\mathbb{N}$ and $N(F)$. Accordingly, for a natural number n we shall often drop the subscript and denote the corresponding element n_F of an ordered field simply by n.

The definition of nub of an ordered field refers to something that may be outside that field, namely, the set $\mathbb{N}$ of natural numbers. An *internal* characterization of the nub—a description referring to elements only of the field itself—is sometimes useful. To suggest such a characterization, we abstract from properties of the natural numbers. We know that the set of natural numbers includes zero[5] and also includes, along with each natural number n, its successor—the next natural number, $n + 1$. Accordingly, we give a name to subsets of the ordered field in question that share this property (the name honors the creator of the Peano Postulates—see page 27).

Definition 3.2.14. Let F be an ordered field. A subset I of F is called a **Peano set**—more precisely, a **Peano subset** of F—if it has the following two properties:

(a) $0 \in I$; and

(b) if $x \in I$, then $x + 1 \in I$.

In an ordered field F the entire set F is a Peano set. So is the set $P \cup \{0\} = \{x \in F : x \geq 0\}$ consisting of all nonnegative elements of F.

Exercises 3.2.15. (1) In the ordered field $\mathbb{R}$, find, if possible:

(a) several Peano subsets besides $\{x \in \mathbb{R} : x \geq 0\}$ and $\mathbb{N}$ that consist only of nonnegative numbers;

(b) several that include negative numbers; and

(c) several that are proper subsets of $\mathbb{N}$.

(2) Let F be an arbitrary ordered field. Prove the following.

(a) The nub $N(F)$ of F is itself a Peano subset of F.

(b) The nub $N(F)$ is a subset of every Peano subset of F.

(c) The nub $N(F)$ is the intersection of the collection of all Peano subsets of F.

The preceding exercise provides the desired internal characterization of nub: The nub $N(F)$ of an ordered field F is the *smallest* Peano subset of F (smallest with respect to set inclusion). In other terms, $N(F)$ consists of exactly those elements $x \in F$ that belong to *every* Peano subset of F.

---

[5]Recall that our convention is that the natural numbers start with 0 rather than 1.

Exercise 3.2.16.   (a) Prove that the field $\mathbb{Q}$ can be ordered in only one way. That is, prove that if P is a subset of $\mathbb{Q}$ satisfying the properties (O1)–(O3) for an ordered field, then P must be the set of rational numbers that are positive in the usual sense.

(*Hint:* Show first that each positive integer belongs to P. From that deduce that the set of rational numbers positive in the usual sense is a subset of P. Then prove the reverse inclusion. To keep things straight, it may help to denote the order relation corresponding to P by $\prec$, and its opposite by $\succ$, while reserving the notation $<$ for the usual order relation in $\mathbb{R}$, and $>$ for its opposite.)

(b) Prove that the field $\mathbb{R}$ can be ordered in only one way.

The nub $N(F)$ plays the same role within an arbitrary ordered field that the set $\mathbb{N}$ of natural numbers does within the particular ordered field $\mathbb{R}$. In our axiomatic approach to the real numbers, the next steps are to construct within any ordered field the analogs of the integers and rational numbers within $\mathbb{R}$. How to accomplish this is easy and fairly obvious. Recall that, in any field F, the quotient a/b of two elements with $b \neq 0$ is defined to be the element ab^{-1} of F.

Definition 3.2.17. Let F be an ordered field. Define $Z(F)$ to be the subset

$$Z(F) = N(F) \cup \{ -n : n \in N(F) \}$$

of F consisting of all elements of the nub of F together with their negatives. We call elements of $Z(F)$ the **integral elements** of F.

Define $Q(F)$ to be the subset

$$Q(F) = \left\{ \frac{m}{n} : m, n \in Z(F), n \neq 0 \right\}$$

of F. We call elements of $Q(F)$ the **rational** elements of F.

Of course, when $F = \mathbb{R}$ (assuming we already know what the real numbers are), then $Z(F) = \mathbb{Z}$ and $Q(F) = \mathbb{Q}$.

Exercise 3.2.18. For an ordered field F, prove the following.

(a) $N(F) \subset Z(F) \subset Q(F)$.

(b) The sum, product, and difference of two elements of $Z(F)$ belong to $Z(F)$.

(c) The sum and product of two elements of $Q(F)$ belong to $Q(F)$, in other words, the subset $Q(F)$ is closed under the operations of addition and multiplication in F. Moreover, these operations, when restricted to just elements of $Q(F)$, make $Q(F)$ a field in its own right. (We may call $Q(F)$ a *subfield* of the field F—see Definition 3.6.9.)

(d) The field $Q(F)$ can be ordered. [Indicate both the set P of positive elements of $Q(F)$ and the corresponding order relation $\le$ in $Q(F)$.]

Both ordered fields $\mathbb{Q}$ and $\mathbb{R}$ have the property named in the following definition.

Definition 3.2.19. An ordered field F is said to be **archimedean** and to have the **Archimedean Ordering Property** when

(AOP) *If* $x \in F$ *and* $y \in F$ *with* $y > 0$, *then there exists some* $n \in \mathbb{N}$ *such that* $ny > x$.

In particular, in an archimedean ordered field, *for each* $x \in F$, *there exists some* $n \in \mathbb{N}$ *such that* $n > x$ (that is, such that $n_F > x$).

Think of the Archimedean Ordering Property like this: Given $0 < y \in F$, no matter how small y is, and given $x \in F$, no matter how large x is, there is some sufficiently large $n \in \mathbb{N}$ for which $ny > x$. Figure 3.2 illustrates this geometrically.

Figure 3.2: Archimedean Ordering Property.

By its recursive definition (see Definition 3.1.16), the product of a positive integer n and an arbitrary element y in an ordered field may be regarded as the n-term sum $y + y + \cdots + y$ obtained by adding y to itself repeatedly. Stated informally, then, the Archimedean Ordering Property promises:

- *If you add a very small positive number to itself enough times, then you can obtain as large a number as you wish.*

Frequently I am pestered to respond to surveys by filling out questionnaires. Nearly every such request promises that responding will "take only a few minutes of your time" (some even enclose a quarter or a dollar bill as enticement). Suppose, indeed, that each questionnaire takes only 5 minutes to complete. That is hardly very long—but with enough surveys the time certainly adds up. The Archimedean Ordering Property in action!

U. S. Senator Everett M. Dirkson is reported to have said once of federal expenditures, "A billion here, a billion there, and pretty soon you're talking about *real* money." The Archimedean Ordering Property in action?

Exercises 3.2.20.   (1) Call an element β in an ordered field F *infinite* when $\beta > n$ for every $n \in \mathbb{N}$; call $\varepsilon \in F$ *infinitesimal* when $0 < \varepsilon < 1/n$ for every $n \in \mathbb{N}$.

    (a) Show that $\beta \in F$ is infinite if and only if $1/\beta$ is infinitesimal, and that $\varepsilon \in F$ is infinitesimal if and only if $1/\varepsilon$ is infinite.

    (b) Prove that an ordered field F is archimedean if and only if F has no infinitesimal elements.

(2) Prove the following generalization of the Division Theorem (Theorem 2.1.1): If a and b are arbitrary elements of an archimedean ordered field F with $b > 0$, then there exists a unique element $q \in \mathbb{N}$ and a unique element $r \in F$ such that $a = qb + r$ and $0 \leq r < b$.

(3) Let F be the ordered field $\mathbb{Q}(\sqrt{2})$ consisting of all real numbers of the form $q + r\sqrt{2}$ with q and r rational [see Exercise 3.1.3 (d)]. Without using the fact that $\mathbb{R}$ is archimedean, prove that the field F is archimedean. (You *may* use the fact that $\mathbb{Q}$ is archimedean.)

Non-archimedean ordered fields do exist, but, unfortunately, constructing examples of them takes some work—see Exercise 3.2.21 below. Of course, since our aim here is to characterize the real number system, we are really interested only in ordered fields that *are* archimedean.

Exercise 3.2.21. By a *rational function* we shall mean a quotient of the form $p(x)/q(x)$, where $p(x)$ and $q(x)$ are polynomials

$$p(x) = \sum_{j=0}^{m} a_j x^j = a_0 + a_1 x + a_2 x^2 + \cdots + a_m x^m,$$

$$q(x) = \sum_{j=0}^{n} b_j x^j = b_0 + b_1 x + b_2 x^2 + \cdots + b_n x^n$$

whose coefficients $a_0, a_1, \ldots, a_m$ and $b_0, b_1, \ldots, b_n$ are all rational and in which the *leading coefficient* b_n of the denominator $q(x)$ is not zero. (The leading coefficient a_m of the numerator $p(x)$ may be zero; that is, the numerator may be identically zero.) We consider two such rational functions $p(x)/q(x)$ and $r(x)/s(x)$ to be the same when $p(x)/q(x) = r(x)/s(x)$ in the usual sense, that is, when $p(x)s(x) = q(x)r(x)$.

In this exercise you are going to verify that, with appropriately defined operations and set of positive elements, the set F of all such rational functions becomes a non-archimedean ordered field.

By definition, the denominator $q(x)$ of such a rational function cannot be the zero polynomial, whereas the numerator $p(x)$ may be. We do not want to worry about the domain of such a quotient—which may exclude some values of x at which the denominator vanishes but the numerator does not. Hence it is useful to consider such quotients $p(x)/q(x)$ purely as formal expressions.

Strictly speaking, we should consider not a quotient $p(x)/q(x)$ but rather a pair $(p(x), q(x))$ of polynomials whose second entry is not identically zero. [Further, we might even consider the polynomial $p(x)$ as the list $(a_0, a_1, \ldots, a_m)$ of its coefficients, and similarly the polynomial $q(x)$.] Then we should use as elements of F not the actual pairs $(p(x), q(x))$ themselves, but rather equivalence classes under the equivalence relation defined by

$$(p(x), q(x)) \sim (r(x), s(x)) \quad \text{if and only if} \quad p(x)s(x) = q(x)r(x).$$

(Compare the treatment of $\mathbb{Z}_m$ as a set of congruence classes in Section 2.5 and see also Appendix C.)

Add and multiply elements of F in the usual way:

$$\frac{p(x)}{q(x)} + \frac{r(x)}{s(x)} = \frac{p(x)s(x) + q(x)r(x)}{q(x)s(x)},$$

$$\frac{p(x)}{q(x)} \cdot \frac{r(x)}{s(x)} = \frac{p(x)r(x)}{q(x)s(x)}.$$

(a) (*Optional*) When F is considered, as indicated above, as a set of equivalence classes, verify that the operations of addition and multiplication are well-defined. (Compare the discussion of defining addition and multiplication in $\mathbb{Z}_m$—see page 129.)

(b) Verify that the operations of addition and multiplication so defined make F a field.

Now let P be the subset of F consisting of those rational functions $p(x)/q(x)$ whose leading coefficients a_m and b_n are both positive or both negative.

(c) (*Optional*) When F is considered as a set of equivalence classes, indicate how to define P. Remember that, since F is the set of equivalence classes, then P also has to be a set of equivalence classes. Hence when you indicate whether an equivalence class belongs to P, you need to be sure this does not depend on which representative of the equivalence class you use.

(d) Verify that P orders F, that is, F becomes an ordered field when P is taken as its set of positive elements.

(e) Determine the nub $N(F)$ of F.

(f) Show that the ordered field F does not have the Archimedean Ordering Property. (*Hint:* Show that—with respect to the order relation corresponding to P—the inequality $n/1 < x/1$ holds for all positive integers n.)

(g) Which elements of F are infinitesimal? Which are infinite? [See Exercise 3.2.20(1) for the necessary definitions.]

(h) What, if anything, above changes substantially if the coefficients of the polynomials are allowed to be real numbers instead of just rational numbers?

Exercise 3.2.22. Let F be the set of all "generalized polynomials" in a variable x that have the form

$$f(x) = \cdots + a_{-2}x^{-2} + a_{-1}x^{-1} + a_0x^0 + a_1x^1 + a_2x^2 + \cdots + a_mx^m$$

$$= \sum_{j=-\infty}^{m} a_jx^j, \tag{*}$$

where m is some integer and the coefficients a_j are rational numbers. Notice that in the expression for $f(x)$ there are dots at the far left but not at the far right. In other words, the terms a_jx^j with positive powers of x in such an $f(x)$ are all zero beyond some value m of j (which varies from one such generalized polynomial to another); by contrast, the terms $a_{-j}x^{-j}$ with negative powers of x may be nonzero for infinitely many, or even all, negative integers $-j$. [When $a_j = 0$ for $j = -1, -2, \ldots$, the generalized polynomial $f(x)$ is just an ordinary polynomial.]

We shall consider generalized polynomials such as (*) as formal expressions involving the symbol x—not as formulas into which we substitute actual numbers for x. If you are uncomfortable with this idea, then you may wish to consider each such formal expression (*) as a list

$$(\ldots, a_{-2}, a_{-1}, a_0, a_1, a_2, \ldots, a_m)$$

—that is, as a family indexed by $\{ j \in \mathbb{Z} : j \leq m \}$ for some integer m. Then you should translate everything said below about operations with generalized polynomials into corresponding statements about operations with such lists.

Addition in F is defined term by term, just as for ordinary polynomials. Suppose $f(x) \in F$ is given by (*) and $g(x) \in F$ is given by

$$g(x) = \cdots + b_{-2}x^{-2} + b_{-1}x^{-1} + b_0x^0 + b_1x^1 + b_2x^2 + \cdots + b_nx^n$$

$$= \sum_{j=-\infty}^{n} b_jx^j. \tag{**}$$

Then the sum $f(x) + g(x)$ is defined by

$$f(x) + g(x) = \cdots + (a_{-2} + b_{-2})\,x^{-2} + (a_{-1} + b_{-1})\,x^{-1} + (a_0 + b_0)\,x^0$$
$$+ (a_1 + b_1)\,x^1 + (a_2 + b_2)\,x^2 + \cdots + (a_k + b_k)\,x^k$$
$$= \sum_{j=-\infty}^{k} \left(a_j + b_j\right) x^j,$$

where $k = \max\{m, n\}$, $a_j = 0$ for $j > m$, and $b_j = 0$ for $j > n$.

To see how to define multiplication in F, recall the way ordinary polynomials are customarily multiplied by collecting like powers of x. For example,

$$\left(a_0 + a_1 x + a_2 x^2\right) \cdot \left(b_0 + b_1 x + b_2 x^2\right)$$
$$= (a_0 b_0) + (a_0 b_1 + a_1 b_0)\,x + (a_0 b_2 + a_1 b_1 + a_2 b_0)\,x^2$$
$$+ (a_1 b_2 + a_2 b_1)\,x^3 + (a_2 b_2)x^4.$$

In such a product, the coefficient of x^j (when j is not too large) is the sum $a_0 b_j + a_1 b_{j-1} + \cdots + a_{j-1} b_1 + a_j b_0$, which may be written as $\sum_{i=0}^{j} a_i b_{j-i}$. We extend this multiplication to typical elements $f(x)$ and $g(x)$—of the forms (*) and (**), respectively—as follows:

$$f(x) \cdot g(x) = \sum_{j=-\infty}^{m+n} \left(\sum_{i=-\infty}^{j} a_i b_{j-i} \right) x^j,$$

where we let $a_i = 0$ for all $i > m$ and $b_i = 0$ for all $j > n$. (Why is the inner sum in the double sum really an ordinary sum of finitely many numbers?)

(a) Verify that the operations of addition and multiplication so defined make F a field.

Now let P be the subset of F consisting of those $f(x)$ of the form (*) whose leading coefficient a_m is positive. Note that m may be negative. For example, the element $(-2/5)x^{-3} + 4x^{-1}$ of F belongs to P, but the element $\cdots + 3x^{-3} + 2x^{-2} + 1x^{-1} + 0x^0 + (-1)x^1 + (-2)x^2$ does not.

(b) Verify that P orders F.

(c) Is F archimedean when ordered by P?

Recall that, for each x in an archimedean ordered field F, there exists $n \in \mathbb{N}$ such that $n > x$ (see page 157). Since we are postulating that the real numbers form a certain archimedean ordered field, then, for each $x \in \mathbb{R}$, there exists $n \in \mathbb{N}$ such that $n > x$. Let us sharpen this result.

Lemma 3.2.23. *Let $x \in \mathbb{R}$. Then there exists a unique integer n such that $n \le x < n + 1$.*

Proof. Existence. First suppose $x \ge 0$. By the Archimedean Ordering Property, there exists a natural number k with $k > x$. By the Well-Ordering Principle, there is a least such k. Let $n = k - 1$. Then $n \le x < n + 1$ (why?).

Now suppose $x < 0$. Since $-x \ge 0$, there exists a natural number k with $k > x$. Then $k - x > 0$. Now apply what we proved in the preceding paragraph to the real number $k - x$.

Uniqueness. Exercise. $\square$

The unique integer n such that $n \le x < n + 1$ is *the greatest integer that is less than or equal to x.* (Why?) Exercise 2.1.4 (2) asked you to prove the existence of such n in the case that x is rational—and dared you to prove it in the more general case that x is real. As you now see, it takes the Archimedean Ordering Property to prove it for arbitrary real x.

Definition 3.2.24. For $x \in \mathbb{R}$, the **floor** of x, denoted by $\lfloor x \rfloor$, is the greatest integer less than or equal to x.

A real number $x > 0$ may now be written uniquely in the form

$$x = n + t, \qquad \text{where } n \in \mathbb{N}, t \in \mathbb{R}, 0 \le t < 1,$$

namely,

$$n = \lfloor x \rfloor, \qquad t = x - \lfloor x \rfloor.$$

Accordingly, n is often called the **integral part** of x, and t the **fractional part**. For example, the integral part of $18/7$ is $\lfloor 18/7 \rfloor = 2$, and the fractional part of $18/7$ is $4/7$. The integral part of $\sqrt{3}$ is $\lfloor \sqrt{3} \rfloor = 1$, and the fractional part is $\sqrt{3} - 1 = 0.7320508\ldots$, a nonterminating, nonrepeating decimal. (Decimal expansions are treated rigorously in Section 3.5.)

Observe that taking the floor of a real number rounds it *down* to the nearest integer but does not necessarily produce the nearest integer to it.

Exercises 3.2.25.   (1) Is $\lfloor x + y \rfloor = \lfloor x \rfloor + \lfloor y \rfloor$? Is $\lfloor x \cdot y \rfloor = \lfloor x \rfloor \cdot \lfloor y \rfloor$?

  (2) For real x, justify the existence of a unique least integer that is greater than or equal to x. This integer is called the *ceiling* of x and is denoted $\lceil x \rceil$.

  (3) What is the relationship of $\lfloor x \rfloor$ to $\lceil x \rceil$?

(4) What is the relationship of $\lfloor -x \rfloor$ and $\lceil -x \rceil$, on the one hand, to $\lfloor x \rfloor$ and $\lceil x \rceil$, on the other hand?

(5) Explain why taking the ceiling of a real number rounds it *up* to the nearest integer but does not necessarily produce the nearest integer to it.

(6) Express in terms of floor and/or ceiling the *nearest integer* to a given real number x, that is, the integer n for which the distance $|x - n|$ is least. Of course, there is no single nearest integer in the case where x is precisely midway between two consecutive integers, and so you should exclude this case. (This case is what gives rise to all the complications of rounding off numbers expressed in decimal form.) See, though, what your expression(s) for the nearest integer actually give when applied in this exceptional case.

(7) Deduce Lemma 3.2.23 from Exercise 3.2.20 (2).

(8) How many digits are there in the decimal representation of a given positive integer? (Count each digit as many times as it appears.) Express your answer in terms of logarithms along with floor or ceiling.

Before looking at the next theorem, try the following exercise.

Exercise 3.2.26. Prove that between any two distinct rational numbers there is some rational number. Prove, in other words, that if $a, b \in \mathbb{Q}$ with $a < b$, then there exists some $c \in \mathbb{Q}$ such that $a < c < b$.

When applied to the field $\mathbb{R}$ of real numbers, the Archimedean Ordering Property implies that the set $\mathbb{Q}$ of rational numbers is **order-dense** in $\mathbb{R}$, that is:

- *Between any two distinct real numbers there is some rational number.*

In our context of archimedean ordered fields, the statement is as follows:

Theorem 3.2.27 (Order Density). *Let F be an archimedean ordered field. If $a, b \in F$ with $a < b$, then there exists some $q \in Q(F)$ such that $a < q < b$.*

Proof. It is enough to consider the case $0 \le a < b$. (Why is that enough?) Choose $c, \varepsilon \in F$ with $c > 0$, $\varepsilon > 0$, and

$$a = c - \varepsilon, \quad b = c + \varepsilon.$$

(How? What is really going on in this step? Draw a diagram!) What we need to show is that $|c - q| < \varepsilon$ for some $q \in Q(F)$.

There exists some $n \in N(F)$ with $n > 0$ and

$$\frac{1}{n} < \varepsilon.$$

(That is the key step; justify it!) Then it suffices to show that $|c - q| \le 1/n$ for some $q \in Q(F)$. Now there exists $m \in N(F)$ with $m > 0$ and

$$m - 1 \le nc < m.$$

(Why?) Then $m/n \in Q(F)$ with

$$\frac{m}{n} - \frac{1}{n} \le c < \frac{m}{n} < \frac{m}{n} + \frac{1}{n}.$$

Hence $|c - m/n| < 1/n$, as required.   □

When applied to $F = \mathbb{R}$, the preceding proof actually shows:

- **Each real number can be approximated as closely as you wish by rational numbers.**

In fact, for $c \in \mathbb{R}$ given, the proof shows that, for each $\varepsilon > 0$, there exists some $q \in \mathbb{Q}$ with $|c - q| < \varepsilon$. Moreover, the proof shows that for each positive integer n, there exists $q_n \in \mathbb{Q}$ with $|c - q_n| < 1/n$; then $\lim_{n \to \infty} q_n = c$.[6] Thus:

- **Each real number is the limit of a sequence of rational numbers.**

Loosely speaking, order density of $\mathbb{Q}$ in $\mathbb{R}$ means that at every place you look among the reals you see rational numbers nearby. So there are a great many rational numbers, thickly strewn among the reals. Are there, then, so many rational numbers that the remaining real numbers—the irrational numbers—are quite sparsely scattered? Are there, in fact, more rational than irrational numbers? See the following exercise as well as Section 4.3, especially Example 4.3.7 (5).

Exercise 3.2.28. Prove that the set $\mathbb{R} \setminus \mathbb{Q}$ of irrational numbers is also order dense in $\mathbb{R}$: between any two distinct real numbers there is some irrational number. (*Hint:* Let $a, b \in \mathbb{R}$ with $a < b$. Choose any positive irrational number c whatsoever—say $c = \sqrt{2}$. By order density applied to a/c and b/c, there is some rational r with $a/c < r < b/c$.)

Note that this result means there are a great many irrational numbers, thickly strewn among the reals!

---

[6]The notion of limit of a sequence is examined rigorously in Section 3.5.

3.3   Nested Intervals and Completeness

To distinguish the archimedean ordered fields $\mathbb{R}$ and $\mathbb{Q}$ from one another, we shall now introduce one more property an ordered field may have. This one involves the notion of closed intervals, which are defined in any ordered field the same way they are in the familiar case of $\mathbb{R}$.

Definition 3.3.1. Let a, b be elements of an ordered field F with $a < b$. Then the **closed interval** in F with endpoints a and b is the subset $[a, b]$ of F defined by

$$[a, b] = \{\, x \in F : a \leq x \leq b \,\}.$$

Figure 3.3: Closed interval.

Such a closed interval is depicted in the usual way, as in Figure 3.3. The corresponding **open interval** (a, b), defined to be $\{\, x \in F : a < x < b \,\}$, is depicted the same way except that the solid dots at the endpoints are replaced by open dots, as in Figure 3.4.

Figure 3.4: Open interval.

Exercise 3.3.2.   (a) Show that the intersection of two closed intervals in an ordered field is itself a closed interval—unless it is empty.

(b) What can you say about the intersection of two open intervals?

(c) Must the union of two closed intervals always be a closed interval? If so, why? If not always, then when?

(d) Let a, b be elements of the field $\mathbb{Q}$ of rational numbers with $a < b$. As above, we may form the closed interval $J = [a, b]$ in $\mathbb{Q}$ with endpoints a and b. But a, b are also elements of the larger field $\mathbb{R}$ of real numbers, and so we may also form the closed interval $K = [a, b]$ in $\mathbb{R}$ with the same endpoints. What is the relationship of K to J?

The final property needed for our axiomatic characterization of the real numbers is the one that follows. This property really involves certain kinds of *sequences* of closed intervals in F. Recall that a sequence

$(J_n)_{n=0,1,2,...}$ of intervals, like any sequence, is just a family $(J_n)_{n \in \mathbb{N}}$ indexed by the set $\mathbb{N}$ of natural numbers. (See Section B.4.)

Definition 3.3.3. An ordered field F is said to have the **Nested Interval Property** when

(NIP) *For each sequence $(J_n)_{n \in \mathbb{N}}$ of closed intervals for which*

$$J_{n+1} \subset J_n \quad \text{for all } n \in \mathbb{N},$$

there exists some c such that

$$c \in J_n \quad \text{for all } n \in \mathbb{N},$$

that is, the intersection $\bigcap_{n \in \mathbb{N}} J_n = \{x : x \in J_n \text{ for all } n \in \mathbb{N}\}$ is nonempty.

In the notation of the preceding definition, for each $n \in \mathbb{N}$, let

$$J_n = [a_n, b_n].$$

The condition that $J_{n+1} \subset J_n$ for all $n \in \mathbb{N}$ can be expressed as

$$a_0 \le a_1 \le a_2 \le \cdots \le a_n \le a_{n+1} \le b_{n+1} \le b_n \le \cdots \le b_2 \le b_1 \le b_0.$$

Then the Nested Interval Property guarantees the existence of some c for which

$$a_n \le c \le b_n \quad \text{for all } n \in \mathbb{N}.$$

(See Figure 3.5, where the successive intervals, which really lie one inside another on the same line, are shown "exploded" away from that line so that they can more readily be seen. Of course, the figure can show only several of the intervals, not the entire infinite sequence of them.) Thus, the Nested Interval Property is one way to say that points that "ought to be there" in F really are there or, in other terms, that *there are no "holes" in F.*

In our axiomatic approach, our basic assumption about the real numbers is:

- *The field $\mathbb{R}$ of real numbers is an archimedean ordered field having the Nested Interval Property.*

Exercise 3.3.4. (a) Does the conclusion of the Nested Interval Property still hold if the intervals J_n are open rather than closed?

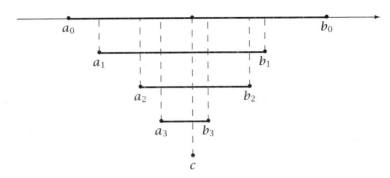

Figure 3.5: Nested Interval Property.

(b) Let the "nested" closed intervals J_n be as in the Nested Interval Property except that there are only finitely many of them, say $J_0, J_1, \ldots, J_k$. Without using the Nested Interval Property, show that there exists some $c \in \bigcap_{n=0}^{k} J_n$.

Does this result for finitely many intervals hold in the field $\mathbb{Q}$ as well as in $\mathbb{R}$? In any ordered field? Does it hold if the intervals are open instead of closed?

The Nested Interval Property does indeed distinguish the archimedean ordered fields $\mathbb{Q}$ and $\mathbb{R}$ from each other.

Example 3.3.5. The field $\mathbb{Q}$ of rational numbers does *not* have the Nested Interval Property. In fact, we may use recursion to obtain a sequence $(J_n)_{n \in \mathbb{N}}$ of closed intervals in $\mathbb{Q}$ such that, for each $n \in \mathbb{N}$,

$$J_n = [a_n, b_n] \cap \mathbb{Q} \subset \left[\sqrt{2} - \frac{1}{n+1}, \sqrt{2} + \frac{1}{n+1} \right] \cap \mathbb{Q},$$

$$J_{n+1} \subset J_n.$$

For example, we may start the recursion by taking

$$J_0 = \left[\frac{1}{2}, 2 \right] \cap \mathbb{Q}.$$

[Justify the existence of the sequence $(J_n)_{n \in \mathbb{N}}$.]

We included the "$\cap \, \mathbb{Q}$" above to emphasize that each J_n is an interval in $\mathbb{Q}$, *not* in $\mathbb{R}$.

If $c \in J_n$ for every $n \in \mathbb{N}$, then $c \in \mathbb{Q}$ with

$$\sqrt{2} - \frac{1}{n+1} \leq c \leq \sqrt{2} + \frac{1}{n+1}$$

for every natural number n. But the only real number c satisfying all these inequalities is $\sqrt{2}$ (why?), and so $c = \sqrt{2}$. This is impossible since, as we know, $\sqrt{2} \notin \mathbb{Q}$ (see Example 1.4.6 or 2.2.20).

There is one gap in the preceding argument showing that $\mathbb{Q}$ does not have the Nested Interval Property: the construction uses the real number $\sqrt{2}$. How do we know $\sqrt{2}$—a real number $x > 0$ for which $x^2 = 2$—actually exists? (In asking this question we are, of course, approaching things strictly from the axiomatic viewpoint.) The existence of such a number is established below, in Proposition 3.3.15, as a consequence of our axiomatic characterization of $\mathbb{R}$.

We would like to say that $\mathbb{R}$ is *the* one and only archimedean ordered field having the Nested Interval Property. This is not exactly true but nearly so. It turns out that any two such fields are essentially the same—in technical terms, there is an isomorphism between the two that preserves the ordering. These ideas are discussed in Section 3.4.

Once we know that $\mathbb{R}$ as an archimedean ordered field is essentially unique, then we may identify within it the natural numbers, integers, and rational numbers by defining $\mathbb{N} = N(\mathbb{R})$, $\mathbb{Z} = Z(\mathbb{R})$, and $\mathbb{Q} = Q(\mathbb{R})$.

Another way, besides using the Nested Interval Property, to express the idea that there are no "holes" in $\mathbb{R}$ is to say that the ordered field of real numbers is *order-complete*—least upper bounds of certain sets exist. Let us introduce the associated terminology and then establish that order completeness of the reals is a consequence of our assumption that $\mathbb{R}$ is an archimedean ordered field having the Nested Interval Property.

Definition 3.3.6. Let A be a subset of an ordered field F. An element b of F is called an **upper bound of** A **in** F when each $a \in A$ satisfies $a \leq b$, that is,

$$(\forall a \in A)\,(a \leq b).$$

When such an upper bound exists, we say that A is **bounded above in** F.

Equivalently, $b \in F$ is an upper bound of $A \subset F$ when no element of A is greater than b, that is,

$$\neg\,(\exists a \in A)\,(b < a),$$

where the symbol $\neg$ denotes logical negation. Moreover, $b \in F$ is *not* an upper bound of A in F when

$$(\exists a \in A)\,(b < a).$$

Examples 3.3.7. (1) The numbers $3, 4, 5, 18/5, \pi$, and $3\sqrt{2}$ are upper bounds of the closed interval $A = [-4, 3]$ in $F = \mathbb{R}$, whereas neither 2 nor -2 is an upper bound. Notice that one of these upper bounds, namely, 3, belongs to A.

(2) The same things are true of the half-open interval

$$A = [-4, 3) = \{\, x \in \mathbb{R} : -4 \le x < 3 \,\}$$

except that now the upper bound 3 of A does *not* belong to A.

(3) In $\mathbb{R}$, the set $\mathbb{N}$ does not have any upper bound, that is, is not bounded above (why?). Likewise, neither $\mathbb{Z}$ nor $\mathbb{Q}$ is bounded above in $\mathbb{R}$.

(4) The entire field $\mathbb{R}$ is not bounded above in $\mathbb{R}$.

(5) In any F, the empty set $\varnothing$ is bounded above. In fact, *every* element of F is an upper bound of $\varnothing$ in F (why?).

(6) Let

$$A = \{\, t \in \mathbb{Q} : t^2 < 2 \,\}.$$

Then $\sqrt{2}$ is an upper bound of A in $\mathbb{R}$, but $\sqrt{2}$ is not an upper bound of A in $\mathbb{Q}$ because $\sqrt{2} \notin \mathbb{Q}$. Of course, A still has many upper bounds in $\mathbb{Q}$—for example, 2. Thus, whether a number is an upper bound of a subset A of an ordered field F depends on which field F is being considered and not just what A is!

Exercise 3.3.8.   (a) Determine the set U of *all* upper bounds of the open interval $(-4, 3) = \{\, x \in \mathbb{R} : -4 < x < 3 \,\}$ in $\mathbb{R}$.

(b) Repeat (a) for the open interval $\{\, x \in \mathbb{Q} : -4 < x < 3 \,\}$ in $\mathbb{Q}$.

(c) Repeat (a) for the set $\{\, x \in \mathbb{Q} : -4 < x < 3 \,\}$ in $\mathbb{R}$.

Among all the upper bounds of the half-open interval $[-4, 3)$ in $\mathbb{R}$, the upper bound 3 is clearly the least.

Definition 3.3.9. Let A be a subset of an ordered field F. An element $s \in F$ is called a **supremum of** A when it is a least element—hence *the* least element—of the set of all upper bounds of A in F. When A has such a supremum, this unique least upper bound is denoted $\sup A$. (Another notation for the least upper bound of A is simply $\operatorname{lub} A$.)

Thus $s \in F$ is the supremum of A in F when:

- s is an upper bound of A in F, that is, $a \le s$ for each $a \in A$; and

- $s \le b$ for each upper bound b of A in F.

The notation "sup" is read "soup." (And "lub" is read like the job you have done when your car needs grease.)

Exercises 3.3.10.   (1) The second requirement just listed for s to be a supremum of A in F can be symbolized as

$$(\forall b \in F) \, ((\forall a \in A) \, (a \le b) \implies s \le b).$$

Is this equivalent to

$$(\forall b \in F) \, (\forall a \in A) \, (a \le b \implies s \le b)?$$

(2) Let F be an ordered field and let $B \subset A \subset F$. If A has a supremum in F, must B have one? If B has a supremum in F, must A have one?

(3) Does the empty set $\varnothing$ have a supremum in $\mathbb{R}$? Why or why not?

(4) Find the least upper bound in $\mathbb{R}$ of each of the following.

  (a) $\{\, 1 + 1/n : n \in \mathbb{N}^* \,\}$.

  (b) $\{\, 1 - 1/n : n \in \mathbb{N}^* \,\}$.

  (c) $\{\, (1 + 1/n)^n : n \in \mathbb{N}^* \,\}$.

(5) Let F be an archimedean ordered field having the Nested Interval Property. Show that, for each $c \in F$,

$$c = \sup\{\, q \in Q(F) : q < c \,\},$$

where $Q(F)$ is, as in Definition 3.2.17, the set of rational elements of F. In particular, $c = \sup\{\, q \in \mathbb{Q} : q < c \,\}$ for each $c \in \mathbb{R}$.

(6) Let A and B be subsets of an ordered field F having least upper bounds a and b, respectively. Show that $a + b$ is the least upper bound of the set $\{\, a + b : a \in A, b \in B \,\}$ of sums. Is the analogous result true for the set of products?

Exercise 3.3.11. The "below" analogs of *upper bound* and *supremum* are **lower bound** and **infimum**. Do the following:

(a) Define the notion of a *lower bound* in an ordered field F of a subset A of F. Give examples of subsets of $\mathbb{R}$ and of $\mathbb{Q}$ having no such lower bounds and having many lower bounds.

(b) Define the notion of an *infimum* in an ordered field F of a subset A of F. Show that such an infimum, if it exists, is necessarily unique. Then we may speak of *the* infimum of A and denote it by $\inf A$ or by $\mathrm{glb}\, A$. What does the notation glb stand for?

(c) Given a subset A of an ordered field F having an infimum g, find a subset B of F with $-g = \sup B$.

(d) Repeat (d) the other way around; that is, start with a supremum s of a subset of the field.

A subset of an ordered field F is said to be **bounded** when it is both bounded below and bounded above.

(e) Give examples of subsets of $\mathbb{R}$ and of $\mathbb{Q}$ that are bounded below but not bounded; bounded above but not bounded; and bounded.

(f) Show that a subset A of an ordered field F is bounded if and only if there exists some $b > 0$ for which $|a| \leq b$ for all $a \in A$.

As the following exercise indicates, the notions of supremum and infimum generalize the notions of maximum and minimum, respectively. Then a good way to regard the supremum of a set that has no actual maximum is as the next best thing to a maximum, and similarly for the infimum.

Exercise 3.3.12.   (a) Prove: If a subset A of an ordered field F has a maximum m, then $m = \sup A$.

(b) Give an example of a subset of $\mathbb{R}$ having no maximum that has a supremum.

(c) Repeat (a) and (b) for minimum and infimum instead of maximum and supremum.

The following theorem is another way of saying that $\mathbb{R}$ does not have any "holes."

Theorem 3.3.13 (Order Completeness). *Let F be an archimedean ordered field having the Nested Interval Property. Then each nonempty subset of F that is bounded above in F has a supremum in F.*

Proof. Let A be a nonempty subset of F that is bounded above in F. Construct recursively, as follows, a family $([a_n, b_n])_{n \in \mathbb{N}}$ of closed intervals in F each of which intersects A.

Choose $a_0 \in A$ and choose some upper bound b_0 of A in F. Form the closed interval $[a_0, b_0]$. Given $[a_n, b_n]$, form the next interval $[a_{n+1}, b_{n+1}]$ as follows. Let

$$c = \frac{a_n + b_n}{2},$$

the midpoint of $[a_n, b_n]$. If the right half $[c_n, b_n]$ of $[a_n, b_n]$ intersects A, let $[a_{n+1}, b_{n+1}] = [c_n, b_n]$; otherwise, let $[a_{n+1}, b_{n+1}] = [a_n, c_n]$. In either case, b_n is an upper bound of A and

$$b_{n+1} - a_{n+1} = \frac{1}{2}(b_n - a_n).$$

By induction,

$$b_n - a_n = \frac{b_0 - a_0}{2^n}$$

for each $n \in \mathbb{N}$.

By the Nested Interval Property, there exists some c with

$$c \in [a_n, b_n]$$

for each $n \in \mathbb{N}$. We shall show that c is the supremum of A in F.

(Drawing diagrams now may help you keep track of the relative positions of the various numbers involved.)

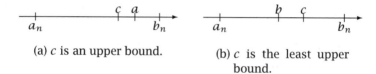

(a) c is an upper bound.

(b) c is the least upper bound.

Figure 3.6: Least upper bound exists.

First, we show that c is an upper bound of A in F. Just suppose there exists some $a \in A$ with $c < a$. Then $c < a \leq b_1$. By the Archimedean Ordering Property (supply details), there is some n with

$$b_n - a_n < a - c.$$

But $a_n \leq c \leq b_n$ and $c < a \leq b_n$, whence $a - c < b_n - a_n$ [see Figure 3.6 (a)]. This is a contradiction.

Second, we show that c is the *least* upper bound of A in F. Just suppose there is some upper bound b of A in F with $b < c$. There exists an n with $b_n - a_n < (c - b)/2$ (why?). But $a_n \leq c \leq b_n$ and $a_n \leq b < c \leq b_n$ [see Figure 3.6 (b)], whence $c - b \leq b_n - a_n$. This is a contradiction. $\square$

In view of the preceding theorem, we may say that the ordered field $\mathbb{R}$ is **order-complete**.

Exercises 3.3.14. (1) Prove or disprove:

(a) Each nonempty subset of the ordered field $\mathbb{Q}$ that is bounded above in $\mathbb{Q}$ has a supremum in $\mathbb{Q}$.

(b) Each nonempty subset of $\mathbb{Q}$ that is bounded above in $\mathbb{Q}$ has a supremum in $\mathbb{R}$.

    (c) Each nonempty subset of $\mathbb{Q}$ that is bounded above in $\mathbb{R}$ has a supremum in $\mathbb{Q}$.

(2) Formulate and prove an analog of order completeness for lower bounds. (See Exercise 3.3.11.)

(3) Let $([a_n, b_n])_{n \in \mathbb{N}}$ be a sequence of closed intervals in $\mathbb{R}$ for which $[a_{n+1}, b_{n+1}] \subset [a_n, b_n]$ for all n. Justify the existence of $a = \sup\{a_n : n \in \mathbb{N}\}$ and $b = \inf\{b_n : n \in \mathbb{N}\}$. Then show that $[a, b] \subset [a_n, b_n]$ for every n.

One important application of order completeness is the existence of square roots of positive real numbers.

Proposition 3.3.15. *Let $c \in \mathbb{R}$ with $c > 0$. Then there exists a unique $x \in \mathbb{R}$ with $x > 0$ and $x^2 = c$.*

Proof. We write the proof only for the case $c = 2$. The proof for arbitrary c is similar.
    Let

$$A = \{t \in \mathbb{R} : t > 0 \text{ and } t^2 < 2\}.$$

Since $2/3 > 0$ and $(2/3)^2 < 2$, then $2/3 \in A$. Thus A is nonempty. If $t > 3$, then $t^2 \geq t > 2$ and $t \notin A$; in other words, if $t \in A$, then $t \leq 3$. Thus A is bounded above in $\mathbb{R}$.
    By the order completeness of $\mathbb{R}$, the set A has a least upper bound x in $\mathbb{R}$. Certainly $x > 0$ (why?). It remains to show that $x^2 = 2$. Suppose not. There are two cases, $x^2 < 2$ and $x^2 > 2$, which we consider separately.
    Case 1: $x^2 < 2$. The strategy is to produce a (rational) number $h > 0$ for which $(x+h)^2 < 2$ also. This will mean that $x+h \in A$ and $x+h > x$, which will contradict the fact that x is an upper bound of A in $\mathbb{R}$.
    To find such h, note that, for any h whatsoever,

$$(x + h)^2 = x^2 + 2xh + h^2 = x^2 + h(2x + h).$$

When $0 < h < 1$, then $2x + h < 2x + 1$ so that

$$(x + h)^2 < x^2 + h(2x + 1).$$

But we want $(x + h)^2 < 2$, that is,

$$x^2 + h(2x + 1) < 2.$$

The preceding inequality is equivalent to

$$h < \frac{2 - x^2}{2x + 1}.$$

Now the fraction on the right-hand side is positive (why?). By the Archimedean Ordering Property, there exists $h \in \mathbb{Q}$ such that

$$0 < h < 1, \qquad 0 < h < \frac{2 - x^2}{2x + 1}$$

(give the details!). This disposes of Case 1.

Case 2: $x^2 > 2$. The strategy is to produce a (rational) number $k > 0$ for which $k < x$ and $(x - k)^2 > 2$ also. Then $t \in \mathbb{R}$ with $t > x - k$ will imply $t^2 \geq 2$ or, equivalently, $t \in \mathbb{R}$ with $t^2 < 2$ implies $t \leq x - k$. This will mean that $x - k$ is an upper bound of A in $\mathbb{R}$ with $x - k < x$, which will contradict the definition of x as the *least* upper bound of A in $\mathbb{R}$.

To find such k, note that, for any k whatsoever,

$$(x - k)^2 = x^2 - 2kx + k^2 = x^2 - k(2x - k).$$

When $0 < k$, then $2x - k < 2x + 1$ so that

$$(x - k)^2 > x^2 - k(2x + 1).$$

There exists rational $k > 0$ such that

$$k(2x + 1) < x^2 - 2$$

(why?), and so

$$(x - k)^2 > x^2 - (x^2 - 2) = 2.$$

This disposes of Case 2.

We conclude that $x^2 = 2$. Thus a square root exists. As in any ordered field, distinct elements of $\mathbb{R}$ have distinct squares (why?), and so the square root is unique. $\square$

The number 0 is the unique square root of 0. Of course, a negative real number does not have a square root in the field $\mathbb{R}$ of real numbers (why?). However, in the larger field $\mathbb{C}$ of complex numbers, every nonzero real number—negative as well as positive—has a square root (in fact, two of them); see Section 3.6.

Exercise 3.3.16.   (a) Carry out the proof of Proposition 3.3.15 for arbitrary $c > 0$ in place of $\sqrt{2}$.

(b) Prove that every $c \in \mathbb{R}$ has a unique cube root.

(c) Generalize to nth roots.

Exercises 3.3.17.   (1) Prove that the archimedean ordered field $\mathbb{Q}$ is not order-complete by showing that its subset $\{ q \in \mathbb{Q} : q < \sqrt{2} \}$ is bounded above in $\mathbb{Q}$ but has no supremum there.

(2) A subset J of $\mathbb{R}$ is said to have the *betweenness property* when

$$(\forall x, y \in J)(\forall z \in \mathbb{R})\,(x < z < y \implies z \in J).$$

(a) Show that any interval in $\mathbb{R}$ (whether open, closed, or half-open and half-closed) and any ray in $\mathbb{R}$ (whether open or closed), as well as $\varnothing$ and $\mathbb{R}$ itself, has the betweenness property.

(b) Prove that each subset of $\mathbb{R}$ having the betweenness property is necessarily one of the kinds of subsets listed in (a).

Exercise 3.3.18. A *cut* in $\mathbb{Q}$ is a pair (A, B) of subsets of $\mathbb{Q}$ such that A and B partition $\mathbb{Q}$, that is,

$$A \ne \varnothing \ne B, \quad A \cup B = \mathbb{Q}, \quad A \cap B = \varnothing,$$

and such that

$$a \in A, b \in B \implies a < b.$$

(a) If $A = \{\, a \in \mathbb{Q} : a \le \sqrt{2}\,\}$ and $B = \{\, b \in \mathbb{Q} : \sqrt{2} < b\,\}$, show that (A, B) is a cut.

(b) Generalize (a) by replacing $\sqrt{2}$ by an arbitrary real number c.

(c) Let (A, B) be a cut in $\mathbb{Q}$. Prove that there is one and only one element $c \in \mathbb{R}$ such that either

$$A = \{\, a \in \mathbb{Q} : a \le c\,\}, \quad B = \{\, b \in \mathbb{Q} : c < b\,\}$$

or else

$$A = \{\, a \in \mathbb{Q} : a < c\,\}, \quad B = \{\, b \in \mathbb{Q} : c \le b\,\}.$$

[Note that, according to (a), such c need not be rational!]

In the preceding exercise, the underlying assumption of our axiomatic approach was that $\mathbb{R}$ was already known—as an archimedean ordered field having the Nested Interval Property. It is easy to see, then, that there is a one-to-one correspondence between the set of all real numbers and the set of all cuts in the rational numbers.

For a constructive approach to the real numbers, one can proceed in the opposite direction. After having already obtained the rational numbers, one defines $\mathbb{R}$ to be the set of all such cuts; defines operations of addition and multiplication on this set making it a field; defines a subset that orders this field; verifies that this ordered field has both the Archimedean Ordering Property and the Nested Interval Property; and, finally, finds an isomorphism between the already-constructed $\mathbb{Q}$ and a subfield of this $\mathbb{R}$.[7] That was exactly the approach to constructing the reals from the rationals taken by Richard Dedekind (1831–1916); in his honor what we termed a "cut" is often referred to as a *Dedekind cut*. As a project, you may wish to carry out Dedekind's construction in detail.

3.4 Isomorphism of Fields

In the preceding section, we used an axiomatic approach to describe the real numbers as an archimedean ordered field having the Nested Interval Property. We suggested how to construct such a field from the simpler field of rational numbers (and pointed to a path that takes one from the natural numbers to the rational numbers). And we raised the question of whether there was only one such field so that we would be justified in speaking of *the* field of real numbers.

In a trivial way we can answer this question of uniqueness in the negative. Look at the following example.

Example 3.4.1. Let $F = \mathbb{R} \times \{0\}$, that is,

$$F = \{\, (x,0) : x \in \mathbb{R} \,\}.$$

In other words, F is the x-axis in the euclidean plane $\mathbb{R} \times \mathbb{R}$. Define operations of addition and multiplication on F by the formulas

$$(x,0) + (y,0) = (x+y,0),$$
$$(x,0) \cdot (y,0) = (x \cdot y,0).$$

These operations make F a field: each field axiom—which we are assuming holds for $\mathbb{R}$—evidently holds for F. Define the subset P of F by

$$P = \{\, (x,0) : x \in \mathbb{R}, x > 0 \,\}.$$

Then P orders F, and the corresponding order relation in F is given by

$$(x,0) < (y,0) \iff x < y.$$

(Check this.) For each $n \in \mathbb{N}$, the corresponding element $n_F \in N(F)$ is just $(n,0)$. Then the ordered field F is archimedean (why?). Since a closed interval $J = [(a,0),(b,0)]$ in F has the form $J = [a,b] \times \{0\} = \{\, (x,0) : x \in \mathbb{R}, x \in [a,b] \,\}$, the ordered field F also has the Nested Interval Property (write out the details).

---

[7]See Section 3.4 for the notions of an isomorphism between two fields and a subfield of a field. See, in particular, Definitions 3.4.2 and 3.6.9.

Thus F has all the same properties as we postulated for $\mathbb{R}$. But certainly $F \neq \mathbb{R}$: for example, $\sqrt{2} \in \mathbb{R}$, but $\sqrt{2} \notin F$; and $(1, 0) \in F$, whereas $(1, 0) \notin \mathbb{R}$.

Although the field F of the preceding example is different from $\mathbb{R}$, it is essentially the same as $\mathbb{R}$. We have, in effect, renamed each element of $\mathbb{R}$ from x to $(x, 0)$—giving each real number x the "tag" 0. The situation is similar to the naming of family members. Within the family you would refer to "Meg," "Jo," "Beth," and "Amy," but outside the family you might refer to the same people as "Meg March," "Jo March," "Beth March," and "Amy March."

When should two fields be considered essentially the same? Let us start with the simplest case possible: fields with only two elements. Compare the addition and multiplication tables for the field $F = \{\alpha, \omega\}$ from Example 3.1.2(5) and the field $\mathbb{Z}_2 = \{[0], [1]\}$ of congruence classes of the integers modulo 2 from Example 3.1.2(6):

$F = \{\alpha, \omega\}$:

+	ω	α
ω	ω	α
α	α	ω

$\cdot$	ω	α
ω	ω	ω
α	ω	α

$\mathbb{Z}_2 = \{[0], [1]\}$:

+	$[0]$	$[1]$
$[0]$	$[0]$	$[1]$
$[1]$	$[1]$	$[0]$

$\cdot$	$[0]$	$[1]$
$[0]$	$[0]$	$[0]$
$[1]$	$[0]$	$[1]$

The two addition tables are the same but for the names of their entries: the zero element ω of F plays the same role for addition in F as the zero element $[0]$ of $\mathbb{Z}_2$ plays in $\mathbb{Z}_2$; the other, nonzero element α of F plays the same role for addition in F as the nonzero element $[1]$ of $\mathbb{Z}_2$. Similarly, the two multiplication tables are the same but for the names of their entries. This is what we mean, loosely, when we say that the two fields are essentially the same or, as we shall soon say, *isomorphic.*

Imagine two math students, Adrian and Zuleika, discussing fields with one another over the telephone or by electronic mail:

> ADRIAN: I have this field consisting of two elements. When I add the first to the second or the second to the first, I get the second; when I add the first to itself or the second to itself, I get the first.
>
> ZULEIKA: That's interesting: I have a field of two elements, too. When I add *its* first to *its* second or the second to the first, I get *its* second; when I add *its* first to itself or *its* second to itself, I get *its* first.
>
> ADRIAN: Of course, I can also multiply the elements of my field. When I multiply the second element of my field by itself, I get as product that same element; when I multiply any other pair of elements, I get the first element of my field.

ZULEIKA: You're not the only one who can do multiplication. When I multiply the second element of *my* field by itself, I get as product that same element; when I multiply any other pair of elements, I get the first element of *my* field.

ADRIAN: Hmm .... We must be talking about the same field. I'll let you in on a secret: my field's elements are named "omega" and "alpha". Omega is the zero element and alpha is the unity element. What are the elements of your field called?

ZULEIKA: My field's elements are definitely not called "omega" and "alpha." They're named "bracket-zero" and "bracket-one," instead. Bracket-zero is the zero element and bracket-one is the unity element. So we're not talking about the same field after all.

ADRIAN: But if I called my field's first element "bracket-zero" instead of "omega" and its second element "bracket-one" instead of "alpha," then no one could tell I'm talking about a different field than yours—unless they were peeking over my shoulder and saw what the elements actually are.

ZULEIKA: Yeah! Hmm .... (*Thinking:*) I wonder if he realizes the same sort of thing is true the other way 'round. If I renamed my field's first element "omega" and its second "alpha," then no one could tell that I'm talking about a different field than his—not even Adrian himself.

You must get the idea by now. To be a little more precise: if you replace ω by $[0]$ and α by $[1]$, you convert the addition and multiplication tables for F into the corresponding tables for $\mathbb{Z}_2$. Such replacement of each element of F by a corresponding element of $\mathbb{Z}_2$ may be described by the function $f: F \to \mathbb{Z}_2$ defined by

$$f(\omega) = [0], \qquad f(\alpha) = [1].$$

Since f maps the distinct elements ω and α of F to distinct elements of $\mathbb{Z}_2$ (rather than mapping both of them to the same element), the function f is injective (one-to-one); since each of the two elements of the codomain $\mathbb{Z}_2$ of f is the image of some element or other of the domain F, the function f is surjective.[8] Thus the function f is bijective, in other words, a one-to-one correspondence between F and $\mathbb{Z}_2$.

That the replacement f converts the addition table for F into that for $\mathbb{Z}_2$ is to say that

$$f(\omega + \alpha) = [0] + [1] = f(\omega) + f(\alpha),$$
$$f(\alpha + \omega) = [1] + [0] = f(\alpha) + f(\omega),$$
$$f(\omega + \omega) = [0] + [0] = f(\omega) + f(\omega),$$
$$f(\alpha + \alpha) = [1] + [1] = f(\alpha) + f(\alpha),$$

---

[8]For the notions of injective, surjective, and bijective functions, see Section B.3.

or, more concisely,

$$f(x + y) = f(x) + f(y) \text{ for all } x, y \in F. \tag{*}$$

Similarly, that the replacement f converts the multiplication table for F into that for $\mathbb{Z}_2$ is to say that

$$f(x \cdot y) = f(x) \cdot f(y) \text{ for all } x, y \in F. \tag{**}$$

Notice that equation (*) involves two quite different operations of addition—one in F and the other in $\mathbb{Z}_2$—both, however, denoted by +. Similarly, (**) involves two different operations of multiplication.

In this example with two two-element fields we have the germ of a general notion.

Definition 3.4.2. Let F and K be fields. An **isomorphism from F to K** is a function $f \colon F \to K$ that is *bijective* and satisfies the equations

$$f(x + y) = f(x) + f(y) \qquad (x, y \in F), \tag{*}$$
$$f(x \cdot y) = f(x) \cdot f(y) \qquad (x, y \in F). \tag{**}$$

We express (*) by saying that the isomorphism f **preserves addition** and express (**) by saying that f **preserves multiplication**.

When an isomorphism from F to K exists, then we say that the field F is **isomorphic to** the field K.

In the language of this definition, the function $f \colon \{\omega, \alpha\} \to \mathbb{Z}_2$ given by $f(\omega) = [0]$, $f(\alpha) = [1]$ is an isomorphism, and the field $\{\omega, \alpha\}$ is isomorphic to the field $\mathbb{Z}_2$.

We shall regard a field F to be essentially the same as a field K when F is isomorphic to K. This is analogous to the way we regarded an integer a to be essentially the same as an integer b with respect to a given modulus m when $a \equiv b \pmod{m}$. Only now the objects being considered essentially the same are not individual numbers but entire sets of such numbers or other elements.

Exercise 3.4.3.   (a) Is the function $f \colon \mathbb{Q} \to \mathbb{Q}$ defined by $f(x) = -x$ an isomorphism? Why or why not?

(b) Let $\mathbb{Q}(\sqrt{2})$ be the field consisting of all real numbers of the form $q + r\sqrt{2}$ for rational numbers q and r, with such numbers being added and multiplied in the usual way as real numbers [see Exercise 3.1.3 (d)]. Show that each element of this field can be written in such a form for *unique* rational q and r.

(c) Is the function $f \colon \mathbb{Q}(\sqrt{2}) \to \mathbb{Q}(\sqrt{2})$ given by $f\left(q + r\sqrt{2}\right) = q - r\sqrt{2}$ an isomorphism? Why or why not?

(d) Does the function $f: \mathbb{Q}(\sqrt{2}) \to \mathbb{Q}$ given by $f\left(q + r\sqrt{2}\right) = q$ preserve addition and multiplication?

(e) Construct an isomorphism between $\mathbb{R}$ and the field described in Example 3.4.1 and verify that it really is an isomorphism.

(f) Construct another field that is isomorphic to $\mathbb{R}$ yet different from the field in Example 3.4.1.

(g) Is the field $\mathbb{Z}_2$ isomorphic to the field $\mathbb{Z}_3$?

(h) Show that every two-element field is isomorphic to $\mathbb{Z}_2$. (Thus, there is essentially only one two-element field. Said otherwise, the two-element field $\mathbb{Z}_2$ is *unique up to isomorphism.*)

(i) Is the field $\mathbb{Q}$ isomorphic to the field $\mathbb{Q}(\sqrt{2})$?

(j) Let $\mathbb{Q}(\sqrt{3})$ be the set of all real numbers of the form $q + r\sqrt{3}$ for rational numbers q and r, with such numbers being added and multiplied in the usual way as real numbers. Then $\mathbb{Q}(\sqrt{3})$ is a field—the verification of the field axioms is essentially the same as for $\mathbb{Q}(\sqrt{2})$. Is the function $f: \mathbb{Q}(\sqrt{3}) \to \mathbb{Q}(\sqrt{3})$ given by $f(q + r\sqrt{3}) = q - r\sqrt{3}$ an isomorphism? Why or why not?

(k) Are the fields $\mathbb{Q}(\sqrt{2})$ and $\mathbb{Q}(\sqrt{3})$ isomorphic? If so, construct an isomorphism; if not, tell why not.

(l) Let $\mathbb{Q}(\sqrt{2}, \sqrt{3})$ be the set of all real numbers of the form $q + r\sqrt{2} + s\sqrt{3} + t\sqrt{6}$ for rational numbers q, r, s, t, with such numbers being added and multiplied in the usual way as real numbers. Verify that $\mathbb{Q}(\sqrt{2}, \sqrt{3})$ is a field. Construct several isomorphisms from this field to itself.

(m) Prove that the field $\mathbb{R}$ is not isomorphic to the field $\mathbb{Q}$. (*Note:* Later, in Chapter 4, it is proved that there is no bijection between $\mathbb{R}$ and $\mathbb{Q}$ whatsoever, and so there certainly cannot be any isomorphism between them. Here, solve the exercise by a different method.)

Fields, which have two operations, are just one type of algebraic structure. (Other types are groups, which have a single operation, and vector spaces, which have a single internal operation and, besides, an external operation of scalars upon vectors.) Such other types of algebraic structures have similar notions of isomorphism—of being essentially the same. Now a given field can also be considered as a group (by disregarding its operation of multiplication) or as vector space (by regarding elements of itself as the scalars). To avoid ambiguity, when we want to refer to an isomorphism between F and K in so far as F and K are fields, we may speak of a *field isomorphism* and say that F is *field isomorphic* to K. Since, however, only fields are being considered in this book, we can generally omit the "field" modifier.

An isomorphism maps zero and unity to zero and unity, respectively, and preserves negatives and inverses as well.

Proposition 3.4.4. *Let $f: F \to K$ be an isomorphism from a field F to a field K. Then:*

1. $f(0) = 0$ *and* $f(1) = 1$.[9]

2. $f(-x) = -f(x)$ *for each $x \in F$.*

3. $f(x^{-1}) = (f(x))^{-1}$ *for each $x \in F$ with $x \neq 0$.*

Proof.   1. Since f preserves addition, $f(0) + f(0) = f(0 + 0) = f(0)$ so that $f(0) + f(0) = f(0)$. Subtracting $f(0)$ from this equation—in the field K—leaves $f(0) = 0$. The proof that $f(1) = 1$ is similar.

2. Exercise. [*Hint:* What is $f(x) + f(-x)$?]

3. Exercise.   □

Note one consequence of part 1 of the preceding proposition: an isomorphism maps nonzero elements to nonzero elements—if $x \neq 0$ in F, then $f(x) \neq 0$ in K. This was used tacitly in the statement of part 3.

Exercise 3.4.5. Let $f: F \to K$ be an isomorphism of fields. Prove that, for all $x, y \in F$:

(a) $f(x - y) = f(x) - f(y)$.

(b) $f(x/y) = f(x)/f(y)$ provided $y \neq 0$.

Everything about a field depends upon its operations of addition and multiplication. An isomorphism between two fields preserves these operations. Hence anything you can say about one field—as a field—translates into a corresponding thing you can say about any field isomorphic to it. For example, if a field F has an element x for which $x^2 + 1 = 0$, and if K is isomorphic to F, then K must have an element y for which $y^2 + 1 = 0$. Consequently, the field $\mathbb{C}$ of complex numbers (discussed in detail in Section 3.6) is not isomorphic to the field $\mathbb{R}$ of real numbers.

---

[9]The first statement here means, of course, that the image under f of the zero element of the field F is the zero element of the field K. (The second statement has a similar meaning.) To avoid ambiguity we ought to denote the zero elements of the two fields differently, say as 0_F and 0_K, as we did in Definition 3.1.16. Since $f(0_F) = 0_K$, there is really no great harm in avoiding this notational distinction.

As already stipulated, a field is to be regarded essentially the same as another when it is isomorphic to the other. If we have reasonably represented the rough idea of being essentially the same by the precise idea of isomorphic, then it is also reasonable to expect that:

1. Any given field is isomorphic to itself (since it actually *is* the same as itself).

2. If one field is isomorphic to a second and the second is isomorphic to a third, then the first is isomorphic to the third.

3. If one field is isomorphic to a second, then the second is isomorphic to the first.

The third of these three properties is just the realization of Zuleika's idea (see page 178) that renaming the elements of two essentially equal fields can be done in either direction.

That the three properties do hold is a consequence of the following proposition about isomorphisms. (For the notions of identity map, composite map, and inverse map, see Appendix B. Note that it is meaningful to form the inverse f^{-1} of an isomorphism f because an isomorphism is, among other things, a bijective map.)

Proposition 3.4.6.   *1. The identity map $i_F\colon F \to F$ of any field F is an isomorphism.*

   2. If $f\colon F \to G$ and $g\colon G \to K$ are isomorphisms, then their composite $g \circ f\colon F \to K$ is an isomorphism.

   3. If $f\colon F \to K$ is an isomorphism, then its inverse $f^{-1}\colon K \to F$ is an isomorphism.

Proof.   1. Obvious. (Why is it obvious?)

   2. Assume $f\colon F \to G$ and $g\colon G \to K$ are isomorphisms. Since f and g are bijections, so is their composite $g \circ f$. Next, $g \circ f$ preserves addition since $x, y \in F$ implies

$$
\begin{aligned}
(g \circ f)(x + y) &= g(f(x + y)) \\
&= g(f(x) + f(y)) &&(f \text{ preserves addition}) \\
&= g(f(x)) + g(f(y)) &&(g \text{ preserves addition}) \\
&= (g \circ f)(f(x)) + (g \circ f)(f(y))
\end{aligned}
$$

Similarly, $g \circ f$ preserves multiplication. (Write out a proof after covering up the one above!).

   3. Exercise.   □

A map $f: F \to K$ from one field to another that preserves addition and multiplication need not be an isomorphism. An example is the inclusion map $\mathbb{Q} \to \mathbb{R}: x \mapsto x$. In general, a map between fields that merely preserves addition and multiplication is called a field **homomorphism**.

Exercise 3.4.7. Suppose $f: F \to K$ is an isomorphism from an ordered field F to an ordered field K. We say that f *preserves order* when $x < y$ in F always implies $f(x) < f(y)$ in K.

(a) Must a field isomorphism between ordered fields always preserve order?

(b) If a field isomorphism between ordered fields does preserve order, show that its inverse isomorphism also preserves order.

The work of showing that a map between fields is an isomorphism can at times be simplified a bit by using the following result.

Lemma 3.4.8. *Let $f: F \to K$ be a surjection from a field F onto a field K that preserves addition and multiplication. Suppose*

$$f(x) = 0 \implies x = 0 \qquad (x \in F). \qquad (*)$$

Then f is an isomorphism.

Proof. Exercise.   □

The displayed condition in the preceding lemma is equivalent to

$$x \neq 0 \implies f(x) \neq 0 \qquad (x \in F).$$

Now $f(0) = 0$ because f preserves addition. Then the condition is equivalent to

$$x \neq 0 \implies f(x) \neq f(0) \qquad (x \in F).$$

This is just a special case of the seemingly more general requirement that f be injective:

$$x \neq y \implies f(x) \neq f(y) \qquad (x \in F, y \in F).$$

Thus to check whether f is injective, you need only compare images of elements with the image of zero rather than images of arbitrary pairs of elements. For an application, see the proof of Theorem 3.4.13.

The work of showing that a map between fields is an isomorphism can also at times be simplified a bit (as in the next example) by using the following result.

Lemma 3.4.9. *Let $f: F \to K$ be a map from a field F to a field K that preserves addition and multiplication. Suppose there exists a map $g: K \to F$ such that $g \circ f$ is the identity map of F and $f \circ g$ is the identity map of K. Then f is an isomorphism and $g = f^{-1}$.*

Proof. The only thing to show is that f is bijective. This has nothing to do with the fact that F and K are fields or that f preserves addition and multiplication. In fact, it is really a general result about maps (see Proposition B.3.12), but we outline the proof here.

First, f is injective. To check this, let $x_1, x_2 \in F$ with $f(x_1) = f(x_2)$. Apply g to conclude that $x_1 = x_2$.

Second, f is surjective. To check this, let $y \in K$. Let $x = g(y)$. Show that $f(x) = y$. $\square$

Here is another, substantial example of a field isomorphism. The example is reminiscent of the definition of the set $\mathbb{Z}_m$ of congruence classes of integers modulo m and the operations of addition and multiplication on this quotient set (see Section 2.5 as well as Appendix C).

Example 3.4.10. We shall construct from the set of integers a field Q that is isomorphic to the field $\mathbb{Q}$ of rational numbers.

The motivating idea. Each rational number has the form m/n for some integers m and n with $n \neq 0$. Of course, a given rational number has such a form for many different pairs of integers—for infinitely many pairs, in fact. For example, $2/3 = 10/15 = (-12)/(-18)$, etc. Given integers m, n, a, and b with $n \neq 0 \neq b$, the equality $m/n = a/b$ of rational numbers holds if and only if the equality $mb = na$ of integers holds.

Very loosely speaking, then, the elements of Q are going to be pairs (m, n) of integers with n nonzero but with two such pairs (m, n) and (a, b) identified with each other—considered *as if* they were the same—whenever $mb = na$. Strictly speaking, the elements of Q are going to be equivalence classes of such pairs of integers with respect to a certain equivalence relation (see Appendix C for the terminology involved here).

Definition of the set Q. To begin construction of the field Q, let

$$\mathbb{Z}^* = \{ n \in \mathbb{Z} : n \neq 0 \},$$

the set of all nonzero integers, and form the product set

$$S = \mathbb{Z} \times \mathbb{Z}^* = \{ (m, n) : m \in \mathbb{Z}, n \in \mathbb{Z}^* \}$$

consisting of all those pairs (m, n) of integers whose second entry n is nonzero. Next, define the relation $\sim$ in S by

$$(m, n) \sim (a, b) \iff mb = na.$$

For example, $(2,3) \sim (10,15)$ and $(2,3) \sim (-12,-18)$, too. This relation is an equivalence relation on the set S. In other words, for all $(m,n), (a,b), (p,q) \in S$,

$$(m,n) \sim (m,n),$$
$$(m,n) \sim (a,b) \implies (a,b) \sim (m,n),$$
$$(m,n) \sim (a,b) \,\&\, (a,b) \sim (p,q) \implies (m,n) \sim (p,q).$$

(Deduce these three properties from corresponding properties of equality of integers—do *not* use nonintegral rational numbers.)

For each pair $(m,n) \in S$, there is a corresponding equivalence class

$$[(m,n)] = \{\, (a,b) \in S : (m,n) \sim (a,b) \,\}$$

consisting of all those pairs in S that are equivalent to (m,n) with respect to the relation $\sim$. For example, $(10,15) \in [(2,3)]$. However, $[(m,n)] = [(a,b)]$ does *not* necessarily mean that $(m,n) = (a,b)$, only that $(m,n) \sim (a,b)$. In fact,

$$(a,b) \in [(m,n)] \iff (m,n) \sim (a,b);$$

that is,

$$(a,b) \in [(m,n)] \iff mb = na.$$

The set that will become the field isomorphic to $\mathbb{Q}$ is the quotient set

$$Q = S/\!\sim$$
$$= \{\, [(m,n)] : (m,n) \in S \,\}$$

consisting of all the equivalence classes with respect to the equivalence relation $\sim$.

Definition of the operations on Q. In Q we wish to add and multiply elements $A = [(m,n)]$ and $B = [(a,b)]$ the way that in $\mathbb{Q}$ we add and multiply the fractions m/n and a/b formed from representatives (m,n) and (a,b) of such equivalence classes. Multiplication is a bit simpler to deal with, and so we start with that. For motivation, note that for $(m,n), (a,b) \in S$, the product of the rational numbers m/n and a/b is $(ma)/(nb)$.

Let $A, B \in Q$. Choose representatives (m,n) and (a,b) of A and B, respectively, that is, $(m,n) \in S$ with $[(m,n)] = A$ and $(a,b) \in S$ with $[(a,b)] = B$. We want to define the product $A \cdot B$ to be

$$A \cdot B = [(ma, nb)], \tag{*}$$

but before doing so we need to know that this is *well-defined,* that is, does not depend on the choice of the particular representatives (m, n) and (a, b) from the equivalence classes A and B. So let $(m', n') \in A$ and $(a', b') \in B$, too. We have

$$(m', n') \sim (m, n) \quad \text{and} \quad (a', b') \sim (a, b),$$

that is,

$$m'n = n'm \quad \text{and} \quad a'b = b'a.$$

Then

$$
\begin{aligned}
(ma)(n'b') &= (n'm)(b'a) \\
&= (m'n)(a'b) \\
&= (nb)(m'a'),
\end{aligned}
$$

which means

$$(ma, nb) \sim (m'a', n'b');$$

in other words,

$$[(ma, nb)] = [(m'a', n'b')].$$

Thus, the product $A \cdot B$, as given by (*), is well-defined.

An operation of multiplication has now been defined on Q. We leave as an exercise definition of an operation of addition. Your definition should reflect the way rational numbers are added (but in the end it should be stated in terms of operations on integers alone). You should check that you have a well-defined operation (but the check should use properties only of operations on integers).

The operations make Q a field. Verification of the field axioms is left as an exercise. (The verification should involve ultimately only properties of addition and multiplication of integers—*not* properties of addition and multiplication of rational numbers.) Here is a start— existence of a unity element. (*Thinking:* For rational fractions, the number $1 = 1/1$ is the multiplicative identity.) The element $[(1, 1)]$ is a unity element because, for $(m, n) \in S$,

$$[(m, n)] \cdot [(1, 1)] = [(m \cdot 1, n \cdot 1)] = [(m, n)].$$

The field Q is isomorphic to $\mathbb{Q}$. The motivation for the construction of Q furnishes us with a map

$$
\begin{aligned}
\varphi \colon S = \mathbb{Z} \times \mathbb{Z}^* &\to \mathbb{Q} \\
(m, n) &\mapsto m/n.
\end{aligned}
$$

Then φ will give rise to a map

$$f: Q \quad\quad \to \mathbb{Q}$$
$$[(m,n)] \mapsto \varphi((m,n)) = m/n$$

provided we can show that this map is well-defined, that is, that the value $f([(m,n)]) = m/n$ does not depend on the choice of the representative of the equivalence class $[(m,n)]$. But this is evident from the definition of the equivalence relation $\sim$. In fact, for $A \in Q$,

$$(m,n),(a,b) \in A \implies (m,n) \sim (a,b)$$
$$\implies mb = na$$
$$\implies \varphi((m,n)) = m/n = a/b = \varphi((a,b)).$$

Thus the map $f: Q \to \mathbb{Q}$ is well-defined.

Similar considerations show that f is injective. In fact, we know that $m/n = a/b$ for rational fractions is equivalent to $mb = na$, and $mb = na$ is equivalent to $(m,n) \sim (a,b)$ by the very definition of $\sim$. Thus the last two of the preceding implications are reversible, and so

$$f([(m,n)]) = f([(a,b)]) \implies \varphi((m,n)) = \varphi((a,b))$$
$$\implies (m,n) \sim (a,b)$$
$$\implies [(m,n)] = [(a,b)].$$

The map f is manifestly surjective (write out the details!). Thus $f: Q \to \mathbb{Q}$ is a bijection.

It remains only to check that f preserves both addition and multiplication. Here is the check for multiplication. Let $A, B \in Q$. Choose representatives $(m,n) \in A$ and $(a,b) \in B$. Then

$$\begin{aligned}
f(A \cdot B) &= f([(m,n)] \cdot [(a,b)]) \\
&= f([(ma,nb)]) \\
&= (ma)/(nb) \\
&= (m/n) \cdot (a/b) \\
&= f([(m,n)]) \cdot f([(a,b)]) \\
&= f(A) \cdot f(B).
\end{aligned}$$

The check that f preserves addition is left as an exercise.

Why bother to construct such a complicated field Q as in Example 3.4.10 only to conclude that it is isomorphic to our familiar old friend $\mathbb{Q}$? Perhaps you have already anticipated the reason. If you had wanted to follow a constructive approach to the real numbers, instead of our axiomatic approach, then one of the steps would have been to construct the rational numbers from the integers. And that is, in effect, precisely what we did. The field Q is constructed using solely the set $\mathbb{Z}$ of integers and the operations of addition and multiplication of integers. As a field, Q is essentially the same as what we expect the field of rational numbers to be.

Well, not quite! You have long believed that $\mathbb{Z} \subset \mathbb{Q}$, in other words, that each integer m *is* a rational number, because $m = m/1$, the quotient of an integer by a nonzero integer. Unfortunately, our Q, as constructed, does *not* have $\mathbb{Z}$ as a subset! It *almost* has $\mathbb{Z}$ as a subset: among the elements of Q are those of the form $[(m,1)]$. In fact, the map

$$J\colon \mathbb{Z} \to Q$$
$$m \mapsto [(m,1)]$$

is an injection that preserves addition and multiplication (check that!). Then J defines a one-to-one correspondence between $\mathbb{Z}$ and its range $Z = \{\,[(m,1)] : m \in \mathbb{Z}\,\}$ that makes Z look essentially the same as the integers (but not as fields—$\mathbb{Z}$ is not a field). We can, therefore, "identify" each integer m with the corresponding element $J(m) = [(m,1)]$ of $Z \subset Q$, that is, pretend m is the same as $J(m)$.

Of course, just because we pretend that each integer m is the same as the corresponding element $J(m)$ does not make it actually the same. (If you call a dog's tail a leg, the dog will still have only four legs!) A way out is indicated in Exercise 3.4.11 (c).

Exercise 3.4.11. This exercise refers to the field Q in Example 3.4.10 and the injection $J\colon \mathbb{Z} \to Q$ with range Z just discussed.

(a) Complete all unfinished or undone steps in the verification that Q is a field and f is an isomorphism from Q to $\mathbb{Q}$.

(b) What is the nub $N(Q)$ of the field Q? What is $Z(Q)$?

(c) Reconstruct Q as follows in order to make $\mathbb{Z}$ actually a subset of $\mathbb{Q}$. "Cut out" the subset Z of Q and "paste in" the set $\mathbb{Z}$ in its place, that is, form the set

$$\widetilde{Q} = (Q \setminus Z) \cup \mathbb{Z}.$$

Define operations of addition and multiplication on $\widetilde{Q}$ that extend the usual addition and multiplication on $\mathbb{Z}$ as well as the addition and multiplication of elements of the subset $Q \setminus Z$. Show that these operations make $\widetilde{Q}$ a field. Finally, show that $\widetilde{Q}$ is isomorphic to Q.

The next exercise outlines a construction of $\mathbb{Z}$ from $\mathbb{N}$. A different—perhaps simpler—construction appears in Appendix D. The one below has the virtue of being quite similar to the construction of $\mathbb{Q}$ from $\mathbb{Z}$ suggested by Example 3.4.10 (even though $\mathbb{Z}$ is, of course, not a field). Together, the next exercise and Example 3.4.10 therefore provide a way to construct $\mathbb{Q}$ from $\mathbb{N}$. Coupled with the construction of $\mathbb{R}$ from $\mathbb{Q}$ by either Dedekind cuts (Exercise 3.3.18 and the following discussion) or Cauchy sequences (Exercise 3.5.46), they provide all the steps needed to construct the real numbers from the natural numbers.

Exercise 3.4.12. Each integer k can be written as the difference $k = m - n$ of a pair of natural numbers m and n. In fact, a given integer k can be written as such a difference for infinitely many different such pairs. Two pairs (m, n) and (i, j) of natural numbers have the same integer as their difference exactly when $m - n = i - j$ (a statement about integers, still) or, equivalently, when $m + j = n + i$ (a statement solely about natural numbers).

(a) Use these observations to define an equivalence relation $\sim$ on the set $S = \mathbb{N} \times \mathbb{N}$ of all pairs of natural numbers. (Try to phrase the definition strictly in terms of natural numbers.)

(b) Define appropriate operations of addition and multiplication on the quotient set $Z = S/\sim$. These operations should have the usual properties of addition and multiplication of integers: all the axioms for a field are valid *except* the existence of reciprocals; moreover, the product of two elements is 0 only when at least one of them is 0.

(c) Construct a bijection from the quotient set Z to the set $\mathbb{Z}$ of integers that preserves addition and multiplication.

(d) Assume you know about $\mathbb{N}$ but not $\mathbb{Z}$. Starting from $\mathbb{N}$ along with its addition and multiplication, construct the integers $\mathbb{Z}$ along with its addition and multiplication. Indicate what you would do to make it turn out that $\mathbb{N} \subset \mathbb{Z}$ and that the addition and multiplication on $\mathbb{Z}$ extend addition and multiplication on $\mathbb{N}$.

You may recall that the question raised at the beginning of this section was whether the there is only one archimedean ordered field having the Nested Interval Property. Although the answer was shown to be no in a superficial sense, we shall now show that the answer is essentially yes—that such a field is **unique up to isomorphism**.

Theorem 3.4.13. (Uniqueness of Real Numbers up to Isomorphism)
Any two archimedean ordered fields having the Nested Interval Property are isomorphic. Moreover, there is a unique *isomorphism between any two such fields.*

Proof. Let the two fields be F and K.

Existence of an isomorphism. We shall define the desired isomorphism $f: F \to K$ in several steps.

 Step 1: Definition of f on the nub $N(F)$ of F. Recall that $N(F) = \{ n_F : n \in \mathbb{N} \}$. For each natural number n, define $f(n_F) = n_K$. Thus $n \in N(F)$ implies $f(n) \in N(K)$, and so we have a map

$$f: N(F) \to N(K).$$

This f is bijective (why?) and preserves both addition and multiplication (why?).

Step 2: Extension of f to the set $Z(F)$ of integral elements of F. Recall that $Z(F) = N(F) \cup \{ -n : n \in N(F) \}$. We extend the previously defined f to $Z(F)$ simply by specifying $f(-n) = -f(n)$ for each $n \in N(F)$. The map

$$f: Z(F) \to Z(K)$$

so defined is bijective (why?) and preserves both addition and multiplication (why?).

Step 3: Extension of f to the set $Q(F)$ of rational elements of F. Recall that $Q(F) = \{ m/n : m, n \in Z(F), n \neq 0 \}$. We extend the previously defined f to $Q(F)$ by specifying $f(m/n) = f(m)/f(n)$ for each $m, n \in Z(F)$ with $n \neq 0$. [Because $f(m/1) = f(m)$ already has a meaning for $m \in Z(F)$, you must check that this new definition is consistent with the preceding definition of f on $Z(F)$, in other words, that this new f actually extends the old f. What else must you check?] Thus $n \in Q(F)$ implies $f(n) \in Q(K)$, and so we have a map

$$f: Q(F) \to Q(K).$$

This f is bijective (why?) and preserves both addition and multiplication (why?), and so it is an isomorphism from the field $Q(F)$ to the field $Q(K)$.

Before proceeding, note that

$$q < r \implies f(q) < f(r) \qquad (q, r \in Q(F))$$

(proof?). Thus the isomorphism $f: Q(F) \to Q(K)$ preserves order.

Step 4: Extension of f to the entire set F. Let $x \in F$. Recall from Exercise 3.3.10(5) that

$$x = \sup F(x),$$

where

$$F(x) = \{ q \in Q(F) : q < x \}.$$

By the Archimedean Ordering Property of F (at last!), there exists some $n \in N(F)$ with $x < n$. Then $q \in Q(F)$ with $q < x$ implies $q < n$, and so $f(q) < f(n)$. This means that the image $\{ f(q) : q \in Q(F), q < x \}$ of $F(x)$ under f is bounded above in K. Moreover, this image is not empty (why?). By order completeness of K (which holds because K has the Nested Interval Property), the image of $F(x)$ has a supremum in K. We define

$$f(x) = \sup\{ f(q) : q \in F(x) \}.$$

This definition is consistent with the preceding definition of $f(x)$ for $x \in Q(F)$ (why?). Finally, we have a map

$$f : F \to K.$$

It remains to show that f is bijective and preserves both addition and multiplication.

The map f preserves addition. Let $x, y \in F$. By definition,

$$f(x + y) = \sup\{ f(s) : s \in F(x + y) \}.$$

Since, as we already know, $f(q + r) = f(q) + f(r)$ for all $q, r \in Q(F)$, then

$$\begin{aligned}
f(x) + f(y) &= \sup\{ f(q) : q \in F(x) \} + \sup\{ f(r) : r \in F(y) \} \\
&= \sup\{ f(q) + f(r) : q \in F(x), r \in F(y) \} \\
&= \sup\{ f(q + r) : q \in F(x), r \in F(y) \}.
\end{aligned}$$

Hence it suffices to show that

$$F(x + y) = \{ q + r : q \in F(x), r \in F(y) \}.$$

If $q \in F(x)$ and $r \in F(y)$, then $q + r \in Q(F)$ with $q + r < x + y$. Thus $\{ q + r : q \in F(x), r \in F(y) \}$ is a subset of $F(x + y)$. To prove the opposite inclusion, let $s \in F(x + y)$. We must show $s = q + r$ for some $q \in F(x)$, $r \in F(y)$. Since $s < x + y$, by archimedean ordering there exists $n \in N(F)$ with $n > 0$ and

$$1/n < (x + y) - s.$$

By order density of the rational elements of F, there exists $q \in Q(F)$ with

$$x - 1/n < q < x.$$

Then $q \in F(x)$. Let $r = s - q$. Then $r \in Q(F)$ also, and $q + r = s$. It remains to show that $r < y$. But since $x - 1/n < q$, then, as desired,

$$r = s - q < s - x + 1/n = (s + 1/n) - x < (x + y) - x = y.$$

The map f preserves multiplication. Since we know now that f preserves addition, we have $f(0) = 0$ and $f(-x) = -f(x)$ for all $x \in F$. Hence to show that f preserves multiplication, it suffices to show that $f(xy) = f(x)f(y)$ for all $x, y \in F$ with $x > 0$ and $y > 0$. (Why does this suffice?)

Notice that by archimedean ordering, for $x \in F$ with $x > 0$, there exists $n \in N(F)$ with $n > 0$ and $1/n \in F(x)$, and consequently,

$$f(x) = \sup\{\, f(q) : q \in F^+(x)\,\},$$

where

$$F^+(x) = \{\, q \in Q(F) : 0 < q < x\,\}.$$

Now let $x, y \in F$ with $x > 0$ and $y > 0$. It suffices to show that

$$F^+(xy) = \{\, qr : q \in F^+(x), r \in F^+(y)\,\}.$$

(Why does that suffice?) If $q \in F^+(x)$ and $r \in F^+(y)$, then $qr \in F^+(xy)$ (why?). Thus $\{\, qr : q \in F^+(x), r \in F^+(y)\,\} \subset F^+(xy)$.

To show the opposite inclusion, let $s \in F^+(xy)$. Then $s \in Q(F)$ with $0 < s < xy$. There exists $n \in N(F)$ with $n > 0$ and $1/n < xy - s$ so that

$$s < xy - 1/n.$$

Next, there exists $m \in N(F)$ with

$$m > nxy - 1.$$

Now $0 < m/(m+1) < 1$, whence $0 < (mx)/(m+1) < x$, and so by order density there exists $q \in Q(F)$ with

$$\frac{mx}{m+1} < q < x.$$

Thus $q \in F^+(x)$. Define $r = s/q$. Then $r \in Q(F)$ also with $s = qr$, $q > 0$, and $r > 0$. It remains to show that $r < y$ so that $r \in F^+(y)$, as needed.

Since $q > (mx)/(m+1)$, then

$$\frac{1}{q} < \frac{m+1}{mx} = \frac{1}{x} + \frac{1}{mx}.$$

Since also $s < xy - 1/n$, then

$$\begin{aligned}
r = \frac{s}{q} &< \left(xy - \frac{1}{n}\right)\left(\frac{1}{x} + \frac{1}{mx}\right) \\
&= y + \frac{y}{m} - \frac{1}{nx} - \frac{1}{mnx} \\
&= y + \frac{nxy - m - 1}{mnx} \\
&< y.
\end{aligned}$$

The map f is bijective. To show that f is injective, by Lemma 3.4.8 it suffices to show that $x \neq 0$ implies $f(x) \neq 0$. Since f preserves addition, $f(-x) = -f(x)$, and so $f(x) = 0$ if and only if $f(-x) = 0$. Hence it suffices to show that $x > 0$ implies $f(x) > 0$. This is immediate for $x \in N(F)$ (why?) and hence for $x \in Q(F)$ (why?). That it then follows for arbitrary $x \in F$ is left as an exercise.

That f is surjective is also left as an exercise.

Uniqueness of an isomorphism. Exercise. [*Hint:* First show that there is only one isomorphism from the field $Q(F)$ to the field $Q(K)$.] □

Exercise 3.4.14. State and prove a characterization of the field $\mathbb{Q}$ of rational numbers, that is, properties with respect to which this field is unique up to isomorphism.

3.5   Null Sequences and Limits

Our axiomatic approach has led us to characterize the field $\mathbb{R}$ of real numbers as an archimedean ordered field having the Nested Interval Property—the unique such field up to isomorphism, as we now know. From this characterization it should be possible to derive the familiar representation of real numbers by their decimal expansions. In this section, we shall do precisely that.

In order to understand what such decimal expansions mean when the decimal digits "go on forever," we need to discuss the general idea of limit of a sequence (and the particular case of sum of a series). For limits, we start with the special case of sequences whose limit is 0—or, as we shall call them, *null sequences.* (Dealing with the precise definition of limit in the special case of limit 0 is technically a bit simpler than dealing with general limits, and the more general case is readily reducible to the case of limit 0.)

One by-product of studying sequences here will be a way to construct the field $\mathbb{R}$ of real numbers from the field $\mathbb{Q}$ of rational numbers. For that reason, our definitions and results will initially be phrased in terms of sequences not just in $\mathbb{R}$ but rather in an arbitrary archimedean ordered field—in $\mathbb{Q}$, for example. Accordingly, in order to avoid repeating hypotheses, we shall use the following standing notation.

Standing Notation 3.5.1. *Unless explicitly indicated to the contrary, F denotes an archimedean ordered field, and all sequences considered are sequences in F.*

Some of the results below will depend in an essential way on the assumption that the archimedean ordered field F has the Nested Interval Property. Since any such field is isomorphic to $\mathbb{R}$ (see Theorem 3.4.13), there will be no loss of generality in stating such results for just $\mathbb{R}$.

Look at the following sequences:

$$1, \frac{1}{2}, \frac{1}{3}, \ldots, \frac{1}{n+1}, \ldots,$$
$$1, \frac{1}{10}, \frac{1}{100}, \ldots, \frac{1}{10^n}, \ldots,$$
$$\frac{1}{2}, \frac{1}{4}, \frac{1}{8}, \ldots, \frac{1}{2^{n+1}}, \ldots,$$
$$1, -\frac{1}{2}, \frac{1}{3}, \ldots, (-1)^n \frac{1}{n+1}, \ldots.$$

The successive entries of each sequence become closer and closer to zero the further along in the sequence you look. More precisely, the entries approach and stay as close to zero as you wish to make them provided you look far enough along in the sequence. But *how close* to zero and *how far* along in the sequence? Carefully and clearly answering these two questions will nail down the idea of what it means to say that such a sequence is a null sequence.

For motivation, start with an arbitrary sequence

$$(x_0, x_1, x_2, \ldots) = (x_n)_{n=0,1,2,\ldots} = (x_n)_{n \in \mathbb{N}}$$

in $\mathbb{R}$ or $\mathbb{Q}$. When saying x_n is close to 0, we do not especially care whether it is above 0 or below 0, and so we look at the distance $|x_n - 0|$, that is, at the absolute value $|x_n|$. Then we should compare $|x_n|$ with small positive numbers. But there are no such things as "small" numbers in an archimedean ordered field—only numbers that are small*er* than other numbers. (Whether you regard a quantity as small depends entirely on your point of reference. For most of us, a penny is certainly a small amount of money, and probably a nickel or a dime as well. In the federal budget, $1000 is definitely small, but if it had to come completely out of your own pocket, you probably would not think it was so small after all.)

We could compare $|x_n|$ with *all* numbers $\varepsilon > 0$ (and, indeed, what we shall actually do will be equivalent to making these comparisons). However, note that

$$(\forall \varepsilon > 0)\, (\exists p \in \mathbb{N}^*)\, \left(\frac{1}{p} < \varepsilon \right).$$

(Why?) In other words, however (relatively) small you might regard an element $\varepsilon \in F$, there is a still smaller element of the form $1/p$ with $p \in \mathbb{N}^*$. This means that we need only compare $|x_n|$ with elements of F of the form $1/p$ for positive integers p. (In this context, think of "p" as standing for "positive integer," not "prime.")

For each $p \in \mathbb{N}^*$, regard $1/p \in F$ as a *tolerance* for how big $|x_n|$ is allowed to be yet still be considered small with reference to $1/p$. To say that the entries of the sequence get within—and stay within—such a tolerance as you move further out in the sequence is to refer to $|x_n|$ for all values of n from some point onward, that is, for all $n \geq m$ where m is some (suitably large) natural number. Now put everything together: each positive integer p provides a tolerance $1/p$ that challenges us to show that there is some m that makes $|x_n| < 1/p$ whenever $n \geq m$.

We are ready to spell out the precise definition of null sequence. At times we shall refer to the criterion in this definition as the $(1/p, m)$-**criterion**.

Definition 3.5.2. A sequence $(x_n)_{n \in \mathbb{N}}$ is said to be **null** if, for every $p \in \mathbb{N}^*$, there exists some $m \in \mathbb{N}$ such that $|x_n| < 1/p$ for every $n \in \mathbb{N}$ with $n \geq m$.

Just so there is no misunderstanding, here is the definition in logical notation instead of words: $(x_n)_{n \in \mathbb{N}}$ is a null sequence in F when

$$(\forall p \in \mathbb{N}^*)\,(\exists m \in \mathbb{N})\,(\forall n \in \mathbb{N})\left(n \geq m \;\Rightarrow\; |x_n| < \frac{1}{p}\right).$$

Sometimes we leave implicit the specifications that p belong to $\mathbb{N}^*$ and n and m belong to $\mathbb{N}$. Then the logical formulation takes the abbreviated form

$$(\forall p)\,(\exists m)\,(\forall m \geq n)\left(|x_n| < \frac{1}{p}\right).$$

Once you have chewed, swallowed, and thoroughly digested the precise definition—and when you can regurgitate it upon demand—then you may formulate it more suggestively as follows:

> For each positive integer p, no matter how large, there exists some sufficiently large m for which $|x_n| < 1/p$ whenever $n \geq m$.

Here are some examples showing how the definition works.

Examples 3.5.3. (1) The sequence $(1/(n+1))_{n=0,1,2,...}$ in $\mathbb{R}$ (or in $\mathbb{Q}$) is a null sequence. In fact, let p be an arbitrary positive integer. We wish to find a nonnegative integer m so large that $|1/(n+1)| < 1/p$ whenever $n \geq m$. Now $1/(n+1) > 0$, and so $|1/(n+1)| = 1/(n+1)$ for all $n \in \mathbb{N}$. Moreover,

$$\frac{1}{n+1} < \frac{1}{p} \;\Longleftrightarrow\; n+1 > p \;\Longleftrightarrow\; n \geq p.$$

Thus, take $m = p$ itself to get $|1/(n + 1)| < 1/p$ whenever $n \geq m$.

Note that, for given p, there is nothing sacred about taking m exactly equal to p. We could just as well have taken $m = 18p$, say, and still we would get $|1/(n + 1)| < 1/p$ whenever $n \geq m$.

(2) The sequence $((-1)^n/2^n)_{n \in \mathbb{N}}$ is a null sequence. In fact, let p be an arbitrary positive integer. We wish to find a nonnegative integer m so large that $|(-1)^n/2^n| < 1/p$ whenever $n \geq m$. Now $|(-1)^n/2^n| = 1/2^n$ for each n. Moreover,

$$\frac{1}{2^n} < \frac{1}{p} \iff 2^n > p.$$

There always exists some $m \in \mathbb{N}^*$ with $2^m > p$ (why?). Then $|(-1)^n/2^n| < 1/p$ whenever $n \geq m$.

(3) The sequence $(n/(n + 1))_{n \in \mathbb{N}}$ is *not* a null sequence. To show this, we seek some $p \in \mathbb{N}^*$ for which there is *no* $m \in \mathbb{N}$ with the property that, for every $n \in \mathbb{N}$, $n \geq m \implies |n/(n + 1)| < 1/p$. Equivalently, we seek $p \in \mathbb{N}^*$ such that, for every $m \in \mathbb{N}$, there exists some $n \in \mathbb{N}$ with $n \geq m$ yet $|n/(n + 1)| \geq 1/p$.

For which $n \in \mathbb{N}$ is $|n/(n + 1)| < 1/p$ for a given positive integer p? For exactly those n satisfying $n(p - 1) < 1$. Take, say, $p = 2$. Then $|n/(n + 1)| = n/(n + 1) < 1/p$ exactly when $n < 1$. If $m = 0$, then $n = 1 > m$ and $n/(n + 1) \geq 1/p$ (for $p = 2$, still); if $m \geq 1$, then $m \geq m$ and $m/(m + 1) \geq 1/p$.

(4) The sequence $(x_n)_{n \in \mathbb{N}} = (3, 1/1, 3, 1/3, 3, 1/5, \dots)$ given by $x_{2j} = 3$ and $x_{2j+1} = 1/(2j + 1)$ for all $j = 0, 1, 2, \dots$ is *not* a null sequence even though it includes values that get as close to 0 as you wish if you look far enough out in the sequence. (What p has no corresponding m?)

(5) A *constant sequence* $(x_n)_{n \in \mathbb{N}}$, that is, one in which $x_n = x_0$ for all $n \in \mathbb{N}$, is null if and only if $x_n = 0$ for all n.

Exercise 3.5.4. Determine whether the sequence $(x_n)_{n \in \mathbb{N}}$ defined by the given formula is null or not and justify your answer.

(a) $x_n = (-1)^n/(n + 1)$.

(b) $x_n = 2^n$.

(c) $x_n = 1/10^n$.

(d) $x_n = c^n$ for a real or rational number c. Consider various possibilities for c separately, including negative values.

(e) $x_n = d_n/10^n$ where, for each n, $d_n \in \{0, 1, 2, \ldots, 9\}$, that is, d_n is a decimal digit.

(f) $x_n = c^n/n!$ for a real or rational number c. (*Hint:* What is the ratio x_{n+1}/x_n?)

Exercise 3.5.5. Let $([a_n, b_n])_{n \in \mathbb{N}}$ be a sequence of closed intervals in $\mathbb{R}$ for which $[a_{n+1}, b_{n+1}] \subset [a_n, b_n]$ for all n. Suppose $(b_n - a_n)_{n \in \mathbb{N}}$ is a null sequence. Show that $\bigcap_{n \in \mathbb{N}} [a_n, b_n]$ consists of exactly one real number.

The $(1/p, m)$-criterion in Definition 3.5.2 is particularly simple (at least, as simple as these things can be) in that it uses for comparison just numbers of the form $1/p$ for positive integers p. Nonetheless, you (or your instructor) might prefer the following equivalent of our definition, using for comparison arbitrary elements $\varepsilon > 0$ in the field.

Lemma 3.5.6 ((ε, m)-criterion). *A sequence $(x_n)_{n \in \mathbb{N}}$ is null if and only if, for every $\varepsilon \in F$ with $\varepsilon > 0$, there exists some $m \in \mathbb{N}$ such that $|x_n| < \varepsilon$ for every $n \in \mathbb{N}$ with $n \geq m$.*

Proof. Exercise. ☐

Exercises 3.5.7. (1) Use logical symbols ($\Rightarrow$, $\forall$, etc.) to express the (ε, m)-criterion for a null sequence.

(2) Starting with the (ε, m)-criterion in Lemma 3.5.6, state in both words and symbols what it means to say that $(x_n)_{n \in \mathbb{N}}$ is *not* a null sequence.

(3) Apply the (ε, m)-criterion to redo Examples 3.5.3.

(4) Apply the (ε, m)-criterion to resolve Exercise 3.5.4.

(5) Let $(\varepsilon_n)_{n \in \mathbb{N}}$ be a fixed null sequence. Prove that then a sequence $(x_n)_{n \in \mathbb{N}}$ is null if and only if, for every $k \in \mathbb{N}$, there exists $m \in \mathbb{N}$ such that $|x_n| < \varepsilon_k$ whenever $n \geq m$.

A diagram can help to understand what it means to say that a sequence in $\mathbb{R}$ is null. A sequence $(x_n)_{n \in \mathbb{N}}$ in $\mathbb{R}$ is just a function $x \colon \mathbb{N} \to \mathbb{R}$ with $x(n) = x_n$ for each $n \in \mathbb{N}$. Then the graph of such a sequence is the subset

$$\{ (n, x_n) : n \in \mathbb{N} \}$$

of $\mathbb{N} \times \mathbb{R}$, and the latter is a subset of the plane $\mathbb{R} \times \mathbb{R}$. Hence the graph of x may be visualized as an array of points in the plane, one vertically above or below the point $(n, 0)$ on the x-axis for each nonnegative integer n.

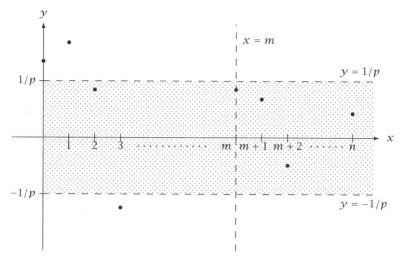

Figure 3.7: Graph of a null sequence.

To say that $(x_n)_{n \in \mathbb{N}}$ is a null sequence means the following, as suggested in Figure 3.7. Suppose you take an arbitrary positive integer p and draw a horizontal band centered around the x-axis and extending from the line $y = -1/p$ below to the line $y = 1/p$ above. Then for a suitably large integer m, all the points (n, x_n) of the graph on and to the right of the vertical line $x = m$ must lie within that horizontal band. (Since you must be able to do this for *every* positive integer p, evidently a single figure cannot capture the entire meaning of when the sequence is null.)

Exercise 3.5.8.   (a) Draw a diagram suggesting a sequence that is not null.

(b) Use a graph to suggest, in terms of the (ε, m)-criterion, when a sequence is null. Do the same for a sequence that is not null.

(c) Draw graphs illustrating each of Examples 3.5.3.

The following criteria help in analyzing some examples more readily.

Proposition 3.5.9.   *1.* (eventual equality criterion) *Let $(x_n)_{n \in \mathbb{N}}$ and $(y_n)_{n \in \mathbb{N}}$ be two sequences for which there is some $k \in \mathbb{N}$ with $x_n = y_n$ for all $n \geq k$. Then $(x_n)_{n \in \mathbb{N}}$ is null if and only if $(y_n)_{n \in \mathbb{N}}$ is null.*

2. (absolute value criterion) *A sequence $(x_n)_{n \in \mathbb{N}}$ is null if and only if the sequence $(|x_n|)_{n \in \mathbb{N}}$ of its absolute values is null.*

3. (comparison criterion) *If $(y_n)_{n \in \mathbb{N}}$ is a null sequence and $(x_n)_{n \in \mathbb{N}}$ is a sequence with $|x_n| \le |y_n|$ for all $n \in \mathbb{N}$, then $(x_n)_{n \in \mathbb{N}}$ is also null.*

Proof. Exercise. □

The hypothesis in the eventual equality criterion that $x_n = y_n$ for all n greater than or equal to some fixed k may be expressed by saying that the sequences $(x_n)_{n \in \mathbb{N}}$ and $(y_n)_{n \in \mathbb{N}}$ are *eventually* the same. In other words, *altering some finite number of entries in a sequence does not affect whether it is a null sequence.* Then the comparison criterion is still valid if its hypothesis that $|x_n| \le |y_n|$ for all n is replaced by the weaker hypothesis that $|x_n| \le |y_n|$ *eventually*, that is, for all n greater than or equal to some fixed k.

Exercise 3.5.10. Rework any parts of Exercise 3.5.4 that may be simplified by applying the preceding criteria.

Recall from Exercise 3.3.11 that a subset S of an ordered field is said to be *bounded* in F when S is bounded below in F and bounded above in F, that is, when there exist $a, b \in F$ such that $a \le s \le b$ for all $s \in S$. It is convenient to call $(x_n)_{n \in \mathbb{N}}$ itself a **bounded** sequence when its set of values $\{ x_n : n \in \mathbb{N} \}$ is bounded in F.

Proposition 3.5.11. *Every null sequence is bounded.*

Proof. Assume $(x_n)_{n \in \mathbb{N}}$ is a null sequence in F. We apply the $(1/p, m)$-criterion with $p = 1$. There exists $m \in \mathbb{N}$ such that $|x_n| < 1/1 = 1$ for all $n \ge m$. Let

$$b = \max \{ 1, |x_0|, |x_1|, \ldots, |x_{m-1}| \}.$$

Then $|x_n| \le b$ for all $n \in \mathbb{N}$. □

Did you notice that the statement of the proposition is actually a quantified implication? It really means: For every sequence $(x_n)_{n \in \mathbb{N}}$, if $(x_n)_{n \in \mathbb{N}}$ is null, *then* $(x_n)_{n \in \mathbb{N}}$ is bounded.

Exercise 3.5.12. Prove or disprove the converse of this implication.

The tools are now at hand to see when various algebraic combinations of null sequences are null.

Proposition 3.5.13. *Suppose the sequences $(x_n)_{n \in \mathbb{N}}$ and $(y_n)_{n \in \mathbb{N}}$ are null. Then so are the following sequences:*

- *The sum $(x_n + y_n)_{n \in \mathbb{N}}$.*

- *The difference $(x_n - y_n)_{n \in \mathbb{N}}$.*

- *For any $c \in F$, the constant multiple $(cx_n)_{n \in \mathbb{N}}$.*

- *The product $(x_n y_n)_{n \in \mathbb{N}}$.*

Proof. We give proofs only for sums and constant multiples and suggest a strategy for products. The proof for differences is left as an exercise.

Sum. (*Thinking:* For each n, we have $|x_n + y_n| \leq |x_n| + |y_n|$. Hence to make $|x_n + y_n|$ small, we need only make each of $|x_n|$ and $|y_n|$ twice as small.)

Let p be a positive integer. Then $2p$ is also a positive integer. Therefore, there are nonnegative integers m_1 and m_2 such that, for all $n \in \mathbb{N}$,

$$n \geq m_1 \implies |x_n| < \frac{1}{2p},$$

$$n \geq m_2 \implies |y_n| < \frac{1}{2p}.$$

Let $m = \max\{m_1, m_2\}$. Then $n \geq m$ implies both $n \geq m_1$ and $n \geq m_2$, whence

$$|x_n + y_n| \leq |x_n| + |y_n| < \frac{1}{2p} + \frac{1}{2p} = \frac{1}{p}.$$

Constant multiple. Fix $c \in F$. If $c = 0$, we are finished, and so we now assume $c \neq 0$. (*Thinking:* $|cx_n| < 1/p \iff |x_n| < 1/|c|p$. Of course, $|c|$ need not be an integer.)

Let p be a positive integer. Choose an integer $k \geq |c|$. There exists $m \in \mathbb{N}$ such that, for all $n \in \mathbb{N}$,

$$n \geq m \implies |x_n| < \frac{1}{kp} \implies |x_n| < \frac{1}{|c|p}$$

$$\implies |cx_n| = |c|\,|x_n| < \frac{1}{p}.$$

Product. [*Thinking:* If $(y_n)_{n \in \mathbb{N}}$ were actually constant, then the product sequence $(x_n y_n)_{n \in \mathbb{N}}$ would certainly be null. Of course, the sequence $(y_n)_{n \in \mathbb{N}}$ need not be constant, but each of its entries is at most the corresponding entry in a constant sequence because .... (Finish the thought and then write the actual proof.)] $\square$

The preceding proposition did not say anything about quotients. To deal with these, we need another notion. Observe that a sequence $(x_n)_{n \in \mathbb{N}}$ is bounded if there is some $b > 0$ for which $|x_n| \leq b$ for all n. If always $x_n \neq 0$, then the sequence $(1/x_n)_{n \in \mathbb{N}}$ of reciprocals is bounded if and only if there is some $c > 0$ for which $|x_n| \geq c$ for all n. (Why?) This situation deserves a definition.

Definition 3.5.14. A sequence $(x_n)_{n\in\mathbb{N}}$ is said to be **bounded away from 0** if there exists some $c > 0$ such that $|x_n| \geq c$ for all $n \in \mathbb{N}$.

Exercise 3.5.15.   (a) Let $(x_n)_{n\in\mathbb{N}}$ and $(y_n)_{n\in\mathbb{N}}$ be sequences where each $y_n \neq 0$. Show that the quotient sequence $(x_n/y_n)_{n\in\mathbb{N}}$ is a null sequence if $(x_n)_{n\in\mathbb{N}}$ is null and $(y_n)_{n\in\mathbb{N}}$ is bounded away from 0.

(b) For a sequence $(x_n)_{n\in\mathbb{N}}$ of nonzero numbers, under what circumstances, if any, is the sequence $(1/x_n)_{n\in\mathbb{N}}$ of reciprocals null?

Convergence of sequences to limits other than 0 is easily expressed in terms of null sequences.

Definition 3.5.16. The sequence $(x_n)_{n\in\mathbb{N}}$ is said to **converge to** an element $L \in F$ when the sequence $(x_n - L)_{n\in\mathbb{N}}$ is null, and then we write $(x_n)_{n\in\mathbb{N}} \to L$ or, more suggestively, write $x_n \to L$ as $n \to \infty$. The sequence is said to **converge in** F if it converges to some $L \in F$ and is said to **diverge** otherwise.

Note that a sequence converges to 0 exactly when it is a null sequence.

By definition, $(x_n)_{n\in\mathbb{N}} \to L$ if and only if the following $(1/p, m)$-**criterion** is met.

For every $p \in \mathbb{N}^*$, there exists some $m \in \mathbb{N}$ such that, for every $n \in \mathbb{N}$,

$$n \geq m \implies |x_n - L| < \frac{1}{p}.$$

Exercise 3.5.17.   (a) Formulate an (ε, m)-*criterion* for $(x_n)_{n\in\mathbb{N}} \to L$.

(b) Express both the $(1/p, m)$-criterion and your (ε, m)-criterion using logical symbols.

(c) Draw diagrams similar to Figure 3.7 illustrating these criteria for convergence of a sequence to a limit.

It is tempting to call a number L such that $(x_n)_{n\in\mathbb{N}} \to L$ *the* limit of the sequence $(x_n)_{n\in\mathbb{N}}$. However, we have no right to do so until we know that such an L is unique.

Lemma 3.5.18 (Uniqueness of Sequential Limits). *Let $(x_n)_{n\in\mathbb{N}}$ be a sequence. If $(x_n)_{n\in\mathbb{N}} \to L$ and $(x_n)_{n\in\mathbb{N}} \to M$, then $L = M$.*

Proof. Assume $(x_n)_{n\in\mathbb{N}} \to L$ and $(x_n)_{n\in\mathbb{N}} \to M$. Then $(x_n - M)_{n\in\mathbb{N}}$ and $(x_n - L)_{n\in\mathbb{N}}$ are both null, hence so is their difference. But the nth entry in the difference is $(x_n - M) - (x_n - L) = L - M$, and so the constant sequence with value $L - M$ for all entries is null. This is impossible unless $L - M = 0$, that is, $L = M$. $\square$

This lemma justifies the definition of limit of a sequence.

Definition 3.5.19. An element $L \in F$ is called **the limit of** a sequence $(x_n)_{n \in \mathbb{N}}$ when $(x_n)_{n \in \mathbb{N}} \to L$, and then we write

$$\lim (x_n)_{n \in \mathbb{N}} = L$$

or, more suggestively,

$$\lim_{n \to \infty} x_n = L.$$

Exercise 3.5.20. Explain what is wrong with the following "proof" that the limit is unique. If $\lim_{n \to \infty} x_n = L$ and $\lim_{n \to \infty} x_n = M$, then by symmetry and transitivity of equality,

$$L = \lim_{n \to \infty} x_n = M.$$

Evidently a constant sequence with value L converges to L. This provides an endless supply of examples of convergent sequences: Take any $L \in F$, take any null sequence $(y_n)_{n \in \mathbb{N}}$, and form the sequence $(L + y_n)_{n \in \mathbb{N}}$, which will converge to L.

Exercise 3.5.21. Show, more generally, that if $(x_n)_{n \in \mathbb{N}}$ is a null sequence and the sequence $(y_n)_{n \in \mathbb{N}} \to L$, then the sum $(x_n + y_n)_{n \in \mathbb{N}} \to L$, too.

Earlier, we saw that any null sequence is bounded. More generally, the following proposition holds.

Proposition 3.5.22. *Every convergent sequence is bounded.*

Proof. Assume $(x_n)_{n \in \mathbb{N}}$ converges. Then $(x_n)_{n \in \mathbb{N}} \to L$ for some (necessarily unique) $L \in F$. This means $(x_n - L)_{n \in \mathbb{N}}$ is a null sequence. But every null sequence is bounded, and so there exists $b \in F$ such that $|x_n - L| \le b$ for all n. Then $|x_n| \le |x_n - L| + |L| \le b + |L|$ for all n. This means $(x_n)_{n \in \mathbb{N}}$ is bounded. $\square$

Exercise 3.5.23. Show that the converse of the preceding proposition is not true—that a bounded sequence need not converge.

The following generalizes Proposition 3.5.13 about algebraic combinations of sequences.

Proposition 3.5.24. *Suppose* $(x_n)_{n \in \mathbb{N}} \to L$ *and* $(y_n)_{n \in \mathbb{N}} \to M$. *Then:*

- *The sum* $(x_n + y_n)_{n \in \mathbb{N}} \to L + M$.

- *The difference* $(x_n - y_n)_{n \in \mathbb{N}} \to L - M$.

- *For any $c \in F$, the constant multiple $(cx_n)_{n\in\mathbb{N}} \to cL$.*

- *The product $(x_ny_n)_{n\in\mathbb{N}} \to LM$.*

Proof. The assertions about the sum, difference, and constant multiple sequences are immediate consequences of the corresponding result for null sequences.

Product. For each $n \in \mathbb{N}$,

$$x_ny_n - LM = x_ny_n - Ly_n + Ly_n - LM = (x_n - L)\,y_n + L\,(y_n - M).$$

Now use the fact that $(y_n)_{n\in\mathbb{N}}$ is bounded. □

Exercise 3.5.25. (a) Suppose $(y_n)_{n\in\mathbb{N}} \to M$ with $M \neq 0$. Explain why $(y_n)_{n\in\mathbb{N}}$ is eventually nonzero, that is, $y_n \neq 0$ for all sufficiently large n. What more can you say in the case where $M > 0$?

(b) Suppose $(x_n)_{n\in\mathbb{N}} \to L$ and $(y_n)_{n\in\mathbb{N}} \to M$ with $M \neq 0$ and $y_n \neq 0$ for all n. Show that $(x_n/y_n)_{n\in\mathbb{N}} \to L/M$.

(c) If $(x_n)_{n\in\mathbb{N}} \to L$, must then $(|x_n|)_{n\in\mathbb{N}} \to |L|$? What about the converse?

The first part of the preceding proposition is sometimes stated in the form

$$\lim_{n\to\infty} (x_n + y_n) = \lim_{n\to\infty} x_n + \lim_{n\to\infty} y_n$$

(and similarly for the other parts). Unfortunately, this terse form fails to include an important part of the complete formulation: *if* the two sequences $(x_n)_{n\in\mathbb{N}}$ and $(y_n)_{n\in\mathbb{N}}$ converge, *then* their sum $(x_n + y_n)_{n\in\mathbb{N}}$ also converges.

In general, an often tougher question than what the limit of a sequence might be ·s the earlier question of whether the sequence has a limit at all—whether the sequence converges. It would be futile to try to find the limit of a sequence if it does not converge. We have already seen a necessary condition for a sequence to converge: the sequence is bounded. But this condition is not in general sufficient for convergence—see Exercise 3.5.23. There is one situation, though, when the condition is sufficient, namely, when the sequence is increasing or decreasing.

Definition 3.5.26. A sequence $(x_n)_{n\in\mathbb{N}}$ is said to be **increasing** if $x_n \leq x_k$ whenever $n < k$, and **decreasing** if $x_n \geq x_k$ whenever $n \leq k$.

Thus a sequence $(x_n)_{n\in\mathbb{N}}$ is increasing when

$$x_0 \leq x_1 \leq x_2 \leq \cdots \leq x_n \leq x_{n+1} \leq \cdots,$$

and it is decreasing when

$$x_0 \geq x_1 \geq x_2 \geq \cdots \geq x_n \geq x_{n+1} \geq \cdots.$$

For example, $(1 - 1/(n + 1))_{n \in \mathbb{N}}$ is increasing, whereas $(1/2^n)_{n \in \mathbb{N}}$ is decreasing. (The sequence may be said to be *strictly* increasing if actually $x_n < x_k$ whenever $n < k$. Similarly, the sequence may be said to be *strictly* decreasing if actually $x_n > x_k$ whenever $n < k$.)

Exercises 3.5.27. (1) Determine whether the sequence $(x_n)_{n \in \mathbb{N}}$ is increasing, decreasing, or neither.

    (a) $x_n = n/(n + 1)$.

    (b) $x_n = 1 + (-1)^n/2^n$.

    (c) $x_n = [1 + 1/(n + 1)]^{n+1}$.

(2) Prove that a sequence $(x_n)_{n \in \mathbb{N}}$ is increasing (respectively, decreasing) if $x_n \leq x_{n+1}$ (respectively, $x_n \geq x_{n+1}$) for all $n \in \mathbb{N}$.

The name of the next theorem comes from the term *monotonic* which is often used to mean "increasing or decreasing."

Theorem 3.5.28. (Bounded Monotonic Convergence) *Every bounded increasing or decreasing sequence in $\mathbb{R}$ converges in $\mathbb{R}$.*

Proof. Let the sequence $(x_n)_{n \in \mathbb{N}}$ in $\mathbb{R}$ be bounded and increasing (the proof for the decreasing case is left as an exercise). Choose an upper bound b of the range $R = \{x_n : n \in \mathbb{N}\}$. We shall successively bisect the closed interval $[x_0, b]$ and apply the Nested Interval Property to find a number to which the sequence $(x_n)_{n \in \mathbb{N}}$ converges.

    Let $a_0 = x_0$ and $b_0 = b$. The interval $[a_0, b_0]$ contains at least one point of R (in fact, it contains the point $a_0 \in R$). Bisect $[a_0, b_0]$ into two closed intervals. If the right half contains a point of R, let $[a_1, b_1]$ be that right half; otherwise, let $[a_1, b_1]$ be the left half. Thus $[a_1, b_1]$ has length $(b_0 - a_0)/2$, the intersection $R \cap [a_1, b_1]$ is nonempty, and b_1 is also an upper bound of R (why?).

    Bisect $[a_1, b_1]$ into two closed intervals. If the right half contains a point of R, let $[a_2, b_2]$ be that right half; otherwise, let $[a_2, b_2]$ be the left half. Thus $[a_2, b_2]$ has length $(b_1 - a_1)/2 = (b_0 - a_0)/2^2$, the intersection $R \cap [a_2, b_2]$ is nonempty, and b_2 is also an upper bound of R.

    Continue this process, that is, use recursion, to obtain a sequence

$([a_n, b_n])_{n \in \mathbb{N}}$ of closed intervals such that

$$[a_0, b_0] \supset [a_1, b_1] \supset [a_2, b_2] \supset \cdots [a_n, b_n] \supset [a_{n+1}, b_{n+1}] \supset \cdots,$$
$$R \cap [a_n, b_n] \neq \varnothing \quad (n \in \mathbb{N}),$$
$$b_n \text{ is an upper bound of } R \quad (n \in \mathbb{N}),$$
$$b_n - a_n = \frac{b_0 - a_0}{2^n} \quad (n \in \mathbb{N}).$$

Since the field $\mathbb{R}$ has the Nested Interval Property, there exists some $L \in \bigcap_{n=0}^{\infty} [a_n, b_n]$. We claim $(x_n)_{n \in \mathbb{N}} \to L$.

Let $p \in \mathbb{N}^*$. There exists $k \in \mathbb{N}$ such that

$$b_k - a_k < \frac{1}{p}.$$

Since $R \cap [a_k, b_k] \neq \varnothing$, there exists some $m \in \mathbb{N}$ with $x_m \in [a_k, b_k]$. Since the sequence $(x_n)_{n \in \mathbb{N}}$ is increasing and b_k is an upper bound of R, then $x_n \in [a_k, b_k]$ for all $n \geq m$. Now $L \in [a_k, b_k]$ also, and so for all $n \in \mathbb{N}$,

$$n \geq m \implies |L - x_n| \leq b_k - a_k < \frac{1}{p}. \quad \square$$

The proof of Theorem 3.5.28 used the Nested Interval Property in an essential way: if the Nested Interval Property were not applicable, then a bounded monotonic sequence would not necessarily converge.

Exercises 3.5.29. (1) Show that the statement of the preceding theorem is false for $\mathbb{Q}$ instead of $\mathbb{R}$.

(2) Give an appropriate definition for a sequence to be eventually increasing or eventually decreasing. Show that every bounded eventually increasing or eventually decreasing sequence in $\mathbb{R}$ converges.

One significant consequence of the theorem on bounded monotonic convergence will be that every infinite decimal $0.b_1 b_2 b_3 \ldots$ represents a real number. Here $b_1, b_2, \ldots$ are decimal digits, that is, elements of $\{0, 1, 2, \ldots, 9\}$. By such a decimal we mean a "sum,"

$$\frac{b_1}{10} + \frac{b_2}{10^2} + \cdots + \frac{b_n}{10^n} + \cdots,$$

of infinitely many terms. Of course, it is really possible to add only finitely many numbers together (and then only by adding two numbers at a time). So an explanation is demanded of what it means to add infinitely many numbers. For a decimal such as the one above, what we mean is to form the finite—or, as we shall call them, *partial*—sums

$$\frac{b_1}{10}, \frac{b_1}{10} + \frac{b_2}{10^2}, \frac{b_1}{10} + \frac{b_2}{10^2} + \frac{b_3}{10^3}, \dots$$

and to take the limit of the sequence so obtained. Naturally, we need to know that a limit really exists, that is, that this sequence converges.

We are going to look, more generally, at the *series* obtained from arbitrary sequences. Our look will be brief—just long enough to handle the matter of infinite decimals.

Definition 3.5.30. Let $(a_n)_{n\in\mathbb{N}}$ be a sequence in $\mathbb{R}$. The corresponding **series** is the sequence $\left(\sum_{j=0}^{n} a_j\right)_{n\in\mathbb{N}}$ of **partial sums of** $(a_n)_{n\in\mathbb{N}}$. The entries in the given sequence $(a_n)_{n\in\mathbb{N}}$ are referred to as the **terms** of this series.

The series is said to **converge to** the number $S \in \mathbb{R}$ if this sequence of partial sums converges to S in $\mathbb{R}$, and then (since limits of sequences are unique) we call S the **sum** of the series and write

$$\sum_{j=0}^{\infty} a_j = S.$$

The series is said to **converge in** $\mathbb{R}$ if it converges to some number in $\mathbb{R}$ and to **diverge** otherwise.

Notice the distinction between the *series* corresponding to a given sequence $(a_n)_{n\in\mathbb{N}}$ and the *sum* of that series (if the latter exists). The series itself is the sequence of partial sums. For each $n \in \mathbb{N}$, the nth partial sum is the number

$$S_n = \sum_{j=0}^{n} a_j = a_0 + a_1 + \cdots + a_n.$$

Thus the series *is* the sequence

$$S_0, S_1, S_2, \dots, S_n, \dots,$$

that is,

$$a_0, a_0 + a_1, a_0 + a_1 + a_2, \dots, a_0 + a_1 + a_2 + \cdots + a_n, \dots.$$

When the series converges to S, then

$$\lim_{n\to\infty} \sum_{j=0}^{n} a_j = S.$$

Just to confuse things, the notation $\sum_{j=0}^{\infty} a_j$—which is the limit of the sequence of partial sums—is sometimes used to denote the series itself!

Example 3.5.31. Let $r \in \mathbb{R}$ with $|r| < 1$. Then the *geometric series* $\sum_{j=0}^{\infty} r^j$ with *ratio r* converges. In fact, for each $n \in \mathbb{N}$, the nth partial sum of the series is

$$S_n = \sum_{j=0}^{n} r^j = \frac{1 - r^{n+1}}{1 - r} = \frac{1}{1 - r} - \frac{r^{n+1}}{1 - r}$$

(see Example 1.1.22). Then the sequence $(S_n)_{n \in \mathbb{N}}$ has limit $1/(1 - r)$. Thus

$$\sum_{j=0}^{\infty} r^j = \frac{1}{1 - r}.$$

When the original sequence uses subscripts starting with 1 instead of 0, then the corresponding series uses these same subscripts and we write $\sum_{j=1}^{\infty} a_j$ for the sum. For example, $\sum_{j=1}^{\infty} r^j = r/(1 - r)$ if $|r| < 1$.

Exercises 3.5.32. (1) (a) Prove that $\sum_{j=0}^{\infty} r^j$ diverges when $|r| \geq 1$.

(b) Show that $\sum_{j=0}^{\infty} c r^n = c/(1 - r)$ if $|r| < 1$.

(2) Show that every sequence of real numbers is a series, that is, is the sequence of partial sums of some—in fact, of a unique— sequence. (*Hint:* What is the difference $S_n - S_{n-1}$ of consecutive partial sums?)

(3) Prove: If $\sum_{j=0}^{\infty} a_j$ converges, then $(a_n)_{n \in \mathbb{N}}$ is null. (*Note:* The converse is false: see Example 3.5.40.)

(4) Given a sequence $(a_n)_{n \in \mathbb{N}}$ and a constant $c \in \mathbb{R}$, what relationship is there between convergence of the series corresponding to $(a_n)_{n \in \mathbb{N}}$ and convergence of the series corresponding to the sequence $(c a_n)_{n \in \mathbb{N}}$?

(5) Given two sequences $(a_n)_{n \in \mathbb{N}}$ and $(b_n)_{n \in \mathbb{N}}$, what relationship is there between convergence of the two series corresponding to these sequences, on the one hand, and the series corresponding to the sum sequence $(a_n + b_n)_{n \in \mathbb{N}}$, on the other hand?

Proposition 3.5.33. *Let $(a_n)_{n \in \mathbb{N}}$ be a sequence with $a_n \geq 0$ for all $n \in \mathbb{N}$. If the sequence of the partial sums of $(a_n)_{n \in \mathbb{N}}$ is bounded, then the series $\sum_{j=0}^{\infty} a_j$ converges in $\mathbb{R}$.*

Proof. Apply the theorem on bounded monotonic convergence. □

We are ready to show that any infinite decimal represents a real number.

Example 3.5.34. Let $(b_n)_{n \in \mathbb{N}}$ be a sequence in $\{0, 1, 2, \ldots, 9\}$. Then the series $\sum_{j=1}^{\infty} b_j / 10^j$ converges in $\mathbb{R}$.

In fact, for each j, the jth term $b_j / 10^j \leq 9/10^j$. Now the series corresponding to the sequence $(9/10^n)_{n \in \mathbb{N}^*}$ is a geometric series that converges to $(9/10)/(1 - 1/10) = 1$. The sum 1 is actually an upper bound of the partial sums of this geometric series (why?). Then 1 is also an upper bound of the partial sums of the given series $\sum_{j=1}^{\infty} b_j / 10^j$. But the partial sums of this series form an increasing sequence (why?). It follows that the series $\sum_{j=1}^{\infty} b_j / 10^j$ converges (why?).

Exercises 3.5.35. (1) What real number is represented by $0.999\ldots$? Explain.

(2) Generalize the method used in the preceding example: Prove that if $(b_n)_{n \in \mathbb{N}}$ and $(a_n)_{n \in \mathbb{N}}$ are sequences with $0 \leq b_n \leq a_n$ for all n and if the series corresponding to $(a_n)_{n \in \mathbb{N}}$ converges, then so does the series corresponding to $(b_n)_{n \in \mathbb{N}}$, and in this case $\sum_{n=0}^{\infty} b_n \leq \sum_{n=0}^{\infty} a_n$.

(3) Why is the condition $0 \leq b_n$ needed in (2)?

We just saw that every infinite decimal represents a real number. Conversely, we are about to see, every real number has a representation in terms of an infinite decimal. More precisely, every real number is the sum of an integer and an infinite decimal. (The decimal representation of integers was discussed in Section 2.1 beginning on page 66.) This will be an application of the Archimedean Ordering Property and will not use the Nested Interval Property.

Example 3.5.36. Each real number x has a decimal expansion of the form

$$x = \pm a_m \ldots a_2 a_1 a_0 . b_1 b_2 \ldots,$$

that is,

$$x = \pm \left(a_m \cdot 10^m + \cdots + a_2 \cdot 10^2 + a_1 \cdot 10 + a_0 + \frac{b_1}{10} + \frac{b_2}{10^2} + \cdots \right) \tag{*}$$

for some nonnegative integer m and some decimal digits $a_m, \ldots, a_2, a_1$, a_0 and $b_1, b_2, \ldots$ all belonging to $\{0, 1, 2, \ldots 9\}$. By (*) we mean

$$x = \pm \left(\sum_{j=0}^{m} a_j \cdot 10^j + \sum_{j=1}^{\infty} \frac{b_j}{10^j} \right) \tag{**}$$

(the quantity in parentheses is the sum of an ordinary finite sum and the sum of a series).

It suffices to handle the case where $x \geq 0$ (why?). To establish such a decimal expansion of $x \geq 0$, we first split x into the sum of its integral part $k = \lfloor x \rfloor$ and its fractional part $t = x - \lfloor x \rfloor$ (see page 162):

$$x = k + t, \qquad \text{where } k \in \mathbb{N} \text{ and } 0 \leq t < 1.$$

In view of the Base Representation Theorem (Theorem 2.1.10), it remains to show that t satisfying $0 \leq t < 1$ has the form

$$t = \lim_{n \to \infty} \sum_{j=1}^{n} \frac{b_j}{10^j} \qquad (\text{***})$$

for some decimal digits $b_1, b_2, \ldots$.

To begin, such a number t lies in the half-closed, half-open interval[10] $[0, 1)$. The numbers $0 = 0/10, 1/10, 2/10, \ldots, 10/10 = 1$ partition $[0, 1)$ into 10 subintervals

$$[0/10, 1/10), [1/10, 2/10), \ldots, [9/10, 10/10)$$

exactly one of which must contain t. Let this subinterval be

$$\left[\frac{b_1}{10}, \frac{b_1 + 1}{10} \right)$$

so that

$$\frac{b_1}{10^1} \leq t < \frac{b_1 + 1}{10^1} = \frac{b_1}{10^1} + \frac{1}{10^1}$$

with $b_1 \in \{0, 1, 2, \ldots, 9\}$.

Next, the difference $t - b_1/10$ lies in the half-closed, half-open interval $[0, 1/10)$. The numbers $0 = 0/10^2, 1/10^2, 2/10^2, \ldots, 10/10^2 = 1/10$ partition $[0, 1/10)$ into 10 subintervals

$$\left[0/10^2, 1/10^2 \right), \left[1/10^2, 2/10^2 \right), \ldots, \left[9/10^2, 10/10^2 \right)$$

exactly one of which must contain $t - b_1/10$. Let this subinterval be

$$\left[\frac{b_2}{10^2}, \frac{b_2 + 1}{10^2} \right)$$

so that

$$\frac{b_2}{10^2} \leq t - \frac{b_1}{10^1} < \frac{b_2 + 1}{10^2} = \frac{b_2}{10^2} + \frac{1}{10^2}$$

---

[10] As usual, the notation $[a, b)$ for a half-closed, half-open interval in $\mathbb{R}$ means the set $\{x \in \mathbb{R} : a \leq x < b\}$.

with $b_2 \in \{0, 1, 2, \ldots, 9\}$. Thus

$$\frac{b_1}{10^1} + \frac{b_2}{10^2} \leq t < \frac{b_1}{10^1} + \frac{b_2}{10^2} + \frac{1}{10^2}$$

with $b_1, b_2 \in \{0, 1, 2, \ldots, 9\}$.

Proceeding in this way, we obtain recursively a sequence $(b_1, b_2, \ldots)$ such that

$$\sum_{j=1}^{n} \frac{b_j}{10^j} \leq t < \sum_{j=1}^{n} \frac{b_j}{10^j} + \frac{1}{10^n} \qquad (n = 1, 2, \ldots)$$

with $b_n \in \{0, 1, 2, \ldots, 9\}$ for all $n = 1, 2, \ldots$.

Since $(1/10^n)_{n=1,2,\ldots}$ is a null sequence, the expansion (***) of t follows at once. $\square$

$$
\begin{array}{r}
.625 \\
8\,)\overline{5.000} \\
\underline{4\,8} \\
20 \\
\underline{16} \\
40 \\
\underline{40} \\
0
\end{array}
$$

Figure 3.8: Terminating decimal by long division.

Years ago you learned how to find the decimal expansion of rational numbers through long division. For example, the calculation in Figure 3.8 shows that

$$\frac{5}{8} = 0.62500\ldots0\ldots.$$

In the case of the decimal expansion of a real number x, where all the digits $b_{k+1}, b_{k+2}, \ldots$ after the kth to the right of the decimal point are 0, we say that the decimal expansion is **terminating** and often drop all the infinitely many trailing zeros. For example, we write

$$\frac{5}{8} = 0.625.$$

(This is *not* the same thing as the common scientific practice of showing only as many digits to the right of the decimal point as is justified by the precision of a measurement or by calculations resulting from such a measurement.)

Exercises 3.5.37.   (1) Find the decimal expansions of $101/8$ and $3/7$ by the method used in Example 3.5.36.

(2) Repeat (1) but for $\sqrt{2}$ and $\sqrt{3}$. (Do not attempt to find all the decimal digits of the fractional part—six or so will do.)

(3) Which real numbers x have terminating decimal expansions?

(4) The decimal expansion $\pm a_m \ldots a_2 a_1 a_0.b_1 b_2 \ldots$ of a real number x is said to be *repeating* if there are positive integers k and p with

$$b_{k+mp+j} = b_{k+j} \qquad (m = 1, 2, \ldots; j = 0, 1, \ldots, p - 1).$$

In other words, the fractional part has the form

$$.b_1 b_2 \ldots b_{k-1} b_k b_{k+1} \ldots b_{k+(p-1)} b_k b_{k+1} \ldots$$
$$b_{k+(p-1)} b_k b_{k+1} \ldots b_{k+(p-1)} \ldots .$$

(For example, $137/1110 = 0.1234234234\ldots$, and this decimal expansion has the above form with $k = 2$ and $p = 3$.) Which real numbers have repeating decimal expansions?

(5) By changing 10 to 2 in Example 3.5.36, show similarly that each real number has a binary (base 2) expansion.

(6) Find the binary expansions of $5/32$ and $1/10$.

(7) Generalize decimal expansion of reals to expansion using any integer base $b > 1$.

(8) Let k be a nonnegative integer. Let x be a positive real number. Write x in the form $x = n + t$, where n is the integral part of x, t is the fractional part of x, and t has the decimal expansion

$$t = 0.b_1 b_2 b_3 \ldots b_k b_{k+1} \ldots .$$

We say that the real number y is x *chopped to k decimal places* to mean

$$y = n + 0.b_1 b_2 \ldots b_k.$$

(All digits after the kth to the right of the decimal point have been dropped because they are 0.) For example, since

$$\sqrt{2} = 1.4142135623\ldots,$$

then $\sqrt{2}$ chopped to 6 decimal places is the (rational) number 1.414213. Show that x chopped to k decimal places is given by the formula $10^{-k}\lfloor 10^k x \rfloor$.

(9) Let k be a nonnegative integer and let x be a positive real number. Write x in the form $x = n + t$ with n and t as in (8). We say that the real number z is x *rounded to k decimal places* to mean

$$|x - z| \leq (1/2) \cdot 10^{-k} = 5 \cdot 10^{-(k+1)}.$$

For example, since $\sqrt{2} = 1.4142135623\ldots$, then $\sqrt{2}$ rounded to 6 decimal places is 1.414214 (rounded up). However, $\sqrt{2}$ rounded to 5 decimal places is 1.41421 (rounded down). (In the special case that the decimal expansion of x terminates with a 5 in its $(k + 1)$st decimal place, that is, when $t = 0.b_1 b_2 b_3 \ldots b_k 5$, then there is ambiguity—there are two different numbers that are x rounded to k decimal places. For example, according to what we said, both 6.1237 and 6.1238 are 6.12375 rounded to 4 decimal places.)

Develop a formula involving floor (or ceiling) for rounding real numbers to k decimal places. What does your formula do in the special ambiguous case?

In a convergent sequence, the entries will be as close as you wish to some fixed number L if you look far enough along in the sequence. Then these entries will be as close as you wish *to each other* if, again, you look far enough along in the sequence. Here is the formal definition of the latter property.

Definition 3.5.38. A sequence $(x_n)_{n \in \mathbb{N}}$ is said to be a **Cauchy** sequence[11] if, for each $p \in \mathbb{N}^*$, there exists $m \in \mathbb{N}$ such that, for all $n, k \in \mathbb{N}$,

$$n \geq m \,\&\, k \geq m \implies |x_n - x_k| < \frac{1}{p}.$$

What led us to the definition is the following result.

Proposition 3.5.39. *Every convergent sequence is a Cauchy sequence.*

Proof. Exercise.   □

Owing to this proposition, we have lots of examples of Cauchy sequences. Here is an example of a sequence that is not a Cauchy sequence.

---

[11] Such sequences are named after the French analyst Augustin-Louis Cauchy (1789–1857), who contributed crucially toward making the notions of limit and derivative rigorous.

Example 3.5.40. For each $n \in \mathbb{N}^*$, let S_n be the nth partial sum of the series corresponding to the sequence $(1/n)_{n \in \mathbb{N}^*}$, that is,

$$S_n = \sum_{j=1}^{n} \frac{1}{j}.$$

Then $(S_n)_{n \in \mathbb{N}^*}$ is *not* a Cauchy sequence because, for each $j \in \mathbb{N}^*$,

$$S_{2^{j+1}} - S_{2^j} = \sum_{i=2^j+1}^{2^{j+1}} \frac{1}{i} \geq \left(2^{j+1} - 2^j\right) \frac{1}{2^{j+1}} = 2^j \frac{1}{2^{j+1}} = \frac{1}{2}.$$

(Write the details of why this means the sequence is not a Cauchy sequence.)

Exercises 3.5.41. (1) Prove: If $(x_n)_{n \in \mathbb{N}}$ is a Cauchy sequence, then, for each $p \in \mathbb{N}^*$, there exist $m \in \mathbb{N}$ such that $n \geq m \implies |x_{n+1} - x_n| < 1/p$. In other words, in a Cauchy sequence, consecutive terms get as close to one another as you wish if you go far enough along in the sequence.

It is tempting to suppose the converse is true, too. Show that it is not.

(2) If $(x_n)_{n \in \mathbb{N}}$ is a Cauchy sequence and $(y_n)_{n \in \mathbb{N}}$ is a sequence such that $(x_n - y_n)_{n \in \mathbb{N}}$ is a null sequence, show that $(y_n)_{n \in \mathbb{N}}$ must be a Cauchy sequence, too.

We already know that a convergent sequence must be bounded. More generally, the following proposition holds.

Proposition 3.5.42. *Every Cauchy sequence is bounded.*

Proof. For the positive integer $p = 1$ there exists $m \in \mathbb{N}$ such that $|x_k - x_n| < 1$ whenever $k, n \geq m$. In particular, $|x_m - x_n| < 1$ whenever $n \geq m$. Let

$$a = \max \{ |x_k - x_n| : 0 \leq k \leq m, 0 \leq n \leq m \}.$$

Then $|x_k - x_m| \leq a$ for all k with $0 \leq k \leq m$. It follows that

$$|x_k - x_n| \leq 1 + a$$

for *all* k, n (why?). $\square$

In a Cauchy sequence, the successive entries get closer and closer to one another as you look further and further along the sequence, and so it is plausible that the entries ought to get closer and closer to some fixed number as you look further and further along the sequence; that is, the sequence ought to converge. Unfortunately, this is not always the case in an archimedean ordered field—the converse of Proposition 3.5.39 is not necessarily true.

Example 3.5.43. Let $F = \mathbb{Q}$. By the order density of $\mathbb{Q}$ in $\mathbb{R}$, there exists a sequence $(x_n)_{n\in\mathbb{N}}$ of *rational* numbers such that $(x_n)_{n\in\mathbb{N}} \to \sqrt{2}$ in the field $\mathbb{R}$ of real numbers (indicate why such a sequence exists). Then $(x_n)_{n\in\mathbb{N}}$ is a Cauchy sequence in $\mathbb{R}$, and so it must be a Cauchy sequence in $\mathbb{Q}$ (explain). But $(x_n)_{n\in\mathbb{N}}$ cannot converge in $\mathbb{Q}$. In fact, just suppose $(x_n)_{n\in\mathbb{N}} \to L$ in $\mathbb{Q}$. Then also $(x_n)_{n\in\mathbb{N}} \to L$ in $\mathbb{R}$ (why?). By the uniqueness of limits in $\mathbb{R}$, necessarily $L = \sqrt{2}$. This is impossible because L is rational, whereas $\sqrt{2}$ is irrational.

What makes the preceding example work—and the converse of Proposition 3.5.42 fail—is that the archimedean ordered field $\mathbb{Q}$ does not have the Nested Interval Property: the very number $\sqrt{2}$ that "ought to be there" is missing from $\mathbb{Q}$ (but present in $\mathbb{R}$). When the field does have the Nested Interval Property, then Cauchy sequences necessarily converge. The proof will use a basic fact about finite and infinite sets: the union of finitely many finite sets is itself finite (see Section 4.1).

Theorem 3.5.44 (Cauchy Completeness). *Every Cauchy sequence in $\mathbb{R}$ converges in $\mathbb{R}$.*

Proof. Let $(x_n)_{n\in\mathbb{N}}$ be a Cauchy sequence in $\mathbb{R}$. Let

$$R = \{x_n : n \in \mathbb{N}\}$$

be the range of the sequence. We consider two cases.

Case 1: The set R is finite. Then there is a number $L \in \mathbb{R}$ such that $x_n = L$ for infinitely many values of n in $\mathbb{N}$ (why?). We claim $(x_n)_{n\in\mathbb{N}} \to L$. Let $p \in \mathbb{N}^*$. Since our sequence is a Cauchy sequence, there is some $j \in \mathbb{N}$ such that

$$n, m \ge j \implies |x_n - x_m| < 1/p.$$

Now the set $\{m \in \mathbb{N} : x_m = L\}$ is infinite, whereas the set $\{m \in \mathbb{N} : m < j\}$ is only finite. Hence there is some integer $m \ge j$ with $x_m = L$. Then $n \ge m$ implies $|x_n - L| < 1/p$.

Case 2: The range R of $(x_n)_{n\in\mathbb{N}}$ is infinite. There exist $a_0, b_0 \in \mathbb{R}$ with $a_0 < b_0$ and $R \subset [a_0, b_0]$ (why?). We shall successively bisect this closed interval and apply the Nested Interval Property to obtain the desired limit.

The interval $[a_0, b_0]$ contains infinitely many points of R (in fact, all of the infinitely many points of R). Bisect $[a_0, b_0]$ into two closed intervals. One or the other of these two halves must contain infinitely many points of R (why?); call such a half $[a_1, b_1]$. Thus $[a_1, b_1]$ has length $(b_0 - a_0)/2$, and $R \cap [a_1, b_1]$ is infinite.

Bisect $[a_1, b_1]$ into two closed intervals. One or the other of these two halves must contain infinitely many points of R; call such a half

$[a_2, b_2]$. Thus $[a_2, b_2]$ has length $(b_1 - a_1)/2 = (b_0 - a_0)/2^2$, and $R \cap [a_2, b_2]$ is infinite.

Continue this process, that is, use recursion, to obtain a sequence $([a_n, b_n])_{n \in \mathbb{N}}$ of closed intervals such that

$$[a_0, b_0] \supset [a_1, b_1] \supset [a_2, b_2] \supset \cdots [a_n, b_n] \supset [a_{n+1}, b_{n+1}] \supset \cdots,$$
$$R \cap [a_n, b_n] \text{ is infinite} \quad (n \in \mathbb{N}),$$
$$b_n - a_n = \frac{b_0 - a_0}{2^n} \quad (n \in \mathbb{N}).$$

Since the field $\mathbb{R}$ has the Nested Interval Property, there exists some $L \in \bigcap_{n=0}^{\infty} [a_n, b_n]$. We claim $(x_n)_{n \in \mathbb{N}} \to L$.

Let $p \in \mathbb{N}^*$ and let $\varepsilon = 1/p$. Since $(x_n)_{n \in \mathbb{N}}$ is a Cauchy sequence, there exists j such that

$$n \geq j \,\&\, k \geq j \implies |x_n - x_k| < \frac{\varepsilon}{2}.$$

Since $\mathbb{R}$ is archimedean, there exists $m \in \mathbb{N}$ such that

$$m \geq j, \quad b_m - a_m < \frac{\varepsilon}{2}$$

(supply missing details). Since $R \cap [a_m, b_m]$ is infinite, there exists $k \in \mathbb{N}$ such that

$$k \geq j, \quad x_k \in [a_m, b_m].$$

But also $L \in [a_m, b_m]$, and so

$$|x_k - L| < \frac{\varepsilon}{2}.$$

Then (see Figure 3.9)

$$n \geq m \implies |x_n - L| \leq |x_n - x_k| + |x_k - L| < \frac{\varepsilon}{2} + \frac{\varepsilon}{2} = \varepsilon. \quad \square$$

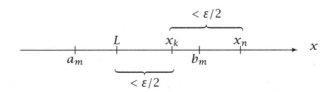

Figure 3.9: The situation in the proof of Cauchy completeness.

Exercise 3.5.45. Reprove Cauchy completeness of $\mathbb{R}$ using order completeness instead of the Nested Interval Property.

The next exercise asks you to use Cauchy sequences of rational numbers to construct a certain field and to prove this field is isomorphic to the field $\mathbb{R}$ (assuming $\mathbb{R}$ is already known axiomatically). The very same construction can be used to define $\mathbb{R}$ in terms of $\mathbb{Q}$ when following, instead, a constructive approach to the reals.

Exercise 3.5.46. Let S be the set of all Cauchy sequences of *rational* numbers. We want to associate with each such Cauchy sequence not happening to have a rational limit the missing irrational real number that should be its limit. Unfortunately, several different Cauchy sequences may have the same limit. Accordingly, we would like to identify any two that have the same limit. But we cannot simply say this directly because their limit need not be there among the rationals (and we are trying to construct the missing irrationals). So we have to find an indirect way of saying the same thing.

To find this indirect way, note that two Cauchy sequences will have the same limit (perhaps in $\mathbb{R}$) precisely when their difference is a null sequence. Hence the thing to do is to identify two Cauchy sequences of rationals when their difference is a null sequence.

Formally, we define a relation $\sim$ in S by the rule

$$(x_n)_{n\in\mathbb{N}} \sim (y_n)_{n\in\mathbb{N}} \iff (x_n - y_n)_{n\in\mathbb{N}} \text{ is a null sequence.}$$

[In particular, $(x_n)_{n\in\mathbb{N}} \sim (0)_{n\in\mathbb{N}}$, the constant zero sequence, exactly when $(x_n)_{n\in\mathbb{N}}$ is a null sequence.]

(a) Show that the relation $\sim$ is an equivalence relation on S.

For an element $(x_n)_{n\in\mathbb{N}} \in S$, denote by $[(x_n)_{n\in\mathbb{N}}]$ its equivalence class under $\sim$, that is, the set of all sequences in $\mathbb{Q}$ that are equivalent to $(x_n)_{n\in\mathbb{N}}$ with respect to $\sim$. Denote by F the quotient set of S under $\sim$, that is, the set of all such equivalence classes.

(b) Show that the operations on F given by

$$[(x_n)_{n\in\mathbb{N}}] + [(y_n)_{n\in\mathbb{N}}] = [(x_n + y_n)_{n\in\mathbb{N}}],$$
$$[(x_n)_{n\in\mathbb{N}}] \cdot [(y_n)_{n\in\mathbb{N}}] = [(x_n \cdot y_n)_{n\in\mathbb{N}}]$$

are well-defined, that is, do not depend on the choice of representatives of the equivalence classes $[(x_n)_{n\in\mathbb{N}}]$ and $[(y_n)_{n\in\mathbb{N}}]$.

(c) Show that these operations make F a field.

(d) Construct an isomorphism from the field F to the field $\mathbb{R}$.

3.6   The Complex Number Field

The sets in the chain $\mathbb{N} \subset \mathbb{Z} \subset \mathbb{Q} \subset \mathbb{R}$ may be regarded as number systems created in order to solve larger and larger classes of algebraic equations. Thus, for given natural numbers a and b, the equation $x + a = b$ cannot be solved in $\mathbb{N}$ unless $a \le b$. In the extension $\mathbb{Z}$ of $\mathbb{N}$, this equation can always be solved no matter what the natural numbers a and b are; indeed, the equation can be solved in $\mathbb{Z}$ when a and b are arbitrary integers.

Next, for given integers a and b with $a \ne 0$, the equation $ax = b$ cannot be solved in $\mathbb{Z}$ unless $a \mid b$. In the extension $\mathbb{Q}$ of $\mathbb{Z}$, this equation can always be solved no matter what the integers a and b are with $a \ne 0$; indeed, the equation can be solved in $\mathbb{Q}$ when a and b are arbitrary rational numbers with $a \ne 0$.

Next, for a given rational number $a \ge 0$, the equation $x^2 = a$ cannot be solved in $\mathbb{Q}$ unless a is a quotient of squares of integers. In the extension $\mathbb{R}$ of $\mathbb{Q}$, this equation can always be solved no matter what the rational number a is with $a \ge 0$; indeed, the equation can be solved in $\mathbb{R}$ when a is an arbitrary real number with $a \ge 0$.

But we do not stop there. The simple equation $x^2 + 1 = 0$ or, equivalently, $x^2 = -1$, cannot be solved in $\mathbb{R}$ (where, as in every ordered field, the square of each element is nonnegative). It is therefore reasonable to seek a larger number system than the real numbers in which this equation can be solved. The field of *complex numbers* is such a system.

The aim of this section is to construct the field of complex numbers from the field of real numbers, to show that the equation $x^2 + 1$ (along with others) has a solution in this number system, and to explore a few further aspects of complex numbers.

The naive way to construct complex numbers would be to invent a symbol i and stipulate that $i^2 = -1$.[12] But that would not give a number *system* yet, just one single new number; even less would it give a number system extending the real numbers. To get an entire number system that extends the real numbers we could naively say that complex numbers are expressions of the form $x + yi$, where x and y are real numbers; expressions of this form with $y = 0$ would be the real numbers. Moreover, we could say, such expressions are to be added and multiplied following the usual rules of algebra but with the added stipulation that $i \cdot i = i^2$ is to be replaced by -1. In other words, for expressions $x + yi$ and $u + vi$ we would agree to write

$$(x + yi) + (u + vi) = (x + u) + (y + v)i, \tag{*}$$

---

[12]Because the symbol i can also stand for current, electrical engineers and others often use j for the complex number we denote by i.

and, because

$$(x + yi) \cdot (u + vi) = xu + xvi + yui + yvi^2$$
$$= xu + xvi + yui + yv(-1)$$
$$= (xu - yv) + (xv + yu)i,$$

we would agree to write

$$(x + yi) \cdot (u + vi) = (xu - yv) + (xv + yu)i. \qquad (**)$$

The preceding "definition" of complex numbers doubtless seems somewhat elusive and unsatisfying. After all, what *is* this i? The tempting answer, "the expression $0 + 0i$," just begs the question. After all, precisely what is meant by an "expression of the form $x + yi$"? (How can we form such an expression if we do not know what i means?) And how do we know that the usual rules of algebra employed to calculate sums and products as in (*) and (**) are valid here?

We finesse these questions by extracting the essence of an expression $x + yi$, namely, the pair of real coefficients x and y.

Definition 3.6.1. A **complex number** is an ordered pair (x, y) of real numbers. The set of all complex numbers is denoted by $\mathbb{C}$.

The first coordinate x of a complex number $z = (x, y)$ is called the **real part of** z and is denoted by $\mathrm{Re}(z)$; the second coordinate y is called the **imaginary part of** z and is denoted by $\mathrm{Im}(z)$.

By definition, the set of all complex numbers is the set of all ordered pairs of real numbers:

$$\boxed{\mathbb{C} = \mathbb{R} \times \mathbb{R}.}$$

Then we can use the usual geometric representation of $\mathbb{R} \times \mathbb{R}$ as a plane to represent complex numbers, too, as in Figure 3.10. Indeed, we often refer to $\mathbb{C}$ as the *complex plane.* Notice that the real and imaginary parts $\mathrm{Re}(z)$ and $\mathrm{Im}(z)$ of a complex number z are then just the usual x- and y-coordinates of the corresponding point in the plane.

What we have done, in effect, is to use ordered pairs (x, y) instead of the mysterious expressions $x + yi$. In view of the formulas (*) and (**) above that we wanted to use for adding and multiplying expressions, we may use corresponding definitions for adding and multiplying ordered pairs of real numbers.

Definition 3.6.2. Let $z = (x, y)$ and $w = (u, v)$ be complex numbers. Then their **sum** is the complex number $z + w = (x + u, y + v)$ and their **product** is the complex number $z \cdot w = (xu - yv, xv + yu)$.

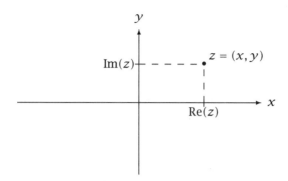

Figure 3.10: The complex plane.

Thus operations of addition and multiplication have been defined on $\mathbb{C}$. (If you have studied vectors, you may recognize this addition as being addition of vectors in the plane. In that case, you should be able to draw a figure showing the geometric meaning of the sum of two complex numbers. Multiplication can also be described geometrically by using trigonometric functions; see Exercise 3.6.3.)

Exercise 3.6.3. Take for granted that any point $(x, y) \in \mathbb{R} \times \mathbb{R}$ can be represented (but not uniquely) in the form $(x, y) = (r \cos \theta, r \sin \theta)$, where $r \geq 0$ is the distance of the point from the origin and θ is a measure of the angle between the positive x-axis and the ray from the origin to the point. We call this a *polar representation* of (x, y).

Let $z = (r \cos \alpha, r \sin \alpha)$ and $w = (s \cos \beta, s \sin \beta)$ be polar representations of complex numbers z and w. Verify the formula

$$z \cdot w = (rs \cos(\alpha + \beta), rs \sin(\alpha + \beta)).$$

Then use this formula to describe multiplication of complex numbers geometrically. Draw a diagram.

Evidently the complex number $(0, 0)$ is a zero element in $\mathbb{C}$. Evidently, also, each $(x, y) \in \mathbb{C}$ has as a negative the element $(-x, -y)$. The complex number $(1, 0)$ is a unity element in $\mathbb{C}$ because, for all $(x, y) \in \mathbb{C}$,

$$(x, y) \cdot (1, 0) = (x \cdot 1 - y \cdot 0, x \cdot 0 + y \cdot 1) = (x, y).$$

We are on our way to proving that $\mathbb{C}$ is a field.

What about reciprocals? Suppose $z = (x, y)$ is a nonzero complex number. We seek a complex number $w = (u, v)$ for which $zw = (1, 0)$, the unity of $\mathbb{C}$. We want

$$(xu - yv, xv + yu) = (1, 0)$$

or, equivalently,

$$\begin{cases} xu - yv = 1 \\ xv + yu = 0. \end{cases}$$

This system of two equations in the two unknowns u and v can readily be solved. Do it! You should get the unique solution $u = x/(x^2 + y^2)$, $v = -y/(x^2 + y^2)$. Thus, (x, y) has the reciprocal

$$(x, y)^{-1} = \left(\frac{x}{x^2 + y^2}, \frac{-y}{x^2 + y^2} \right).$$

The right-hand side of the preceding formula can be simplified by another definition.

Definition 3.6.4. For a complex number $z = (x, y)$, its **conjugate** is the complex number $\bar{z} = (x, -y)$ and its **magnitude** is the real number $|z| = \sqrt{x^2 + y^2}$.

Geometrically, the conjugate of a complex number $z = (x, y)$ is obtained by reflecting the corresponding point in the complex plane through the x-axis, and the magnitude of z is the distance between the origin and that point. (See Figure 3.11.)

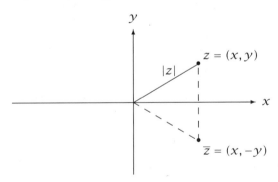

Figure 3.11: Conjugate and magnitude.

The formula above for the reciprocal of $z \in \mathbb{C}$ can now be written

$$z^{-1} = (1/|z|^2, 0)\bar{z}.$$

You should verify the remaining field axioms to complete the proof of the following result.

Proposition 3.6.5. *With the operations of addition and multiplication defined above, the set $\mathbb{C}$ of all complex numbers is a field.*

Exercises 3.6.6.   (1) Verify that $z^{-1} = (1/|z|^2, 0)\bar{z}$ for each nonzero $z \in \mathbb{C}$.

(2) Develop formulas for the difference $z - w$ and quotient z/w of complex numbers (with w nonzero to form the quotient). In particular, give a formula for z/i.

(3) Show that conjugation—forming conjugates—of complex numbers defines an isomorphism of the field $\mathbb{C}$ with itself.

(4) Suppose that $f: \mathbb{C} \to \mathbb{C}$ is an isomorphism that "fixes" $\mathbb{R}$ in the sense that $f(x) = x$ for each $x \in \mathbb{R}$. Show that f is either the identity map or conjugation.

(5) Prove that magnitude of complex numbers has the properties that, for all $z, w \in \mathbb{C}$: $|z| \geq 0$; $|z| = 0 \iff z = $ zero; $|zw| = |z| \cdot |w|$; and $|z + w| \leq |z| + |w|$.

The complex number $(1, 0)$ is the unity of the field $\mathbb{C}$. If this number is so special, what about $(0, 1)$? First, let us give the latter a name.

Definition 3.6.7. The complex number i is defined to be the element $(0, 1)$ of $\mathbb{C}$.

What is special about i is that, for any complex number $z = (x, y)$,

$$z \cdot i = (x, y) \cdot (0, 1) = (x \cdot 0 - y \cdot 1, x \cdot 1 + y \cdot 0) = (-y, x). \quad (***)$$

In particular, the product of i with itself is

$$i \cdot i = (0, 1) \cdot (0, 1) = (-1, 0) = -(1, 0).$$

The complex number i is thus an element of $\mathbb{C}$ whose square is the negative of unity—a solution of the equation $z^2 = -$unity.

Exercise 3.6.8. Using formula $(***)$, explain why the effect of multiplying $z \in \mathbb{C}$ by i is to rotate the point z counterclockwise around the origin through an angle of $\pi/2$. Draw a diagram to illustrate this.

We have almost, but not quite, succeeded in our quest for a "number" whose square is the *real* number -1. In $\mathbb{C}$, we have $-$unity $= (-1, 0)$, not -1. What we need to do is to *identify* each real number x with the corresponding complex number $(x, 0)$. What is involved in such an identification?

We are accustomed to speaking of the x-axis in the xy-plane $\mathbb{R} \times \mathbb{R}$ *as if* it were the set of real numbers. [Of course, the x-axis is not really the set of real numbers. Points on the x-axis, like those anywhere in the xy-plane, are ordered pairs of real numbers—ordered pairs that happen to be of the form $(x, 0)$.] Then it is tempting to speak of complex numbers of the form $(x, 0)$ *as if* they were real numbers. And, indeed, complex numbers of this form are added and multiplied in the same way as the corresponding real numbers that are their first coordinates:

$$(x, 0) + (y, 0) = (x + y, 0),$$
$$(x, 0) \cdot (y, 0) = (x \cdot y, 0).$$

(Check these identities.)

Form the subset

$$R = \{ (x, 0) : x \in \mathbb{R} \}$$

of $\mathbb{C}$. Then the preceding formulas show that the complex sum and product of elements of R are again elements of R. Moreover, the addition and multiplication operations on $\mathbb{C}$, restricted just to elements of R, make R itself into a field. More precisely, in accordance with the following definition they make R into a *subfield* of the field $\mathbb{C}$.

Definition 3.6.9. A **subfield** of a field F is a field K such that K is a subset of F and the operations of addition and multiplication on K are obtained by restricting the addition and multiplication of F to elements of K.

Let us make this definition perfectly clear. Suppose F and K are fields. Denote addition and multiplication in F by $+_F$ and $\cdot_F$, respectively; similarly, denote addition and multiplication in K by $+_K$ and $\cdot_K$, respectively. To say that K is a subfield of F means, first, that $K \subset F$, and second, that for arbitrary elements $x, y \in K$ (which, therefore, are also elements of F), $x +_K y = x +_F y$ and $x \cdot_K y = x \cdot_F y$.[13]

Another example is that the field $\mathbb{Q}$ of rational numbers is a subfield of the field $\mathbb{R}$ of real numbers.

If F and K are fields with K a subset of F, then K may or may not be a subfield of F—see the following exercise.

Exercise 3.6.10. Let K be the *set* of rational numbers. Define operations denoted $\oplus$ and $\odot$ on K by

$$x \oplus y = x + y + 1,$$
$$x \odot y = x \cdot y + x + y,$$

where $+$ and $\cdot$ denote the usual addition and multiplication of rational numbers.

---

[13]Technically speaking, addition $+_F$ and multiplication $\cdot_F$ on F are functions $F \times F \to F$; addition $+_K$ and multiplication $\cdot_K$ on K are functions $K \times K \to K$. The first requirement for K to be a subfield of F is that $K \subset F$. In this case, $K \times K \subset F \times F$. Then the second requirement for K to be a subfield of F is that the function $+_K$ is the restriction $(+_F)|_{K \times K} : K \times K \to K$ of $+_F : F \times F \to F$ (so that necessarily $x, y \in K$ implies $x +_F y \in K$), and similarly for multiplication.

(a) Verify that these operations make K a field. [For example, to verify the associative law for addition you would begin with $(x \oplus y) \oplus z = (x + y + 1) \oplus z = (x + y + 1) + z + 1$ and manipulate this to get $x \oplus (y \oplus z)$.]

Observe that always $x \oplus y \neq x + y$, and usually $x \odot y \neq x \cdot y$. Thus the field K is different from the field $\mathbb{Q}$ because their operations are different, even though the underlying sets of elements are the same. Thus it may be possible to make a given set into a field in more than one way.

(b) Show that K is not a subfield of $\mathbb{R}$. (Thus a field that is a subset of another field need not be a subfield of the latter.)

If you start with a subset of a given field, then it is often easy to tell whether that subset can be made into a subfield.

Lemma 3.6.11. *Let F be a field and let K be a subset of F. Suppose:*

(a) $0_F \in K$ *and* $1_F \in K$.

(b) *(closure) For every $x, y \in K$, the sum $x + y \in K$ and the product $x \cdot y \in K$.*

(c) *For every $x \in K$, its negative $-x \in K$; for every $x \in K$ with $x \neq 0_F$, its inverse $x^{-1} \in K$.*

Then K becomes a subfield of F when the addition and multiplication on F are restricted to elements of K.

Proof. We have $x + 0_F = x$ for all $x \in K$ so that the zero 0_F of F is a zero element of K; similarly, 1_F is a unity element of K. If $x \in K$, then $x + (-x) = 0$ in F and hence in K so that the negative $-x$ of x in F is a negative of x in K; similarly, if $x \in K$ with $x \neq 0$, then its inverse x^{-1} in F is its inverse in K.

The remaining field axioms are identities such as $x + y = y + x$ and $(xy)z = x(yz)$ that are supposed to hold for all elements of the field. These do hold for all elements of K because they hold, more generally, for all elements of the larger set F. $\square$

In view of the preceding, if F is a field and K is a subset of F with the properties (a)-(c), then we often say that K is a subfield of F to mean that K becomes a field when the operations on F are restricted to K.

Exercises 3.6.12. (1) Use Lemma 3.6.11 to verify that $R = \{ (x, 0) : x \in \mathbb{R} \}$ is a subfield of $\mathbb{C}$.

(2) Show that a subset K of a field F becomes a subfield of F, by restricting the operations of F to K, if the following are true: (a) $0_F \in K$ and $1_F \in K$; (b) for all $x, y \in K$, their difference $x - y \in K$; and (c) for all $x, y \in K$ with $y \neq 0$, their quotient $x/y \in K$.

Let us return to the field $\mathbb{C}$ and its subfield $R = \{(x, 0) : x \in \mathbb{R}\}$. What we have accomplished so far is to construct a field $\mathbb{C}$ such that:

- The field $\mathbb{C}$ has a subfield R that is isomorphic to the field $\mathbb{R}$ of real numbers.

- The field $\mathbb{C}$ includes an element i whose square is the negative of the unity.

- Each element of $\mathbb{C}$ can be represented in the form $(x, 0) + (y, 0)i$ for unique $(x, 0), (y, 0) \in R$.

As indicated earlier, we want to identify the field $\mathbb{R}$ with the subfield R of $\mathbb{C}$. The map

$$\varphi : \mathbb{R} \to R$$
$$x \mapsto (x, 0)$$

makes such an identification possible, for it is not just a bijection but also a field isomorphism (why?).

For an arbitrary complex number $z = (x, y)$,

$$(x, y) = (x, 0) + (y, 0)i$$

(why?), in other words,

$$(x, y) = \varphi(x) + \varphi(y)i.$$

The quantity on the right-hand side looks suspiciously like the expression $x + yi$. It would look exactly like it if we were to elide both occurrences of the isomorphism φ, for then we would have $(x, y) = x + yi$. Still, x is *not* $\varphi(x)$, and y is *not* $\varphi(y)$.

To make $(x, y) = x + yi$ true literally, what we do is cut out from $\mathbb{C}$ each complex number of the form $(x, 0)$ and paste in, in its place, the corresponding real number x. That is, we *redefine*

$$\mathbb{C} = ((\mathbb{R} \times \mathbb{R}) \setminus (\mathbb{R} \times \{0\})) \cup \mathbb{R}.$$

(Compare the cut-and-paste method used in Exercise 3.4.11 (c) to construct $\mathbb{Q}$ from $\mathbb{Z}$.) Notice that

$$\mathbb{C} \setminus \mathbb{R} = \{(x, y) \in \mathbb{R} \times \mathbb{R} : y \neq 0\}.$$

We call elements of $\mathbb{C} \setminus \mathbb{R}$ **nonreal**, in contrast to elements of $\mathbb{R}$, which of course we call **real**. Further, we call elements of $\{0\} \times \mathbb{R}$ **imaginary**.[14]

Exercise 3.6.13. (a) Show directly, in detail, how to define addition and multiplication on the redefined $\mathbb{C}$ that extend addition and multiplication operations on the subset $\mathbb{C} \setminus \mathbb{R}$ and those on the subset $\mathbb{R}$. Consider separately the cases of two real numbers, a real number and a nonreal number, and two nonreal numbers.

(b) Let $\psi \colon \mathbb{R} \times \mathbb{R} \to \mathbb{C}$ be the map defined by

$$\psi(x,y) = \begin{cases} (x,y) & \text{if } y \neq 0, \\ x & \text{if } y = 0. \end{cases}$$

Verify that ψ is a bijection. Then show that the addition and multiplication you defined in (a) are given by the formulas

$$z + w = \psi\left(\psi^{-1}(z) + \psi^{-1}(w)\right),$$
$$z \cdot w = \psi\left(\psi^{-1}(z) \cdot \psi^{-1}(w)\right),$$

where the addition and multiplication on the right-hand sides of these formulas are as originally defined on the complex plane $\mathbb{R} \times \mathbb{R}$.

Express informally in words what these formulas say.

(c) Regarding the just-defined addition on the redefined $\mathbb{C}$ as a function $\mathbb{C} \times \mathbb{C} \to \mathbb{C}$ and the previously defined addition on the complex plane as a function $(\mathbb{R} \times \mathbb{R}) \times (\mathbb{R} \times \mathbb{R}) \to \mathbb{R} \times \mathbb{R}$, express the preceding formula for addition as a statement about function composition. Do the same for multiplication.

After addition and multiplication are defined as in the preceding exercise, the situation with the redefined set $\mathbb{C}$ is:

- $\mathbb{C}$ is a field.

- $\mathbb{R}$ is a subfield of $\mathbb{C}$.

- The element $i \in \mathbb{C}$ satisfies $i^2 = -1$.

- Each element of $\mathbb{C}$ can be represented in the form $x + yi$ for unique real x and y.

---

[14]The name "imaginary" is an unfortunate relic from a time when complex numbers were new and abstruse. In view of the preceding construction of $\mathbb{C}$, for us today imaginary numbers are quite familiar and comprehensible—as "real" as real numbers.

At long last it is meaningful and correct to say that complex addition and multiplication are given by the formulas

$$(x + yi) + (u + vi) = (x + u) + (y + v)i,$$
$$(x + yi) \cdot (u + vi) = (xu - yv) + (xv + yu)i.$$

Exercises 3.6.14. (1) Complete verification of the preceding statements.

(2) Show that $z^{-1} = (1/|z|^2)\bar{z}$ for all nonzero $z \in \mathbb{C}$.

(3) Rewrite your formulas from Exercise 3.6.6 (2) for difference and quotient in terms of the $x + yi$ representation of complex numbers.

(4) Show that $\text{Re}(z) = (z + \bar{z})/2$ for each complex number z. Establish a similar formula for $\text{Im}(z)$.

Exercise 3.6.15. Let K be the subset $\{ q + ri : q, r \in \mathbb{Q} \}$ of $\mathbb{C}$.

(a) Show that K is a subfield of $\mathbb{C}$ and $\mathbb{Q}$ is a subfield of K.

(b) Another field having $\mathbb{Q}$ as a subfield is the field $\mathbb{Q}(\sqrt{2})$ in Exercise 3.1.3 (d). Is $\mathbb{Q}(\sqrt{2})$ isomorphic to K?

The equation $z^2 = -1$ has not just one solution in the field $\mathbb{C}$, but two: i and $-i$. We may say that i and $-i$ are square roots of -1 in $\mathbb{C}$. It should not be a surprise that some real numbers have several cube roots in $\mathbb{C}$.

Exercise 3.6.16. (a) Show that, for every real $c \neq 0$, the equation $z^2 = c$ has exactly two solutions in $\mathbb{C}$.

(b) Show that, for every real number $c > 0$, the equation $z^3 = -c$ has as solutions the three complex numbers $r(1/2 + i\sqrt{3}/2)$, $-r$, and $r(1/2 - i\sqrt{3}/2)$, where $r = \sqrt[3]{c}$. (You are *not* asked here to show that these are the only solutions, just that they are solutions.)

By enlarging the field $\mathbb{R}$ to include a square root of -1, we get, in fact, a field $\mathbb{C}$ that includes roots of every order, and not just of -1 but of every nonzero real number. In fact, for every positive integer n and every nonzero complex number w, the equation $z^n + w = 0$ has exactly n distinct solutions in $\mathbb{C}$. (Seeing this requires a little trigonometry; see any book on complex analysis or almost any book on advanced calculus or differential equations.) But even more is true.

Theorem 3.6.17 (Fundamental Theorem of Algebra). *Every polynomial with real or complex coefficients has a root in the field of complex numbers.*

Most of the proofs of this theorem require some nonalgebraic apparatus, namely, use of the geometry—more specifically, the *topology*—of the complex plane. Hence a proof is beyond the scope of this book. (See any text on complex analysis.)

Exercise 3.6.18. (*Prerequisite:* a little matrix algebra.) Let K be the set of all 2×2 matrices of the form $\begin{bmatrix} x & y \\ -y & x \end{bmatrix}$, where x and y are arbitrary real numbers. Show that the usual addition and multiplication of matrices make K a field that is isomorphic to $\mathbb{C}$. What element of K corresponds to i under such an isomorphism?

Exercise 3.6.19. The magnitude $|z|$ of a complex number z is a real number. Moreover, magnitude on $\mathbb{C}$ has the same basic properties as ordinary absolute value in $\mathbb{R}$ (or in any ordered field)—see Exercise 3.6.6 (5). Then convergence of sequences can be defined in $\mathbb{C}$ in essentially the same way as in $\mathbb{R}$. Do it.

Let $(z_n)_{n \in \mathbb{N}}$ be a sequence in $\mathbb{C}$ and, for each n, let x_n and y_n be the real and imaginary parts of z_n, respectively. Prove that $(z_n)_{n \in \mathbb{N}}$ converges in $\mathbb{C}$ if and only if the sequences $(x_n)_{n \in \mathbb{N}}$ and $(y_n)_{n \in \mathbb{N}}$ converge in $\mathbb{R}$, and that in this case,

$$\lim_{n \to \infty} z_n = \lim_{n \to \infty} x_n + i \lim_{n \to \infty} y_n.$$

Chapter 4

Cardinality

Infinity is the land of mathematical hocus pocus.
— Paul Carus

How many elements does a set have? And how do you tell whether two sets have the same number of elements, or whether one of the sets instead has a smaller number of elements than the other? In fact, what is meant by "number of elements" in a set? How big can sets be? Such questions are the subject of this chapter.

4.1   Finite and Infinite Sets

What is an infinite set? Are some infinite sets larger than others? These are two of the questions we shall answer in this section. Such questions were first addressed by the German mathematician Georg Cantor (1845–1918), who in 1874 initiated an epoch-making series of papers about infinite sets.

To get started, we need a precise notion of when two sets are of the same size, in the sense that they have the same number of elements.

The ancient shepherd could tell that his flock had not lost any sheep during the day—nor gained any from his neighbor's flock—by dropping a pebble into a pile as each sheep trotted past him in the morning on the way to the meadow. In the evening, as each sheep passed him on the way back to the fold, he moved one pebble from the morning's pile to a new pile. If he dropped the last pebble just as the last sheep walked by, then there were still the same number of sheep as there had been in the morning. Likewise, we can say that two sets have the same size when we can pair up all their elements in a one-to-one manner.

229

The mathematical tool for studying sizes of sets is thus going to be one-to-one correspondences—*bijections*, as such maps are named in Appendix B.[1] Recall that a **bijection** $f\colon X \to Y$ from a set X to a set Y is a map f with domain (set of inputs) X and codomain (set including outputs) Y with the two properties:

- The map f is injective; that is, for every $x_1, x_2 \in X$, if $x_1 \neq x_2$, then $f(x_1) \neq f(x_2)$.

- The map f is surjective; that is, for each $y \in Y$, there is at least one $x \in X$ such that $f(x) = y$.

Since such a bijection f is, in particular, injective, then for each $y \in Y$, there is exactly one $x \in X$ such that $f(x) = y$. And, of course, merely since f is a function, for each $x \in X$, there is exactly one $y \in Y$ such that $f(x) = y$.

Definition 4.1.1. Let X and Y be sets. To indicate that a map $f\colon X \to Y$ is a bijection, we write

$$f\colon X \approx Y.$$

We say that X **matches** Y and write $X \approx Y$ to mean there exists some such $f\colon X \approx Y$. When X does not match Y, we write $X \not\approx Y$.

Thus we think of a set as having the same size as another set when the first set matches the second in the preceding sense.

Examples 4.1.2. (1) The two-element set $\{0,1\} \approx \{3,8\}$ because the map $f\colon \{0,1\} \to \{3,8\}$ defined by $f(0) = 3$ and $f(1) = 8$ is a bijection.

This map f is not the only bijection from $\{0,1\}$ to $\{3,8\}$. In fact, the map $g\colon \{0,1\} \to \{3,8\}$ defined by $g(0) = 8$ and $g(1) = 3$ is another bijection.

(2) The singleton (that is, the one-element set) $\{0\}$ matches $\{1\}$. Moreover, there is exactly one bijection $f\colon \{0\} \to \{1\}$.

(3) The empty set $\varnothing$ matches set Y if and only if $Y = \varnothing$, and in this case there is exactly one bijection $f\colon \varnothing \to Y$, namely, the unique map from $\varnothing$ to itself. (What is that map?)

(4) $\{1,2\} \not\approx \{1\}$. (Why not?)

Matching is a way of classifying sets as being like or unlike other sets with respect to size. Just like any scheme of classifying things as being alike, it has the same basic properties as actual equality.

---

[1] See the discussion about bijections beginning on page A36 in Appendix B.3.

Proposition 4.1.3.   *1.* (reflexivity) *For each set A: $A \approx A$.*

   2. (symmetry) *For all sets A and B: $A \approx B \implies B \approx A$.*

   3. (transitivity) *For all sets A, B, C: $A \approx B$ and $B \approx C \implies A \approx C$.*

Proof. Exercise. $\square$

According to the preceding proposition, the "matches" relation $\approx$ forms an equivalence relation[2] on any given collection of sets—for example, on the collection of all subsets of a given set.

Definition 4.1.4. If X is a set, then its **power set** $\mathcal{P}(X)$ is the set of all subsets of X. In symbols,

$$\mathcal{P}(X) = \{A : A \subset X\}.$$

Examples 4.1.5.   (1) If $X = \{1, 2\}$, then

$$\mathcal{P}(X) = \{\emptyset, \{1\}, \{2\}, \{1, 2\}\}.$$

(2) $\mathcal{P}(\{1\}) = \{\emptyset, \{1\}\}$.

(3) $\mathcal{P}(\emptyset) = \{\emptyset\}$.

Exercises 4.1.6.   (1) What is $\mathcal{P}(X)$ if $X = \{1, 2, 3\}$?

(2) According to the evidence from (1) and the above examples, how many elements does $\mathcal{P}(X)$ seem to have when $X = \{1, 2, \ldots, n\}$ for a positive integer n?

Why didn't we formulate Proposition 4.1.3 simply to say that the relation $\approx$ is an equivalence relation on the set of all sets? The reason is that there is a problem in forming the set of all sets—the *Russell paradox*, so named after the logician, philosopher, and peace activist Bertrand Russell (1872-1970). Rather than trying to form such an all-encompassing set, let's try something more modest.

Example 4.1.7. Most sets you come across are not elements of themselves. For example, the set $\mathbb{N}$ of all natural numbers is not itself a natural number, and so $\mathbb{N} \notin \mathbb{N}$. Also, $\emptyset \notin \emptyset$ because the empty set $\emptyset$ has no elements whatsoever. (Can you think of any set that *is* an element of itself?) Let $\mathcal{R}$ be the set of all sets that are *not* elements of themselves. In symbols,

$$\mathcal{R} = \{A : A \notin A\}.$$

Is $\mathcal{R} \in \mathcal{R}$? If $\mathcal{R} \in \mathcal{R}$, then from the definition of $\mathcal{R}$ we would have $\mathcal{R} \notin \mathcal{R}$, a contradiction. Hence $\mathcal{R} \notin \mathcal{R}$. But then, again by the definition of $\mathcal{R}$, we would have $\mathcal{R} \in \mathcal{R}$. Thus $\mathcal{R} \in \mathcal{R} \iff \mathcal{R} \notin \mathcal{R}$, which is a logical impossibility (why?).

---

[2]For generalities about equivalence relations, see Appendix C.

Example 4.1.7 indicates that paradoxes arise if you wantonly try to form "large" collections of sets. To resolve such paradoxes requires the sort of careful axiomatic development of set theory that is beyond the scope of this book.

By using the definition of "matches," we can say in a precise fashion what it means for a set to be finite.

Definition 4.1.8. A set X is said to be **finite** if $X = \varnothing$ or $X \approx \{1, 2, \ldots, n\}$ for some positive integer n. Otherwise, X is said to be **infinite**.

For a given n, the set $\{1, 2, \ldots, n\}$ is just like the shepherd's pile of pebbles.

It is easy to give examples of finite sets and to see that they are finite by exhibiting the requisite matches. It is also easy to give examples of infinite sets—for example, we shall see that $\mathbb{N}$ and $\mathbb{R}$ are infinite—but to establish that they are, in fact, infinite requires some preparation.

Examples 4.1.9. (1) If a is any set, then the singleton $\{a\}$ is finite.

(2) If a and b are any two distinct sets, then the doubleton $\{a, b\}$ is finite.

We shall use repeatedly, usually without explicit mention, the following criteria.

Proposition 4.1.10. *1. If F is finite and $X \approx F$, then X itself is finite.*

2. If I is infinite and $X \approx I$, then X itself is infinite.

Proof. 1. Assume F is finite and $X \approx F$. If F is empty, then X is also empty. Otherwise, if F is nonempty, then $F \approx \{1, 2, \ldots, n\}$ for some positive integer n. By transitivity, $X \approx \{1, 2, \ldots, n\}$ also.
2. Exercise. □

When $X \approx \{1, 2, \ldots, n\}$, we would like to call n the number of elements in X. But before doing so, we have to know there is only one such n. The next lemma and its corollary will guarantee this.

Exercise 4.1.11. This exercise asks you to work out some simple cases of the next lemma and its corollary.

(a) Show that $\{1, 2\} \not\approx \{1\}$.

(b) Show that $\{1, 2, 3\} \not\approx \{1, 2\}$. Deduce that $\{1, 2, 3\} \not\approx \{1\}$.

(c) Show that $\{1, 2, 3, 4\} \not\approx \{1, 2, 3\}$. Deduce that $\{1, 2, 3, 4\} \not\approx \{1, 2\}$ and $\{1, 2, 3, 4\} \not\approx \{1\}$.

Lemma 4.1.12. *For every positive integer n,*

$$\{1, 2, \ldots, n+1\} \not\approx \{1, 2, \ldots, n\}.$$

Proof. We use induction on n. Obviously $\{1, 2\} \not\approx \{1\}$. (Why not?)
  Let $n \in \mathbb{N}^*$ and assume $\{1, 2, \ldots, n+1\} \not\approx \{1, 2, \ldots, n\}$. Just suppose $\{1, 2, \ldots, n+2\} \approx \{1, 2, \ldots, n+1\}$. Then there is some bijection

$$h\colon \{1, 2, \ldots, n+1, n+2\} \to \{1, 2, \ldots, n, n+1\}.$$

We modify h to make $n+2 \mapsto n+1$. That is, define

$$h'\colon \{1, 2, \ldots, n+1, n+2\} \to \{1, 2, \ldots, n, n+1\}$$

by

$$h'(i) = \begin{cases} h(i) & \text{if } i \neq n+2 \text{ and } h(i) \neq n+1, \\ h(n+2) & \text{if } h(i) = n+1, \\ n+1 & \text{if } i = n+2. \end{cases}$$

Then h' is bijective (why?) with $h'(n+2) = n+1$. Hence we may restrict the domain of h' to $\{1, 2, \ldots, n+1\}$ and the codomain of h' to $\{1, 2, \ldots, n\}$ to get a bijection $f\colon \{1, 2, \ldots, n+1\} \approx \{1, 2, \ldots, n\}$. The existence of such an f contradicts the induction hypothesis. □

Corollary 4.1.13. *If m and n are positive integers with $m \neq n$, then*

$$\{1, 2, \ldots, m\} \not\approx \{1, 2, \ldots, n\}.$$

Proof. Exercise. □

  Thus, if a set X matches $\{1, 2, \ldots, n\}$ for some positive integer n, then that n is unique. Hence we are justified in making the following definition.

Definition 4.1.14. Let X be a finite set. If $X = \varnothing$, then let $\#(X) = 0$. If $X \neq \varnothing$, then $X \approx \{1, 2, \ldots, n\}$ for a unique positive integer n; in this case, let $\#(X) = n$. In either case, the nonnegative integer $\#(X)$ is called the **number of elements in X**.

  For example, $\#(\{0, 1\}) = 2$ and $\#(\{0\}) = 1$.
  Corollary 4.1.13 also allows us to establish our first examples of infinite sets.

Examples 4.1.15.   (1) *The set $\mathbb{N}$ of natural numbers is infinite.* In fact, just suppose $\mathbb{N}$ is finite. Now $\mathbb{N} \neq \varnothing$, and so there is some positive integer n with $\{1, 2, \ldots, n\} \approx \mathbb{N}$ and hence some bijection

$$f\colon \{1, 2, \ldots, n\} \to \mathbb{N}.$$

We shall show that also $\{1, 2, \ldots, n + 1\} \approx \mathbb{N}$, which will contradict Corollary 4.1.13. What we shall do is shift the image of each $i \in \{1, 2, \ldots, n\}$ to the right by 1, thereby leaving 0 no longer an image, and then map the extra number $n + 1$ to 0. More precisely, define

$$g: \{1, 2, \ldots, n, n + 1\} \to \mathbb{N}$$

by

$$g(i) = \begin{cases} f(i) + 1 & \text{if } i < n + 1, \text{ that is, } i \leq n, \\ 0 & \text{if } i = n + 1. \end{cases}$$

Then g is also a bijection (why?). Hence the composite

$$f^{-1} \circ g: \{1, 2, \ldots, n, n + 1\} \to \{1, 2, \ldots, n\}$$

is a bijection. This contradicts Corollary 4.1.13 (and its special case, Lemma 4.1.12).

(2) The set $\mathbb{N}^*$ of positive integers is infinite because $\mathbb{N}^* \approx \mathbb{N}$ (why?).

(3) The set $\{2, 3, \ldots\}$ of all integers greater than or equal to 2 is infinite. In fact,

$$\mathbb{N}^* = \{1, 2, \ldots\} \approx \{2, 3, \ldots\}$$

under the map $n \mapsto n + 1$.

The bijection just exhibited has an amusing interpretation. Imagine a hotel—named "Hilbert's hotel" after the German mathematician and logician David Hilbert (1862-1943)—with infinitely many rooms, numbered $1, 2, \ldots$, each room accommodating only one person. A very large meeting of mathematicians is held at the hotel—so large that every room is filled. Then one more mathematician arrives. How can she be accommodated? The reservations clerk, being himself an amateur mathematician, finds an answer. He moves the mathematician in room 1 to room 2, the mathematician in room 2 to room 3, etc.; this frees up room 1, where the clerk puts the new arrival.

Exercise 4.1.16. Generalize the last example above.

In each of the last two examples, one element is removed from a set, yet what remains has just as many elements as the original set! Does that disturb you? Can that happen with a finite set? What do you think happens if infinitely many elements are removed from a set—can the set that remains have just as many elements as the original set?

The next few results concern the number of elements in sets built up in various ways from given finite sets.

Lemma 4.1.17. *The union of any two* disjoint *finite sets is itself finite. More precisely, let A and B be* disjoint *finite sets. Then $A \cup B$ is also finite, and $\#(A \cup B) = \#(A) + \#(B)$.*

Proof. The result is clear if $A = \varnothing$ or $B = \varnothing$. Suppose now that $A \neq \varnothing \neq B$. Then there exist positive integers m and n and bijections

$$f: \{1, 2, \ldots, m\} \to A, \qquad g: \{1, 2, \ldots, n\} \to B.$$

Define

$$h: \{1, \ldots, m, m+1, \ldots, m+n\} \to A \cup B$$

so that h does the same thing as f on $\{1, 2, \ldots, m\}$ and, after a shift by m, the same thing as g on $\{m+1, \ldots, m+n\}$. That is, let

$$h(i) = \begin{cases} f(i) & \text{if } 1 \leq i \leq m, \\ g(i-m) & \text{if } m+1 \leq i \leq m+n. \end{cases}$$

Then h is bijective (why?).   □

It is important in the preceding lemma that the sets A and B be disjoint. For example,

$$\{1, 2, 3, 4\} \cup \{3, 4, 5\} = \{1, 2, 3, 4, 5\}$$

so that this union, although finite, has

$$\#(\{1, 2, 3, 4\} \cup \{3, 4, 5\}) = \#(\{1, 2, 3, 4, 5\}) = 5,$$

whereas

$$\#(\{1, 2, 3, 4\}) + \#(\{3, 4, 5\}) = 4 + 3 = 7.$$

Proposition 4.1.18. *Any subset of a finite set is itself finite. More precisely, let $A \subset B$ with B finite. Then A is finite, and $\#(A) \leq \#(B)$.*

Proof. If $B = \varnothing$, we are done (why?). So suppose $B \neq \varnothing$. Without loss of generality, we may assume $B = \{1, 2, \ldots, n\}$ for some positive integer n. (Why?)

We use induction on n to show that each subset of $\{1, 2, \ldots, n\}$ is finite and has at most n elements. The case $n = 1$ is clear since the only subsets of $\{1\}$ are $\varnothing$ and $\{1\}$.

Now let n be a positive integer and assume that each subset of $\{1, 2, \ldots, n\}$ is finite and has at most n elements. Let A be a subset of $\{1, 2, \ldots, n+1\}$. We want to deduce that A is finite and that $\#(A) \leq n + 1$. If, actually, $A \subset \{1, 2, \ldots, n\}$, then in view of the induction assumption we are done.

Suppose now that $A \not\subset \{1, 2, \ldots, n\}$. Then $n + 1 \in A$. Look at the complement

$$D = A \setminus \{n + 1\}$$

obtained by removing $n + 1$ from A. Since $D \subset \{1, 2, \ldots, n\}$, then by the induction assumption D is finite and $\#(D) \leq n$. Now the singleton $\{n + 1\}$ is also finite, and it is disjoint from D; moreover,

$$A = D \cup \{n + 1\}.$$

From Lemma 4.1.17, A is finite, and

$$\#(A) = \#(D) + \#(\{n + 1\}) \leq n + 1. \quad \square$$

Now we know that a subset A of a finite set B is itself finite. What about the rest of the set B, in other words, the complement

$$B \setminus A = \{x \in B : x \notin A\}$$

of A in B?

Corollary 4.1.19. *If $A \subset B$ and B is finite, then the complement $B \setminus A$ is also finite, and $\#(B \setminus A) = \#(B) - \#(A)$.*

Proof. Exercise. $\square$

Corollary 4.1.20. *The union of any two finite sets is again finite. More precisely, let A and B be finite sets. Then $A \cup B$ is finite, and*

$$\#(A \cup B) = \#(A) + \#(B) - \#(A \cap B).$$

Proof. Start with

$$A \cup B = (A \setminus (A \cap B)) \cup (B \setminus (A \cap B)) \cup (A \cap B).$$

(Why is this expression for $A \cup B$ correct?) Then apply Lemma 4.1.17 and Corollary 4.1.20. $\square$

The idea of the formula in the preceding corollary is that in counting the elements in A and then the elements in B, some elements are counted twice, namely, those belonging to both A and B.

Exercises 4.1.21. (1) Explain why Lemma 4.1.17 is a special case of the preceding corollary.

(2) For the case of the union of three finite sets A, B, and C, find a formula analogous to that in the preceding corollary.

When dealing with three or more sets, it is often convenient to index the sets by positive integers (for example, A_1, A_2, A_3) or other sets of indices.

Definition 4.1.22. A **family** of sets $(A_i)_{i \in I}$ is just a map $i \mapsto A_i$ assigning to each element of the family's **index set** I a set A_i.

When the index set I of a family has a form such as $I = \{1, 2, \ldots, n\}$ for a positive integer n, we denote the family by $(A_i)_{i=1,\ldots,n}$ or even by $(A_i)_{i=1}^n$. And, when $I = \mathbb{N}^*$, then the family is simply a sequence of sets indexed by $\mathbb{N}^*$, and we denote it by $(A_i)_{i=1,2,\ldots}$ or even by $(A_i)_{i=1}^\infty$. Similar notations are used for index set $\{0, 1, \ldots, n\}$, $\mathbb{N}$, etc. Strictly speaking, such sequences and more general families are just functions whose values are sets; see Section B.4.

The union of a family of sets is defined to be just the collection of all elements that belong to at least one of the family's sets. This can be formally stated as follows.

Definition 4.1.23. Given a family $(A_i)_{i \in I}$ of sets, its **union** is the set

$$\bigcup_{i \in I} A_i = \{ x : (\exists i \in I)(x \in A_i) \}.$$

When the index set $I = \{1, 2, \ldots, n\}$ or $I = \mathbb{N}^*$, then we use the notation $\bigcup_{i=1}^n A_i$ or $\bigcup_{i=1}^\infty A_i$, respectively. Similarly for other special index sets.

Evidently, this notion of the union of any family of sets generalizes the idea of a union of two sets. For example, given a family $(A_i)_{i=1,2,3}$, we have $\bigcup_{i=1}^3 A_i = A_1 \cup A_2 \cup A_3$. (Write out all the details, just this once!)

Exercises 4.1.24. (1) What is $\bigcup_{i=1}^\infty [1/i, 1 - 1/i]$?

(2) Repeat (1) for the union of the corresponding open intervals.

Proposition 4.1.25. *The union of finitely many finite sets is itself finite. That is, the union $\bigcup_{i \in I} A_i$ of a family $(A_i)_{i \in I}$ of finite sets A_i indexed by a finite set I is finite.*

Proof. If $I = \varnothing$, we are done (why?). So suppose $I \neq \varnothing$. Without loss of generality we may assume $I = \{1, 2, \ldots, n\}$ for some positive integer n. In fact, if $n = \#(I)$, then there exists some $\sigma \colon \{1, 2, \ldots, n\} \approx I$. For each $j \in \{1, 2, \ldots, n\}$, define $B_j = A_{\sigma(j)}$. Then $\bigcup_{i \in I} A_i = \bigcup_{j=1}^n B_j$.

Now use mathematical induction. (What, precisely, is the statement to be proved for each n?) $\square$

Recall the following definition. (For further explanation, see Section A.2.)

Definition 4.1.26. For sets A and B, their **(cartesian) product** is the set

$$A \times B = \{ (a,b) : a \in A, b \in B \}$$

of all ordered pairs whose first coordinate belongs to A and second coordinate belongs to B.

For example,

$$\{1,2\} \times \{1,2,5\} = \{(1,1),(1,2),(1,5),(2,1),(2,2),(2,5)\}.$$

As another example, the plane $\mathbb{R}^2 = \mathbb{R} \times \mathbb{R}$.

Proposition 4.1.27. *The product of two finite sets is itself finite. More precisely, let A and B be finite. Then $A \times B$ is finite, and $\#(A \times B) = \#(A) \cdot \#(B)$.*

Proof. For each $a \in A$, we have $\{a\} \times B \approx B$ (what is the obvious bijection?). Then the result follows from the equality

$$A \times B = \bigcup_{a \in A} (\{a\} \times B).$$

(Why does this equality hold? How does the result follow from it?) □

Here is an interpretation of the fact that the number of elements in the product of two finite sets is the product of the numbers of elements in the two sets:

> If you can do a first thing in any one of m ways and then do a second thing—regardless of how you did the first thing—in any one of n ways, then you can do the first followed by the second in any one of $m \cdot n$ ways in all.

For example, if you can choose any one of 4 vegetables and any one of 3 meats for a pizza, then you can choose in all $4 \cdot 3 = 12$ vegetable-meat combinations for the pizza.

Exercise 4.1.28. Suppose you can do a first thing in any one of m ways and then do a second thing in any one of n ways, but the choice of the second way depends on the choice of the first: for various choices of the first way, you might have different sets of choices for the second (but always still n choices in each case).

Identify or invent a concrete example of this situation. Explain why Proposition 4.1.27 need not apply to such a situation. Formulate and prove a mathematical statement that does apply.

Each element (a_1, a_2) of the product $A_1 \times A_2$ of two sets is an ordered pair consisting of a first coordinate a_1 and a second coordinate a_2. Such an ordered pair (a_1, a_2) may be regarded as a family $(a_i)_{i=1,2}$ where $a_1 \in A_1$ and $a_2 \in A_2$, that is, where $a_i \in A_i$ for each index $i \in \{1,2\}$. This generalizes to any index set.

Definition 4.1.29. The **(cartesian) product** $\prod_{i \in I} A_i$ of an arbitrary family $(A_i)_{i \in I}$ of sets is the set of all those families $(a_i)_{i \in I}$ for which $a_i \in A_i$ for each index $i \in I$.

For example, when $I = \{1, 2, 3\}$, then

$$\prod_{i \in \{1,2,3\}} A_i = \prod_{i=1}^{3} A_i$$
$$= \{(a_1, a_2, a_3) : a_1 \in A_1, a_2 \in A_2, a_3 \in A_3\}.$$

Exercise 4.1.30. Describe the product $\prod_{i=1}^{3} K_i$ of intervals geometrically if $K_1 = [0, 1]$, $K_2 = [-2, 2]$, and $K_3 = [1, 3]$.

Corollary 4.1.31. *The product of finitely many finite sets is itself finite. More precisely, let $(A_i)_{i \in I}$ be a family of finite sets indexed by a finite set I. Then $\prod_{i \in I} A_i$ is finite, and*

$$\#\left(\prod_{i \in I} A_i\right) = \prod_{i \in I} \#(A_i).$$

Proof. Exercise. □

An especially interesting case of the product of a family of sets arises when all the sets A_i are one and the same set A. Then a typical element of $\prod_{i \in I} A_i$ assigns to each $i \in I$ an element of A. Thus $\prod_{i \in I} A_i$ is just the set of *all* functions from I to A. We use a special notation for this case.

Definition 4.1.32. Let A and I be sets. Then A^I denotes the set of all maps $f : I \to A$. When $A = \{0, 1\}$, we also denote A^I by 2^I.

In particular, the set $2^{\mathbb{N}}$ is just the set of all sequences of 0s and 1s.

Corollary 4.1.33. *Let I be a finite set. Then 2^I is also finite, and $\#(2^I) = 2^{\#(I)}$.*

Proof. Apply Corollary 4.1.31. □

Exercises 4.1.34. (1) What does Corollary 4.1.33 say in the special case where $I = \varnothing$?

(2) How many ordered triples (a_1, a_2, a_3) are there whose entries belong to a finite set of n elements? Exhibit all these triples in the case where $n = 2$.

(3) How many injective functions are there from a finite set I to another finite set A?

(4) How many surjective functions are there from a finite set I to another finite set A?

One of the reasons for looking at product sets is to provide a tool for counting the number of subsets of a set. Suppose you know a certain subset A of a set X. Imagine that each element x of X is tagged with a 1 in the case where x belongs to A, but with a 0 in the contrary case. Then you can tell whether or not an element of X belongs to A by looking at its tag, 1 or 0. Let us formalize these ideas.

Definition 4.1.35. Let X be a given set. For $A \subset X$, the **characteristic function of A in X** is the map

$$c_A \colon X \to \{0, 1\}$$

defined for all $x \in X$ by

$$c_A(x) = \begin{cases} 1 & \text{if } x \in A, \\ 0 & \text{if } x \in X \setminus A. \end{cases}$$

For example, when $X = \{1, 2, \ldots, 5\}$ and $A = \{2, 4\}$, then $c_A(2) = c_A(4) = 1$ but $c_A(1) = c_A(3) = c_A(5) = 0$.

For a given $A \subset X$, the characteristic function c_A does the tagging described above. In general, $c_A \in 2^X$ for each subset A of X. Since each subset A of a given set X gives rise to the corresponding characteristic function c_A, we therefore have a map

$$\begin{aligned} \varphi \colon \mathcal{P}(X) &\to 2^X \\ A &\mapsto c_A. \end{aligned}$$

A tagging in X, that is, an element c_A of 2^X, is just another way of describing a subset A of X. Hence the map φ should be a one-to-one correspondence between taggings and subsets of X.

Lemma 4.1.36. *Let X be any set. Then the map $\varphi \colon \mathcal{P}(X) \to 2^X$ defined above is a bijection.*

Proof. Rather than show separately that φ is injective and surjective, instead we construct the map "going backward" from 2^X to $\mathcal{P}(X)$ that ought to be φ^{-1}. For $f \in 2^X$, that is, $f \colon X \to \{0, 1\}$, we can form the subset $\{x \in X : f(x) = 1\}$ of X, and we denote this subset by $\psi(f)$. Thus we have defined a map

$$\begin{aligned} \psi \colon 2^X &\to \mathcal{P}(X) \\ f &\mapsto \{x \in X : f(x) = 1\}. \end{aligned}$$

Then $\psi \circ \varphi$ is the identity map of $\mathcal{P}(X)$, and $\varphi \circ \psi$ is the identity map of 2^X. (Verify this.) It follows that φ is, in fact, bijective (and ψ is its inverse). $\square$

Exercises 4.1.37.   (1) For a set X, what is the characteristic function $c_\varnothing$ of the empty subset of X? What is c_X?

(2) For $A \subset X$, what is the relationship of $c_{X \setminus A}$ to c_A?

(3) For $A, B \subset X$, express $c_{A \cup B}$ and $c_{A \cap B}$ in terms of c_A and c_B.

(4) Suppose, in the preceding proof, we had instead defined $\psi(f) = \{x \in X : f(x) = 0\}$ for each $f \in 2^X$. Then what would the composite $\psi \circ \varphi$ do?

For any set X, we now know

$$\boxed{\mathcal{P}(X) \approx 2^X.}$$

Corollary 4.1.38. *If X is a finite set, then its power set $\mathcal{P}(X)$ is also finite, and $\#(\mathcal{P}(X)) = 2^{\#(X)}$.*

Exercise 4.1.39. Prove the preceding corollary directly, without using the above machinery. [*Hint:* Show that $\#(\mathcal{P}(\{1, 2, \ldots, n\})) = 2^n$ for each positive integer n.]

We have one last basic result about finite sets to discuss.

Proposition 4.1.40. *Let X be a finite set, let Y be a set, and suppose there exists some* surjection *$f : X \to Y$. Then Y is also finite, and $\#(Y) \leq \#(X)$.*

Proof. There is nothing to prove in case $X = \varnothing$, for in that case we must also have $Y = \varnothing$. So assume $X \neq \varnothing$. As usual, without loss of generality we may assume that $X = \{1, 2, \ldots, n\}$ for some positive integer n. (Why may we so assume?)

Define $g : Y \to X$ as follows: Since f is surjective, for each $y \in Y$ there is at least one $x \in X$ such that $f(x) = y$, and we let

$$g(y) = \min\{x \in X : f(x) = y\}.$$

(What justifies the existence of this minimum?) Then g is injective (why?), and, since range$(g) = \{g(y) : y \in Y\} \subset X$, range$(g)$ is finite. It now follows (why?) that Y is finite. Moreover, $\#(Y) = \#(\text{range}(g)) \leq \#(X)$. $\square$

4.2   Countable and Uncountable Sets

We are going to distinguish the sizes of various infinite sets.

Definition 4.2.1. Call a set X **denumerable** when $\mathbb{N}^* \approx X$, **countable** when X is either finite or denumerable, and **uncountable** when X is not countable.

Naturally, $\mathbb{N}^*$ itself is denumerable, as is $\mathbb{N}$ (since, as you should recall, $\mathbb{N}^* \approx \mathbb{N}$). Here is a less obvious example.

Example 4.2.2. The set $\{2, 4, 6, \dots\}$ of all even positive integers is denumerable. In fact, the map

$$
\begin{array}{rcl}
\mathbb{N}^* & \to & \{2, 4, 6, \dots\} \\
n & \mapsto & 2 \cdot n
\end{array}
$$

is a bijection.

Some people find this sort of example, and even the earlier example of Hilbert's hotel (see page 234), hard to stomach. After all, how can it be possible to remove half of a set and still have left a set just as big as the one you started with? Of course, we are dealing with infinite sets here, whereas our everyday experience is limited to finite sets. To avoid such curious examples, we would have little choice but to expunge from mathematics the very notions of set matching and infinite sets. And, without these notions, mathematics would be impoverished, indeed.

Exercises 4.2.3.   (1) Show that the set of odd positive integers is denumerable.

(2) If m is an arbitrary positive integer, show that $\{1, 2, \dots, 2m\} \not\approx \{2, 4, 6, \dots, 2m\}$.

We are going to see that $\mathbb{Z}$ and $\mathbb{Q}$ are also denumerable—so there are exactly as many integers as natural numbers, and as many rational numbers as natural numbers. And, with somewhat more work, we shall see that $\mathbb{R}$ is uncountable.

Since $\mathbb{N}^*$ is infinite, then each denumerable set is infinite. Hence *a set is infinite if and only if it is either denumerable or uncountable.*

Since $\mathbb{N}^* \approx \mathbb{N}$, then a set X is denumerable if and only if $\mathbb{N} \approx X$. Moreover, a set X is denumerable (respectively, countable, uncountable) if and only if X matches *some* set that is denumerable (respectively, countable, uncountable).[3] (Why?)

Note that a bijection $f \colon \mathbb{N} \to X$ is simply a sequence $(x_n)_{n=0,1,\dots}$ in X such that

$$
\begin{aligned}
& X = \{x_n : n \in \mathbb{N}\}, \\
& x_i \neq x_j \text{ whenever } i \neq j.
\end{aligned}
$$

Hence a set X is denumerable exactly when such a sequence exists.

---

[3]Did you understand that sentence? The "respectively" yields three assertions: a set X is denumerable if and only if X matches some denumerable set; X is countable if and only if X matches some countable set; and X is uncountable if and only if X matches some uncountable set.

Proposition 4.2.4. *An infinite subset of a countable set is denumerable.*

Proof. Let $A \subset X$ with A infinite and X countable. Then X is also infinite, and so X is denumerable. Hence we may assume without loss of generality that $X = \mathbb{N}$.

Construct recursively a sequence $(x_n)_{n=0,1,\dots}$ as follows. Let

$$x_0 = \min A.$$

(Why does this minimum, the least element of A, exist?) Next, since $A \setminus \{x_0\}$ is still infinite (why?), it is nonempty; let

$$x_1 = \min (A \setminus \{x_0\}).$$

Next, $A \setminus \{x_0, x_1\}$ is still infinite; let

$$x_2 = \min (A \setminus \{x_0, x_1\}).$$

Continuing in this way, we obtain a sequence $(x_n)_{n=0,1,\dots}$ in A which, by construction, satisfies $x_i \neq x_j$ when $i \neq j$. Thus the map $n \mapsto x_n$ of $\mathbb{N} \to A$ is injective.

By an induction on m, we see that

$$m \in A \implies m = x_n \text{ for some } n \in \mathbb{N}.$$

Then $A = \{x_n : n \in \mathbb{N}\}$. Thus the map $n \mapsto x_n$ of $\mathbb{N} \to A$ is also surjective. $\square$

Corollary 4.2.5. *A subset of a countable set is itself countable.*

Proposition 4.2.6. *Each infinite set has a denumerable subset.*

Proof. Let X be infinite. Then $X \neq \varnothing$, and so we may choose some $x_0 \in X$. Next, $X \setminus \{x_0\}$ is still infinite, hence nonempty, and we may choose some $x_1 \in X \setminus \{x_0\}$. Next, $X \setminus \{x_0, x_1\}$ is still infinite, hence nonempty, and we may choose some $x_2 \in X \setminus \{x_0, x_1\}$. Continuing in this way, we construct a sequence $(x_n)_{n=0,1,\dots}$ of *distinct* elements of X. The range $\{x_n : n \in \mathbb{N}\}$ of this sequence is a denumerable subset of X. $\square$

Aside from the informal use of recursion in it, the above proof is not really complete—however convincing it might seem. A guarantee is needed that *all* the choices of elements x_n can be made "simultaneously." Such a guarantee is provided by the **Axiom of Choice**, one version of which is the following.

If $\mathcal{A}$ is any collection of nonempty sets, then there exists some "choice function" for $\mathcal{A}$, that is, a map c with domain $\mathcal{A}$ such that $c(A) \in \mathcal{A}$ for each $A \in \mathcal{A}$.

Exercise 4.2.7. Prove that a set is infinite if and only if it matches some proper subset of itself. (A *proper* subset of X is any subset other than X itself.)

Proposition 4.2.8. *If X is countable and $f: X \to Y$ is a surjection, then Y is countable.*

Proof. Similar to the proof of Proposition 4.1.40.   □

It follows from the above that a nonempty set Y is countable if and only if there exists some sequence $(y_n)_{n=0,1,\dots}$ in Y, not necessarily of distinct elements, whose range $\{\, y_n : n \in \mathbb{N} \,\}$ is Y.

Exercise 4.2.9. In the notation of the preceding proposition, if X is denumerable, must Y be denumerable?

Examples 4.2.10.   (1) *The set $\mathbb{N} \times \mathbb{N}$ of all pairs of natural numbers is denumerable.*

The idea is to list in order $0, 1, 2, \dots$ all such pairs in a single sequence, with no such pair being listed twice. To construct this list, imagine these pairs arrayed in two dimensions as follows:

$$
\begin{array}{cccccc}
(0,0) & (1,0) & (2,0) & (3,0) & \dots \\
(0,1) & (1,1) & (2,1) & (3,1) & \dots \\
(0,2) & (1,2) & (2,2) & (3,2) & \dots \\
(0,3) & (1,3) & (2,3) & (3,3) & \dots \\
\vdots & \vdots & \vdots & \vdots
\end{array}
$$

Then traverse this array in the order

$$(0,0), (1,0), (0,1), (2,0), (1,1), (0,2), (3,0), (2,1), (1,2), (0,3), \dots$$

obtained by starting at the northwest (top left) corner and going down successive diagonals in a northeast to southwest direction. The order in which the pairs are met is their order in the requisite sequence.

(2) *The set $\mathbb{Z}$ is denumerable.* In fact, $\mathbb{N} \approx \mathbb{Z}$ because the map $f: \mathbb{N} \to \mathbb{Z}$ given by

$$
f(n) = \begin{cases} n/2 & \text{if } n \text{ is even,} \\ -(n+1)/2 & \text{if } n \text{ is odd} \end{cases}
$$

is a bijection.

Exercises 4.2.11.   (1) Derive an explicit formula for the bijection $\mathbb{N} \times \mathbb{N} \to \mathbb{N}$ described in the preceding example.

(2) Give the details for the following alternate proof that $\mathbb{N} \times \mathbb{N}$ is denumerable. First, $\mathbb{N} \times \mathbb{N}$ is infinite because $\mathbb{N} \times \{0\} \subset \mathbb{N} \times \mathbb{N}$. Hence it suffices to show that $\mathbb{N} \times \mathbb{N}$ is countable. To do this, we need only find some injection $\mathbb{N} \times \mathbb{N} \to \mathbb{N}$. But the rule $(m, n) \mapsto 2^m \cdot 3^n$ defines such an injection. (Why is it an injection? Why is the fact that it is an injection enough?)

Since $\mathbb{N} \times \mathbb{N}$ is denumerable, we obtain the following.

Proposition 4.2.12. *The product of any two denumerable sets is denumerable. The product of any two countable sets is countable.*

Example 4.2.13. *The set $\mathbb{Q}$ of all rational numbers is denumerable.* In fact, we have an obvious surjection $(m, n) \mapsto m/n$ from $\mathbb{Z} \times \mathbb{Z}^* \to \mathbb{Q}$. (Here $\mathbb{Z}^*$ is, of course, the set of all nonzero integers.) Since $\mathbb{Z}$ and $\mathbb{Z}^*$ are both denumerable, so is their product $\mathbb{Z} \times \mathbb{Z}^*$. Then the image $\mathbb{Q}$ of this product is countable. But $\mathbb{Q}$ is infinite since it contains the infinite set $\mathbb{N}$. Hence $\mathbb{Q}$ is denumerable.

Proposition 4.2.14. *The union of two denumerable sets is itself denumerable.*

Proof. The proof in the case of two *disjoint* denumerable sets is like the proof above that $\mathbb{Z}$ is denumerable. The proof of the general case is left as an exercise.   □

Of course, it follows immediately from the preceding proposition that the union of finitely many (at least one) denumerable sets is itself denumerable. (Write out the requisite proof!) But even more is true.

Proposition 4.2.15. *The union of countably many denumerable sets is denumerable.*

Proof. Let $(A_i)_{i \in I}$ be a family of denumerable sets with countable index set I. Let $A = \bigcup_{i \in I} A_i$. If I is finite and nonempty, then A is denumerable according to the preceding discussion. So assume that I is denumerable. Without loss of generality, we may assume that $I = \mathbb{N}$. Since $A \supset A_0$ and A_0 is infinite, then A, too, is infinite. Hence it now suffices to show that A is countable.

Our inspiration for what follows is that $\mathbb{N} \times \mathbb{N}$ is denumerable and can be partitioned into the union

$$\mathbb{N} \times \mathbb{N} = \bigcup_{n=0}^{\infty} \mathbb{N} \times \{n\}.$$

For each $n \in \mathbb{N}$, there is some bijection $f_n : \mathbb{N} \approx A_n$. Then the map

$$g_n : \mathbb{N} \times \{n\} \to A_n$$
$$(m, n) \mapsto f_n(m)$$

is also a bijection. Hence the map

$$G: \mathbb{N} \times \mathbb{N} \rightarrow A$$
$$(m, n) \mapsto g_n(m)$$

is a surjection. (Why might G not be injective as well?) But $\mathbb{N} \times \mathbb{N}$ is countable, and so its image A is countable, too. $\square$

Exercises 4.2.16.   (1) What can you say—and prove—about the union of countably many countable sets?

(2) Find the flaw in the following "proof" that the set $\mathbb{R}$ of all real numbers is countable:

Proof. For each $x \in \mathbb{R}$, there is some sequence of rational numbers that converges to x, and we denote by A_x the range of some such sequence. Each A_x is a nonempty subset of $\mathbb{Q}$; for each pair x and y of distinct reals, $A_x \cap A_y$ is finite (and possibly empty). If the index set $\mathbb{R}$ of the family $(A_x)_{x \in \mathbb{R}}$ were uncountable, then the union $\bigcup_{x \in \mathbb{R}} A_x$ of this family would also be uncountable. But $\bigcup_{x \in \mathbb{R}} A_x \subset \mathbb{Q}$ (because each $A_x \subset \mathbb{Q}$), and so this union is, to the contrary, countable. $\square$

Exercise 4.2.17. An *algebraic number* is a real number that is a solution of some polynomial equation

$$p(x) = 0$$

where the polynomial

$$p(x) = a_n x^n + a_{n-1} x^{n-1} + \cdots + a_2 x^2 + a_1 x + a_0 \qquad (*)$$

has *integers* as its coefficients $a_n, a_{n-1}, \ldots, a_0$. For example, each integer k is itself an algebraic number since it is the root of the equation $x - k = 0$.

(a) Show that each rational number is algebraic.

(b) Show that $\sqrt[3]{5}$ is algebraic but not rational.

(c) Show that the number

$$\sqrt{5 + 2\sqrt{6}}$$

is algebraic but not rational.

(d) Give some additional examples of irrational algebraic numbers.

(e) Explain why each real root of a polynomial equation whose coefficients are rational must be algebraic.

(f) Define the "height" of a polynomial p given by (*) to be the positive integer

$$h(p) = n + |a_n| + |a_{n-1}| + \cdots + |a_0|.$$

[Thus $h(p)$ measures both the polynomial's degree n and the sizes of all its coefficients.] How many polynomials p are there having a fixed height m?

(g) Prove that the set of all algebraic numbers is denumerable. (*Hint:* You may use the fact that a polynomial of degree n has at most n real roots.)

4.3   Uncountable Sets

At last we are ready to look at the size of $\mathbb{R}$. It is convenient to begin by examining the subset $[0, 1]$ of $\mathbb{R}$. The conclusion we obtain is important enough to be called a theorem.

Theorem 4.3.1. *The closed interval $[0, 1]$ in $\mathbb{R}$ is uncountable.*

Proof. Just suppose $[0, 1]$ is countable. (Of course, we know that it is infinite, and so then it is actually denumerable.) Then we may write

$$[0, 1] = \{x_n : n \in \mathbb{N}\}$$

for some sequence $(x_n)_{n=0,1,\dots}$ in $[0, 1]$ (actually for some sequence of *distinct* numbers—but we do not need that).

Choose a closed interval $[a_0, b_0]$ with

$$[a_0, b_0] \subset [0, 1], \qquad x_0 \notin [a_0, b_0].$$

(Why is such a choice possible? Can you give some explicit formula for such a_0 and b_0?) Next, choose a closed interval $[a_1, b_1]$ with

$$[a_1, b_1] \subset [a_0, b_0], \qquad x_1 \notin [a_1, b_1].$$

Proceeding in this way, we obtain recursively a decreasing sequence $([a_n, b_n])_{n=0,1,\dots}$ of closed subintervals of $[0, 1]$ such that, for each $n \in \mathbb{N}$,

$$x_n \notin [a_n, b_n].$$

By the Nested Interval Property of $\mathbb{R}$ (see Definition 3.3.3 and page 166), there exists some

$$x \in \bigcap_{n=0}^{\infty} [a_n, b_n].$$

Then $x \in [0, 1]$, and so $x = x_n$ for some n. For this n, we have $x \in [a_n, b_n]$. But by construction of $[a_n, b_n]$ we have $x_n \notin [a_n, b_n]$. This is a contradiction.   □

From the theorem we obtain other interesting examples of uncountable sets.

Examples 4.3.2.   (1) Since the initial interval $[a_0, b_0]$ in the proof of the theorem could have been constructed so that $0 < a_0$ and $b_0 < 1$, then the same construction allows us to conclude that the open interval $(0, 1)$ is uncountable, as is each of the half-open, half-closed intervals $[0, 1)$ and $(0, 1]$.

(2) There is nothing special about the endpoints 0 and 1 in the theorem. We could have started with any two endpoints a and b where $a < b$ to conclude that the various kinds of intervals with these endpoints are uncountable.

Something more is true: Any two closed intervals match one another. To see this, it suffices to show that any closed interval $[a, b]$ (with $a < b$, of course) matches $[0, 1]$. In fact, there is a *linear* function

$$f: [0, 1] \to [a, b]$$
$$x \mapsto f(x) = \alpha \cdot x + \beta$$

with f bijective. We construct such a function by requiring

$$f(0) = a \quad \text{and} \quad f(1) = b,$$

in other words,

$$\begin{cases} \alpha \cdot 0 + \beta & = & a \\ \alpha \cdot 1 + \beta & = & b. \end{cases}$$

This pair of simultaneous equations in unknowns α and β has solution $\alpha = b - a$, $\beta = a$, and so the function f is given by the formula

$$f(x) = (b - a)x + a.$$

Since the coefficient $b - a$ of x is nonzero, evidently f is injective. Why is it surjective as well?

(3) *The set $\mathbb{R}$ of all real numbers is uncountable.* For if $\mathbb{R}$ were count-able, then its subset $[0, 1]$ would also be countable.

(4) The set $\mathbb{R} \setminus \mathbb{Q}$ of all irrational numbers is uncountable. The proof is left as an exercise.

Exercises 4.3.3. (1) Modify the proof of Theorem 4.3.1 to show that if $(x_n)_{n \in \mathbb{N}}$ is an arbitrary sequence in $[0, 1]$, then there is some $x \in [0, 1]$ with $x \neq x_n$ for all n.

(2) Construct an explicit bijection between any two given closed in-tervals $[a, b]$ and $[c, d]$ in $\mathbb{R}$. Draw a diagram illustrating this bijection.

(3) Repeat (2) for open intervals.

(4) Obtain another proof that $\mathbb{R}$ is uncountable by constructing a bi-jection $(0, 1) \approx \mathbb{R}$. [*Hint:* Obtain a bijection $(-\pi/2, \pi/2) \approx \mathbb{R}$. Think calculus functions!]

(5) (*Somewhat more challenging*) Construct a bijection from the half-open, half-closed interval $[0, 1)$ to the open interval $(0, 1)$. Thus, adding one more point to the open interval $(0, 1)$ does not change its number of elements!

(6) Show that $[0, 1] \approx (0, 1)$.

Exercises 4.3.4. (1) Using the fact that the real line $\mathbb{R}$ is uncountable, show that the plane $\mathbb{R}^2 = \mathbb{R} \times \mathbb{R}$ is uncountable. [*Note:* A more precise result—which is considerably more difficult to prove—is that the plane matches the real line (see Example 4.3.18).]

(2) Discover the flaw in the following "proof" that the plane $\mathbb{R}^2$ is *not* the union of countably many lines.

Proof. Just suppose that the plane is the union of a sequence $(L_n)_{n \in \mathbb{N}}$ of lines. For each point $p \in \mathbb{R}^2$, let i_p be the least natural number i such that $p \in L_i$.

For each n, let L'_n be the line obtained by rotating L_n counter-clockwise an angle of $\pi/2$ about the origin. For each $p \in \mathbb{R}^2$, let j_p be the least natural number j such that $p \in L'_j$. Note that each point p in the plane is the unique point of intersection of the perpendicular lines L_{i_p} and L'_{j_p}.

The map $f: \mathbb{R}^2 \to \mathbb{N} \times \mathbb{N}$ defined by $f(p) = (i_p, j_p)$ is clearly injective, and its codomain $\mathbb{N} \times \mathbb{N}$ is denumerable; hence its do-main $\mathbb{R}^2$ is countable. But [see (1)] the set $\mathbb{R}^2$ is, to the contrary, uncountable. $\square$

Exercise 4.3.5. A real number is said to be *transcendental* if it is not algebraic (see Exercise 4.2.17). For example, π and e are transcendental—but the proofs are deep and well beyond the scope of this book! Prove that the set of all transcendental numbers is uncountable.

We saw above that the set $\mathbb{R} \setminus \mathbb{Q}$ of all irrational numbers is uncountable. But how many irrationals are there, exactly? Are there fewer than the number of reals, or just as many? To answer such questions, we will need to obtain a more precise result about how many real numbers there are.

We already know what it means for two sets to have the same number of elements, namely, that the two sets match (in the sense that there is some bijection between them). For finite sets X and Y, we can say that X has fewer elements than Y to mean that $\#(X) < \#(Y)$. Let us formulate a careful definition of what it means for an arbitrary set to have fewer elements than another; of course, the definition should generalize the meaning for finite sets.

Definition 4.3.6. We say that set X is **dominated by** set Y, or that Y **dominates** X, and write $X \preceq Y$ to mean there is some $Z \subset Y$ such that $X \approx Z$. And we say that X is **strictly dominated by** Y, or that Y **strictly dominates** X, and write $X \prec Y$ when $X \preceq Y$ but $X \not\approx Y$.

Of course, we use the notations $X \not\preceq Y$ and $X \not\prec Y$ to negate $X \preceq Y$ and $X \prec Y$, respectively.

Note that $X \preceq Y$ exactly when there is some injection $X \to Y$. Also,

$$X \preceq Y \iff X \prec Y \text{ or } X \approx Y.$$

Examples 4.3.7.   (1) If $X \subset Y$, then $X \preceq Y$.

(2) If $X \subset Y$, it does *not* necessarily follow that $X \prec Y$ just because $X \neq Y$. For example, $\mathbb{N}^* \subset \mathbb{N}$, so that

$$\mathbb{N}^* \preceq \mathbb{N}$$

with $\mathbb{N}^* \neq \mathbb{N}$. Yet

$$\mathbb{N}^* \not\prec \mathbb{N}$$

because, as we know, $\mathbb{N}^* \approx \mathbb{N}$.

(3) If X and Y are any two *finite* sets, then $X \prec Y$ if and only if $\#(X) < \#(Y)$. (Why?)

(4) If X is finite and Y is denumerable, then $X \prec Y$. For example, $\{1, 2, \ldots, n\} \prec \mathbb{N}^*$ for each $n \in \mathbb{N}^*$.

(5) If X is countable and Y is uncountable, then $X \prec Y$. For example, $\mathbb{Q} \prec \mathbb{R}$.

Exercises 4.3.8. (1) Prove:

    (a) $X \preceq X$ for every set X.

    (b) If $X \preceq Y$ and $Y \preceq Z$, then $X \preceq Z$.

    (c) $X \npreceq X$ for each set X.

    (d) If $A \approx X$, if $B \approx Y$, and if $X \preceq Y$, then $A \preceq B$.

    (e) Same as (d) except with $\preceq$ replaced by $\prec$.

(2) In (1), properties (a) and (b) of the relation $\preceq$ between sets are analogs of familiar properties of the relation $\leq$ between real numbers. [And property (c) above is an analog of the property $x \nless y$ of numbers.] What analog of a property of $\leq$ is missing from (1)?

According to the following theorem, for any given set whatsoever, there is a set that is still bigger in size.

Theorem 4.3.9 (Cantor). *Let X be any set. Then*

$$X \prec \mathcal{P}(X).$$

Proof. The map

$$
\begin{aligned}
X &\to \mathcal{P}(X) \\
x &\mapsto \{x\}
\end{aligned}
$$

is an injection, and so already $X \preceq \mathcal{P}(X)$.

Just suppose $X \approx \mathcal{P}(X)$. Then there is some bijection

$$f: X \approx \mathcal{P}(X).$$

[Caution: This map f need not be the one $x \mapsto \{x\}$.] Since $f(x) \subset X$ for each $x \in X$, it makes sense to form the set

$$A = \{x : x \in X \ \& \ x \notin f(x)\}.$$

(This sort of thing looks familiar, doesn't it?) Then $A \in \mathcal{P}(X)$. Since f is surjective, there is some $x \in X$ with $f(x) = A$. Complete the proof by obtaining a contradiction. (*Hint:* Is $x \in A$?) □

Cantor's Theorem has a somewhat surprising consequence.

Example 4.3.10. By using Cantor's Theorem, we can obtain a sequence of infinitely many infinite sets each of which strictly dominates its predecessors in the sequence, namely:

$$\mathbb{N} \prec \mathcal{P}(\mathbb{N}) \prec \mathcal{P}(\mathcal{P}(\mathbb{N})) \prec \mathcal{P}(\mathcal{P}(\mathcal{P}(\mathbb{N}))) \prec \cdots$$

Exercise 4.3.11. Find a set strictly dominating all the sets in the preceding sequence. Then find a whole sequence of such sets, each strictly dominating its predecessors. And then find a set strictly dominating all those in the sequence you found.

Examples like these show that there are "infinitely many different infinities"—what Cantor called *cardinal numbers*. (It was just such examples that at the time made many mathematicians distrust Cantor's investigations into infinite sets.)

Exercises 4.3.12. (1) Use Cantor's Theorem (4.3.9) to show that the set $\mathcal{P}(\mathbb{N})$ is uncountable. Then deduce that $2^{\mathbb{N}}$ is uncountable.

(2) Without using Cantor's Theorem, prove that $2^{\mathbb{N}}$ is uncountable. (*Hint:* If, to the contrary, $2^{\mathbb{N}}$ is uncountable, then it is the range of some sequence $(x_n)_{n=0,1,\ldots}$. Each x_n is itself a sequence $x_n = (x_{n,j})_{j=0,1,\ldots}$. Construct a new sequence x in $\mathbb{N}$ that, for each n, differs in its nth entry from the nth entry $x_{n,n}$ of x_n.)

We are now ready to begin answering the question of exactly how many elements $\mathbb{R}$ has. We will actually work with the interval $[0, 1)$, which we know matches $\mathbb{R}$. The idea in the next example is to express each $x \in [0, 1)$ in a binary (base 2) expansion

$$x = (0.b_1 b_2 \ldots)_2,$$

that is,

$$x = \frac{b_1}{2} + \frac{b_2}{2^2} + \cdots$$

for suitable binary digits $b_1, b_2, \cdots \in \{0, 1\}$. Such a expansion is the analog, for the fractional part of a real number, of the binary representation of the integral part of the number. (See pages 66–67 for binary representation of nonnegative integers.)

Example 4.3.13. We claim

$$[0, 1) \preceq 2^{\mathbb{N}^*}.$$

To see this, we define a map

$$F \colon [0, 1) \to 2^{\mathbb{N}^*}$$

as follows.

Let $0 \leq x < 1$. Successively define

$$b_1 = \begin{cases} 0 & \text{if } x < \frac{1}{2}, \\ 1 & \text{otherwise,} \end{cases}$$

$$b_2 = \begin{cases} 0 & \text{if } x - \frac{b_1}{2} < \frac{1}{2^2}, \\ 1 & \text{otherwise,} \end{cases}$$

$$b_3 = \begin{cases} 0 & \text{if } x - \left(\frac{b_1}{2} + \frac{b_2}{2^2}\right) < \frac{1}{2^3}, \\ 1 & \text{otherwise,} \end{cases}$$

etc.

Let

$$F(x) = (b_n)_{n=1,2,\ldots}.$$

Thus $F(x)$ is, for each $0 \leq x < 1$, just a sequence $(b_n)_{n=1,2,\ldots}$ in the set $\{0, 1\}$. We do not get every possible such sequence, however. In fact, the only sequences we get as values of F are those that do *not* eventually become 1. [This is related to the fact that we are, in effect, consistently choosing only one of two possible ways of representing the dyadic rationals in $[0, 1)$—those rationals of the form $m/2^n$. For example, $F(5/8) = (1, 0, 1, 0, 0, \ldots, 0, \ldots)$ and $F(5/8) \neq (1, 0, 0, 1, 1, \ldots, 1, \ldots)$.]

Let S be the range of F, that is, the set of all sequences in $\{0, 1\}$ that are *not* eventually 1. By restricting the codomain $2^{\mathbb{N}^*}$ of F to the range S, we thus get a new map

$$F \colon [0, 1) \to S,$$

which we denote by the same name as the original one.

To see that F is, in fact, a bijection, define the map

$$G \colon S \to [0, 1)$$

by

$$G\left((b_n)_{n=1,2,\ldots}\right) = \sum_{n=1}^{\infty} \frac{b_n}{2^n}.$$

Then (as you should verify as an exercise!) $G \circ F$ is the identity map of $[0, 1)$ and $F \circ G$ is the identity map of S. Hence

$$[0, 1) \approx S.$$

Exercises 4.3.14. These exercises refer to the preceding proof.

(1) Why did we not include 1 in the domain of F?

(2) What is the binary expansion of $1/10$ as given by the construction above?

(3) Without using the previously proved fact that $[0, 1)$ is uncountable, show that S is uncountable by showing that its complement in $2^{\mathbb{N}^*}$ is denumerable. (This provides another proof that $[0, 1)$ is uncountable, and hence that $\mathbb{R}$ is uncountable.)

In Example 4.3.13, we showed that

$$[0, 1) \preceq 2^{\mathbb{N}^*}$$

by constructing a match

$$[0, 1) \approx S_2,$$

where $S_2 \subset 2^{\mathbb{N}^*}$ consists of only those sequences that are not eventually 1 (we denoted S_2 simply by S before).

Instead of binary (base 2) expansion, we could have used decimal (base 10) expansion (see Example 3.5.36) to obtain a match

$$S_{10} \approx [0, 1),$$

where S_{10} is the set of all sequences in $\{0, 1, \ldots, 9\}$ that are not eventually 9. Now each sequence $(b_n)_{n=1,2,\ldots}$ in $\{0, 1\}$ may be regarded as a sequence in $\{0, 1, \ldots, 9\}$ that is not eventually 9, and so we have an injection

$$2^{\mathbb{N}^*} \to S_{10}.$$

It follows that

$$2^{\mathbb{N}^*} \preceq [0, 1).$$

Since $\mathbb{R} \approx [0, 1)$, we consequently have the relations

$$\begin{aligned}
2^{\mathbb{N}^*} &\preceq \mathbb{R}, \\
\mathbb{R} &\preceq 2^{\mathbb{N}^*}.
\end{aligned}$$

Together these say that each of the two sets $2^{\mathbb{N}^*}$ and $\mathbb{R}$ has at most as many elements as the other. Doesn't it follow that two such sets have exactly the same number of elements as each other? Cantor certainly thought so, but he was unable to prove the general theorem. The theorem he needed was proved only later by E. Schröder and F. Bernstein.

Theorem 4.3.15 (Schröder-Bernstein). *Let X and Y be sets for which*

$$X \preceq Y, \quad Y \preceq X.$$

Then

$$X \approx Y.$$

Proof. (You may want to skip this proof on a first reading.) Throughout the proof we use the notation $\varphi(E)$ for the image under a map φ of a subset E of the domain of φ, that is, $\varphi(E) = \{\varphi(x) : x \in E\}$.

There exist injections

$$f : X \to Y, \quad g : Y \to X.$$

The strategy is to restrict f to a subset of X and then extend the restriction to all of X so that the resulting map is the desired bijection from X to Y. Define a sequence $(A_n)_{n \in \mathbb{N}}$ of subsets of X by

$$A_0 = X \setminus g(Y),$$
$$A_{n+1} = g(f(A_n)).$$

Let A be the subset of X defined by

$$A = \bigcup_{n \in \mathbb{N}} A_n.$$

Observe that

$$X \setminus A \subset \operatorname{range}(g).$$

(Why?) Moreover, since g is injective, for each $x \in X \setminus A$ there exists a *unique* $y \in Y$ such that $x = g(y)$. Define

$$h : X \to Y$$

by the rule that, for each $x \in X$,

$$h(x) = \begin{cases} f(x) & \text{if } x \in A, \\ \text{the } y \in Y \text{ such that } x = g(y) & \text{if } x \in X \setminus A. \end{cases}$$

Certainly the map h is injective (why?).

It remains only to show that h is surjective. Let $y \in Y$. Just suppose $y \notin h(X \setminus A)$. We claim that $y \in h(A)$. By supposition, $g(y) \notin X \setminus A$, that is, $g(y) \in A$. Then $g(y) \in A_n$ for some $n \in \mathbb{N}$. Now $n > 0$ (why?), and so $g(y) = g(f(x))$ for some $x \in A_{n-1}$. Since g is injective, then $y = f(x)$. Thus $x \in A$, and so $y = f(x) = h(x)$. $\square$

Recalling that $\mathbb{N}^* \approx \mathbb{N}$, then with the aid of the Schröder-Bernstein Theorem, we obtain the following famous result.

Theorem 4.3.16.

$$\mathbb{R} \approx 2^{\mathbb{N}}.$$

Exercise 4.3.17. Show that there are exactly as many irrational numbers as real numbers. That is, prove

$$\mathbb{R} \setminus \mathbb{Q} \approx \mathbb{R}.$$

(*Hint:* Under the bijection $2^{\mathbb{N}^*} \supset S_2 \approx \mathbb{R}$ obtained above, which sequences map to rational numbers?)

We already know that there are sets having a greater number of elements than $\mathbb{R}$. For example, from Cantor's Theorem (Theorem 4.3.9),

$$\mathbb{R} \prec \mathcal{P}(\mathbb{R}).$$

But what about some other familiar sets that we can build from $\mathbb{R}$, such as $\mathbb{R} \times \mathbb{R}$? Do they also dominate $\mathbb{R}$? You may find the answers surprising.

Example 4.3.18. The plane $\mathbb{R}^2 = \mathbb{R} \times \mathbb{R}$ matches the line $\mathbb{R}$.
In fact, we already know there is a bijection

$$f: 2^{\mathbb{N}} \to \mathbb{R}.$$

But the map

$$2^{\mathbb{N}} \times 2^{\mathbb{N}} \;\to\; 2^{\mathbb{N}}$$
$$\left((s_n)_{n=0,1,\ldots}, (t_n)_{n=0,1,\ldots} \right) \;\mapsto\; (s_0, t_0, s_1, t_1, \ldots)$$

that alternates entries is bijective, for it has the inverse

$$2^{\mathbb{N}} \;\to\; 2^{\mathbb{N}} \times 2^{\mathbb{N}}$$
$$(u_n)_{n=0,1,\ldots} \;\mapsto\; \left((u_0, u_2, \ldots), (u_1, u_3, \ldots) \right).$$

Exercise 4.3.19. Prove that there are exactly as many points in three-dimensional space $\mathbb{R}^3$ as there are points on the real line $\mathbb{R}$.

Cantor used the language of cardinal numbers to describe the sizes of sets. In his language, we say that two sets X and Y **have the same cardinality** and write

$$\text{card}(X) = \text{card}(Y)$$

when $X \approx Y$. Then we write

$$\operatorname{card}(X) \leq \operatorname{card}(Y)$$

when $X \preceq Y$, and

$$\operatorname{card}(X) < \operatorname{card}(Y)$$

when $X \prec Y$. In this language, Cantor's Theorem (Theorem 4.3.9) takes the form

$$\operatorname{card}(X) < \operatorname{card}(\mathcal{P}(X)).$$

Cantor further used the notation[4]

$$\operatorname{card}(X) = \aleph_0 \iff X \approx \mathbb{N}$$

so that $\operatorname{card}(X) = \aleph_0$ exactly when X is denumerable.

Example 4.3.10 gives a sequence $(A_n)_{n \in \mathbb{N}}$ of sets such that

$$\aleph_0 = \operatorname{card}(A_0) < \operatorname{card}(A_1) < \operatorname{card}(A_2) < \cdots.$$

Exercises 4.3.20.   (1) For which sets X is $\operatorname{card}(X) < \aleph_0$?

(2) Express in Cantor's notation of cardinality the fact that a set X is countable.

(3) If $\mathcal{F}$ is the set of all functions from $\mathbb{R}$ to $\mathbb{R}$, how does $\operatorname{card}(\mathcal{F})$ compare to $\operatorname{card}(\mathbb{R})$?

Cantor also used the notation

$$\operatorname{card}(Y) = 2^{\operatorname{card}(X)} \iff Y \approx 2^X.$$

Then for any set X,

$$\operatorname{card}(\mathcal{P}(X)) = 2^{\operatorname{card}(X)}.$$

Since $\mathbb{R} \approx 2^{\mathbb{N}} \approx \mathcal{P}(\mathbb{N})$, it follows that

$$\operatorname{card}(\mathbb{R}) = 2^{\aleph_0}.$$

We know $\mathbb{N} \prec \mathbb{R}$, that is,

$$\aleph_0 < \operatorname{card}(\mathbb{R}).$$

The question then arises: *Is there any set that is uncountable yet has fewer elements than $\mathbb{R}$?* In other words, is there a set X for which

$$\aleph_0 < \operatorname{card}(X) < 2^{\aleph_0}$$

---

[4]The letter $\aleph$ here is *aleph*, the first letter of the Hebrew alphabet.

holds? In Cantor's terms: Is there an uncountable cardinal less than $2^{\aleph_0}$? The assertion that there is no such set is known as the *Continuum Hypothesis—CH*, for short. (The set $\mathbb{R}$ of real numbers is sometimes called the *continuum*.)

Exercise 4.3.21. Explain why CH may be formulated, equivalently, as follows: Each subset of $\mathbb{R}$ matches either $\mathbb{N}$ or else $\mathbb{R}$ itself.

Cantor conjectured that CH is true, but he was unable to prove it. And for good reason: In 1940 Kurt Gödel proved that CH is consistent with set theory in the sense that if assuming it would lead to a contradiction, then so would the ordinary axioms of set theory. (Thus CH is not false.) In 1963 P. J. Cohen proved that the negation of CH is consistent with set theory in the same sense. (Thus the negation of CH is not false.) In summary, the Continuum Hypothesis is *independent* of the other axioms of set theory.

The Continuum Hypothesis bears the same relation to set theory as Euclid's Parallel Postulate does to plane geometry. (One version of the Parallel Postulate is that, through a given point not on a given line, there is exactly one line parallel to the given line.) Roughly a century before Gödel and Cohen, the work of Gauss, J. Bòlyai, and N. I. Lobačevskiĭ showed that the negation of the Parallel Postulate is consistent with the rest of plane geometry. In 1899, after giving the first wholly adequate set of axioms for euclidean geometry, David Hilbert showed that the Parallel Postulate is consistent with the other axioms. Just as assuming the Parallel Postulate, on the one hand, or its negation, on the other, leads to different kinds of geometries, so assuming CH or its negation leads to different kinds of set theories.

Appendix A

Sets

By a set we mean a collection into a whole of definite, distinct objects ... of our perception or thought.

— Georg Cantor

In this appendix we collect basic definitions, notation, and facts about sets that are used throughout the book. The approach is informal rather than axiomatic.

A.1 Elements and Sets

A **set** is a collection of mathematical objects. If x is one of the objects comprising a set X, then we write $x \in X$ and say that x is an **element, member,** or **point of** X and that x **belongs to** X; in the contrary case we write $x \notin X$.

We did *not* just define the term "set"! (Saying that a set is a collection leads to the question of how to define "collection," and so forth.) It is not defined anywhere in this book. Rather, the notion of a set is taken to be *primitive*—incapable of being defined in terms of anything simpler or prior. What you will find here is an indication of how we allow the term "set" to be used. The presentation is informal and incomplete, rather than formal and complete the way it would be were we to present a list of axioms—fundamental assumptions—about sets.

Two sets are declared to be **equal to** one another when they have the same elements, that is, when

$$(\forall x)\,(x \in X \iff x \in Y),$$

and then we write $X = Y$. In the contrary case, we say that X is **unequal to** or **distinct from** Y and write $X \ne Y$.

A1

The preceding paragraph is an instance of what we meant by saying that our presentation is informal rather than axiomatic. In a formal, axiomatic treatment of set theory, the statement

$$(\forall X)\,(\forall Y)\,(X = Y \iff (\forall x)\,(x \in X \iff x \in Y))$$

would be listed explicitly as one of a number of axioms.

Two related general notational devices are used to specify particular sets. The first lists or indicates the elements of the set between braces ("curly brackets"). For example,

$$\{0, 1\}$$

is the set having the two elements 0 and 1 and no others;

$$\{1, 2, 3, \ldots, 10\}$$

is the set consisting of the first 10 positive integers 1, 2, 3, 4, 5, 6, 7, 8, 9, 10; and

$$\{2, 4, 6, \ldots\}$$

is the set of all even positive integers.

The ellipsis dots (...) were used differently in the last two examples. In $\{1, 2, 3, \ldots, 10\}$ the ellipsis serves to abbreviate the list. In $\{2, 4, 6, \ldots\}$, the ellipsis dots represent infinitely many unlisted numbers whose identity is presumed to be implicit in the few that are actually shown; the notation

$$\{2, 4, 6, \ldots, 2n, \ldots\}$$

perhaps makes the identity of these numbers more explicit.

Some particular sets for which we reserve special notation are

$\mathbb{N}^* =$ the set of all **positive integers** $= \{1, 2, 3, \ldots\}$,

$\mathbb{N} =$ the set of all **natural numbers** $= \{0, 1, 2, \ldots\}$,

$\mathbb{Z} =$ the set of all **integers** $= \{\ldots, -2, -1, 0, 1, 2, \ldots\}$,

$\mathbb{Q} =$ the set of all **rational numbers**,

$\mathbb{R} =$ the set of all **real numbers**,

$\mathbb{C} =$ the set of all **complex numbers**.

The second device for specifying particular sets—*set-builder notation*—uses

$$\{x : P(x)\},$$

where $P(x)$ is a predicate in the variable x, to designate the set consisting of those objects x having the property $P(x)$. For example, if $\mathbb{R}$ denotes the set of all real numbers, then

$$\{\, x : x \in \mathbb{R} \,\&\, x^2 = 1 \,\} = \{-1, 1\}.$$

This set may also be denoted

$$\{\, x \in \mathbb{R} : x^2 = 1 \,\} = \{-1, 1\}.$$

In general, if we already have a set X at hand, then $\{\, x \in X : P(x) \,\}$ denotes the set $\{\, x : x \in X \,\&\, P(x) \,\}$ consisting of those elements that belong to X and have the property $P(x)$.

Explicitly or implicitly listing the elements of a set between braces is really a special case of set-builder notation. For example,

$$\{0, 1\} = \{\, x : x = 0 \text{ or } x = 1 \,\},$$
$$\{1, 2, 3, \ldots, 10\} = \{\, x : x = 1 \text{ or } x = 2 \text{ or } x = 3 \text{ or } \ldots \text{ or } x = 10 \,\}.$$

We must not use the notation $\{\, x : P(x) \,\}$ recklessly, for otherwise we may find ourselves neck deep in paradox. For example, suppose we could form the set

$$\mathcal{R} = \{\, x : x \notin x \,\}.$$

Then we ask, "Is $\mathcal{R} \in \mathcal{R}$?" Whichever way we attempt to answer this question, positively or negatively, our answer immediately implies the opposite—see Example 4.1.7. This paradox, discovered by Bertrand Russell in 1900, led to the realization that Cantor's naive definition of "set" would hardly suffice and that a careful, rigorous foundation for set theory is required in order to avoid such paradoxes.

As already indicated, such a rigorous foundation is not attempted here. Rather, to avoid the formation of paradoxical sets such as Russell's $\mathcal{R}$, we shall generally use set-builder notation only in the form $\{\, x : x \in X \,\&\, P(x) \,\}$, and its abbreviation $\{\, x \in X : P(x) \,\}$, where X is a set that already exists.[1]

The simplest example of using set-builder notation is the **empty set** $\varnothing$, defined by

$$\varnothing = \{\, x : x \neq x \,\}.$$

We shall take it as axiomatic that $x = x$ for every mathematical object so that for every x we have $x \notin \varnothing$. Thus $\varnothing$ is the set that has no

---

[1] Another way to avoid paradoxes is to allow relatively unfettered use of set-builder notation and to call objects formed with it *classes* but to designate as a *set* only a class that is an element of some class.

elements whatsoever! We say that a set X is **empty** when $X = \varnothing$ and **nonempty** when $X \neq \varnothing$, that is, when there exists some $x \in X$.

The empty set is not "nothing," but something! And it is unavoidable if, for example, you want to be able to form the set of all solutions of an equation in a given number system but the equation turns out to have no solutions there. For example, $\{ x \in \mathbb{R} : x^2 + 1 = 0 \} = \varnothing$.

The simplest example of a set formed by explicitly listing its elements is the **singleton**, or **one-element**, set $\{x\}$, that is, the set whose one and only element is x: for arbitrary y, we have $y \in \{x\}$ if and only if $y = x$.

Do not confuse x with $\{x\}$. The difference is like that between a lion uncaged and a lion caged! For example, the set $\varnothing$ has no elements whatsoever, whereas the set $\{\varnothing\}$ has exactly one element, namely, $\varnothing$. More generally, always $\{x\} \neq x$ because otherwise $x \in x$ and, in order to avoid various unpleasantries in set theory, it is customary to assume that no object can be an element of itself.

For objects x and y, we may form the set $\{x, y\}$ having x and y as elements and no others. Of course, if $x = y$, then $\{x, y\} = \{x\}$. When $x \neq y$, we call $\{x, y\}$ a **doubleton** or **two-element** set.

Exercise A.1.1. Prove that $\{x, y\} = \{a, b\}$ if and only if

$$(x = a \text{ and } y = b) \quad \text{or} \quad (x = b \text{ and } y = a).$$

We call a set A a **subset of** a set X, say that A is **contained in** or **included in** X, and write $A \subset X$ to mean each element of A is an element of X, that is,

$$(\forall x)\, (x \in A \implies x \in X).$$

The statement $A \subset X$ is called an **inclusion**. Sometimes we write $X \supset A$ to mean the same thing, and then we call X a **superset of** A and say that X **contains** A. To deny that $A \subset X$ we write $A \not\subset X$. For example, $\mathbb{N} \subset \mathbb{Z} \subset \mathbb{Q} \subset \mathbb{R} \subset \mathbb{C}$, but $\mathbb{R} \not\subset \mathbb{Q}$ (because some real numbers, such as $\sqrt{2}$, are irrational).

The relationship between a set X and a subset A of X may be visualized by a *Venn diagram* as in Figure A.1. The entire shaded rectangular region represents the set X, and the more heavily shaded circular region represents the subset A of X. Of course, even when X is actually a subset of the plane, it need not be rectangular (and A need not be circular), and X need not be a subset of the plane at all.

A Venn diagram is merely suggestive of relationships among sets. It can convince us of the validity of statements about sets, but actual proofs are still required. For example, drawing another, still smaller circular region B inside A in Figure A.1 suggests

$$B \subset A \,\&\, A \subset X \implies B \subset X.$$

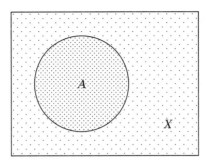

Figure A.1: Venn diagram of set inclusion.

Here is a (rather simple) proof: Assume $B \subset A$ and $A \subset X$. For each x, if $x \in B$, then $x \in A$ (because $B \subset A$) and therefore $x \in X$ (because $A \subset X$).

In view of the meaning of equality of sets, we have

$$X = Y \iff X \subset Y \,\&\, Y \subset X;$$

that is, two sets are equal if and only if each is a subset of the other (proof?). Then the inclusion $X \subset Y$ does not preclude the possibility that $X = Y$. When $X \subset Y$ but $X \neq Y$, we call X a **proper** subset of Y.

If x is an element, then

$$\{x\} \subset X \iff x \in X.$$

Try not to write $x \subset X$ when you mean $x \in X$, and vice versa!

The empty set is a subset of every set X:

$$\varnothing \subset X.$$

Here is the proof in informal terms: Since $\varnothing$ has no elements at all, it does not have any element that fails to be an element of X. Here is the proof in more formal terms: For every x, the implication

$$x \in \varnothing \implies x \in X$$

is true because its hypothesis $x \in \varnothing$ is false.

The **power set** of a given set X is the collection

$$\mathcal{P}(X) = \{A : A \subset X\}$$

consisting of all the subsets of X. For example,

$$\mathcal{P}(\{0, 1\}) = \{\varnothing, \{0\}, \{1\}, \{0, 1\}\}.$$

Always $\varnothing \in \mathcal{P}(X)$ and $X \in \mathcal{P}(X)$.

Exercises A.1.2.   (1) For what nonempty set X, if any, is $\mathcal{P}(X) = \{\varnothing, X\}$?

(2) If X and Y are sets for which $\mathcal{P}(X) = \mathcal{P}(Y)$, show that $X = Y$.

Let A and B be two sets. The **union of A and B** is the set

$$A \cup B = \{x : x \in A \text{ or } x \in B\}$$

of those elements that belong to at least one of the sets A and B (recall that in mathematics "or" is used in its exclusive sense). The **intersection of A and B** is the set

$$A \cap B = \{x : x \in A \text{ and } x \in B\}$$

of those elements that belong to both of the sets A and B. The set A is said to be **disjoint from** the set B when $A \cap B = \varnothing$, that is, when A and B have no elements in common; A is said to **intersect** or **meet** B in the contrary case, that is, when there exists some x with $x \in A$ and $x \in B$.

For example, let $A = \{0, 1, 2\}$, $B = \{1, 2, 3, 4\}$, and $C = \{3, 4\}$. Then $A \cup B = \{0, 1, 2, 3, 4\}$ and $A \cap B = \{1, 2\}$. The sets A and B intersect, whereas A and C are disjoint.

The Venn diagram in Figure A.2 illustrates the relationship between sets A and B and their union and intersection. The union is (represented by) the entire shaded region shaped like a reclining figure eight with a thick "waist"; the intersection is (represented by) the more heavily shaded region common to both of the circular regions (representing) A and B. In such a diagram, it is important not to place the individual sets in a special position. If the circular regions representing A and B had been drawn without any overlap, then they would have suggested that the intersection is empty (of course, the diagram, as drawn, disguises the possibility that the sets A and B might be disjoint after all).

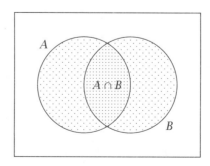

Figure A.2: Union and intersection of sets.

The Venn diagram suggests that, for arbitrary sets A and B,

$$A \cap B \subset A, \qquad A \cap B \subset B,$$
$$A \subset A \cup B, \qquad B \subset A \cup B.$$

You should write the proofs.

Exercise A.1.3. Prove that, for sets A and B:

(a) (*idempotent laws*) $A \cup A = A$ and $A \cap A = A$.

(b) $A \cup \varnothing = A$ and $A \cap \varnothing = \varnothing$.

(c) (*commutative laws*) $A \cup B = B \cup A$ and $A \cap B = B \cap A$.

(d) $A \cup B = B \iff A \subset B$; $A \cap B = A \iff A \subset B$.

Things get interesting when three sets A, B, C are involved. Look at the Venn diagram in Figure A.3; see also Figure A.4. Convince yourself that

$$(A \cup B) \cup C = A \cup (B \cup C), \qquad (A \cap B) \cap C = A \cap (B \cap C)$$

and then write proofs. Because union and intersection are thus *associative*, we may remove the parentheses and unambiguously write $A \cup B \cup C$ and $A \cap B \cap C$.

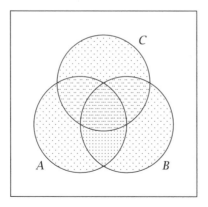

Figure A.3: Venn diagram for three sets.

By shading appropriate regions in copies of this diagram, convince yourself of the *distributive laws*

$$A \cup (B \cap C) = (A \cup B) \cap (A \cup C), \qquad A \cap (B \cup C) = (A \cap B) \cup (A \cap C).$$

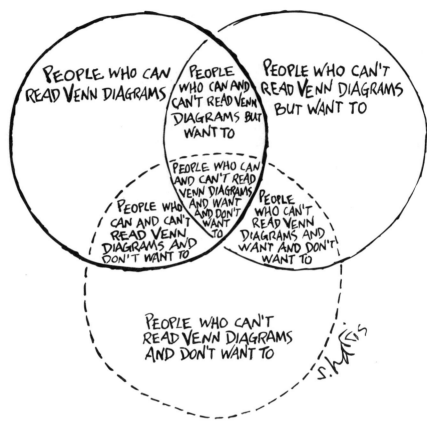

Figure A.4: Another Venn diagram for three sets.

Here is a proof of the first. For every x,

$$x \in A \cup (B \cap C) \iff (x \in A) \text{ or } (x \in B \cap C)$$
$$\iff (x \in A) \text{ or } ((x \in B) \text{ and } (x \in C))$$
$$\iff ((x \in A) \text{ or } (x \in B)) \text{ and } ((x \in A) \text{ or } (x \in C))$$
$$\iff (x \in A \cup B) \text{ and } (x \in A \cup C)$$
$$\iff x \in (A \cup B) \cap (A \cup C).$$

For sets A and X, the **complement of A in X** is the set

$$X \setminus A = \{x \in X : x \notin A\}$$

of those elements of X that do not belong to A. In Figure A.1, the lightly shaded part of the rectangle outside the circle represents $X \setminus A$.

If, again, $A = \{0, 1, 2\}$, $B = \{1, 2, 3, 4\}$, and $C = \{3, 4\}$, then $B \setminus A = C$ and $B \setminus C = \{1, 2\}$. Thus it is permissible to form the complement of A in X even when A is not a subset of X. Nonetheless, the most typical occasion for taking the complement is when A is a subset of X, and then

$$A \cup (X \setminus A) = X, \qquad X \setminus (X \setminus A) = A.$$

Always, even when A is not a subset of X,

$$X \setminus \varnothing = X, \qquad X \setminus X = \varnothing, \qquad A \cap (X \setminus A) = \varnothing.$$

The complement of a set is always taken with respect to some other set; for a set A, we do *not* form an absolute complement $\{x : x \notin A\}$. The reason for not doing so is the usual one of avoiding paradox.

By looking again at Figure A.2, you may discern the following relationships.

Proposition A.1.4 (De Morgan's Laws). *For all sets X, A, and B,*

$$X \setminus (A \cup B) = (X \setminus A) \cap (X \setminus B),$$
$$X \setminus (A \cap B) = (X \setminus A) \cup (X \setminus B).$$

Proof. We prove the first equality and leave the second as an exercise. Let X, A, and B be sets. For each x,

$$
\begin{aligned}
x \in X \setminus (A \cup B) &\iff x \in X \text{ and } x \notin A \cup B \\
&\iff (x \in X) \text{ and } (\text{not } (x \in A \cup B)) \\
&\iff (x \in X) \text{ and } (\text{not } ((x \in A) \text{ or } (x \in B))) \\
&\iff (x \in X) \text{ and } ((\text{not } x \in A) \text{ and } (\text{not } x \in B)) \\
&\iff (x \in X) \text{ and } ((x \notin A) \text{ and } (x \notin B)) \\
&\iff ((x \in X) \text{ and } (x \notin A)) \text{ and } \\
&\qquad ((x \in X) \text{ and } (x \notin B)) \\
&\iff (x \in X \setminus A) \text{ and } (x \in X \setminus B) \\
&\iff x \in (X \setminus A) \cap (X \setminus B). \quad \square
\end{aligned}
$$

Exercises A.1.5.   (1) Prove that $(A \cup B) \cap B = B$ and $(A \cap B) \cup B = B$ for arbitrary sets A and B.

(2) Construct subsets A and B of a set X for which $X \setminus (A \cup B) \neq (X \setminus A) \cup (X \setminus B)$. For which $X \setminus (A \cap B) \neq (X \setminus A) \cap (X \setminus B)$. Illustrate these situations with Venn diagrams.

(3) Let A and B be subsets of a set X. Prove that $A = B$ if and only if $X \setminus A = X \setminus B$.

(4) Let A and B be subsets of a set X. Prove that $A \subset B$ if and only if A and $X \setminus B$ are disjoint.

(5) For sets A and B, their *symmetric difference* is the set $A \triangle B$ defined by

$$A \triangle B = \{\, x : (x \in A \text{ and } x \notin B) \text{ or } (x \in B \text{ and } x \notin A) \,\}.$$

(a) Draw a Venn diagram showing the relationship of $A \triangle B$ to A and B.

(b) Express symmetric difference in terms of union, intersection, and complement.

(c) For an arbitrary set A, what is $A \triangle \varnothing$?

(d) Show that symmetric difference is commutative: $A \triangle B = B \triangle A$ for all sets A and B.

(e) Prove that symmetric difference is associative: $(A \triangle B) \triangle C = A \triangle (B \triangle C)$ for all sets A, B, C.

The associative laws for union and intersection extend to more than three sets. For example, all ways of parenthesizing $A \cup B \cup C \cup D$, such as $(A \cup (B \cup C)) \cup D$, yield the same set, and so we may speak unambiguously of the union $A \cup B \cup C \cup D$ of four sets. But denoting union or intersection when more than a few sets are involved begins to be awkward, and another approach is needed.

Even four sets A, B, C, and D can be collected together into a single set

$$\mathcal{A} = \{A, B, C, D\}.$$

For example, we might have $A = [0, 1]$, the closed interval; $B = (1/2, 2)$, the open interval; $C = \{2\}$, the singleton; and $D = (2, 3)$, the open interval. In this case,

$$\mathcal{A} = \{[0, 1], (1/2, 2), \{2\}, (2, 3)\}.$$

The members of $\mathcal{A}$ are themselves sets of elements. If $a = 1/2$, then

$$a \in A \in \mathcal{A}.$$

Notice the different typographic treatments of different levels of this hierarchy of sets: lowercase letter (a) for an individual element, uppercase letter (A) for a set of such elements, and script uppercase letter ($\mathcal{A}$) for a set of such sets. In such a hierarchy, a set such as $\mathcal{A}$ may be called a **collection** of sets. (A collection is still just a set, but the term "collection" has a useful connotation to remind us where in the hierarchy we are.)

In general, if $\mathcal{A}$ is a collection of sets, then the **union of** $\mathcal{A}$ is the set

$$\bigcup \mathcal{A} = \{\, x : (\exists A \in \mathcal{A})\, (x \in A) \,\}$$

and the **intersection of** $\mathcal{A}$ is the set

$$\bigcap \mathcal{A} = \{\, x : (\forall A \in \mathcal{A})\, (x \in A) \,\}.$$

In the preceding example $\mathcal{A} = \{A, B, C, D\}$, we have $\bigcup \mathcal{A} = [0, 3)$ and $\bigcap \mathcal{A} = \varnothing$.

We are tacitly assuming that the union of any collection of sets may legitimately be formed as a set.

The definition of intersection prohibited intersecting the empty collection. The reason is that otherwise, if $\mathcal{A} = \varnothing$, then for every x it would be the case that $x \in \bigcap \mathcal{A}$ (why?). Thus $\bigcap \mathcal{A}$ would be the set of all sets—one of those paradoxical objects we must avoid.

A collection $\mathcal{A}$ of sets is said to be **pairwise disjoint** when $A \cap B = \varnothing$ for every two distinct members A, B of $\mathcal{A}$. For example, the collection $\{[0, 1], \{2, 3\}, (3, 4)\}$ is pairwise disjoint.

Exercises A.1.6.   (1) Determine $\bigcap \mathcal{A}$ and $\bigcup \mathcal{A}$ if:

(a) $\mathcal{A} = \{\, A : A = [-n, n] \text{ for some } n \in \mathbb{N}^* \,\}$.

(b) $\mathcal{A} = \{\, A : A = (-1/n, 1/n) \text{ for some } n \in \mathbb{N}^* \,\}$.

(2) What is $\bigcup \varnothing$?

(3) Explain carefully why a collection $\mathcal{A}$ of sets is pairwise disjoint if and only if, for all $A, B \in \mathcal{A}$, either $A = B$ or $A \cap B = \varnothing$.

(4) If a collection of sets has empty intersection, must it be pairwise disjoint? What about the converse?

Another approach to forming the union and intersection of several sets together appears in Section B.4.

A.2 Pairs and Products

Let x and y be mathematical objects—sets.[2] We would like to define an object (x, y) determined by x and y together that assigns the order "first x, second y" to the two sets. In other words, $(x, y) = (a, b)$ should imply $x = a$ and $y = b$, but (x, y) should be different from

---

[2] From our viewpoint, *every* mathematical object that qualifies to be an element of some set is itself a set. Even a single natural number is a set; see Section D.2 and, in particular, page A67.

(y, x) unless $x = y$. One set determined by x and y together is $\{x, y\}$. Unfortunately, $\{x, y\} = \{y, x\}$, and so this does not give any preference to x as first and y as second. Now the only subsets of $\{x, y\}$ are $\varnothing$, $\{x\}$, $\{y\}$, and $\{x, y\}$ itself. Then a way to distinguish x as first is to use the singleton $\{x\}$ along with $\{x, y\}$ itself.

Definition A.2.1. For sets x and y, the **ordered pair** (x, y) is the set $\{\{x\}, \{x, y\}\}$. The element x is called the **first coordinate** and y the **second coordinate** of the ordered pair.

It is an annoying fact of mathematical life that sometimes the same notation is overloaded with several different meanings. Surrounding parentheses were just used to denote an ordered pair; if x and y are real numbers, then the very same notation (x, y) is commonly used for the open interval $\{ t \in \mathbb{R} : x < t < y \}$. When reading and writing mathematics, we all must learn to live with these apparent ambiguities and hope the context makes clear what is intended. (In computer algebra systems such as MATHEMATICA, such ambiguity cannot be tolerated.)

Definition A.2.1 does actually accomplish our purpose.

Proposition A.2.2. *For elements x, y, a, and b,*

$$(x, y) = (a, b) \iff x = a \ \& \ y = b.$$

Proof. Assume $x = a$ and $y = b$. Then $\{x\} = \{a\}$ and $\{x, y\} = \{a, b\}$, and so $\{\{x\}, \{x, y\}\} = \{\{a\}, \{a, b\}\}$. This means $(x, y) = (a, b)$.

Conversely, assume $(x, y) = (a, b)$, that is,

$$\{\{x\}, \{x, y\}\} = \{\{a\}, \{a, b\}\}.$$

Then either

$$\{x\} = \{a\} \quad \text{and} \quad \{x, y\} = \{a, b\}$$

or

$$\{x\} = \{a, b\} \quad \text{and} \quad \{x, y\} = \{a\}.$$

Consider each of the cases separately and show that $x = a$ and $y = b$ in each case. □

One consequence of this proposition is that

$$x \neq y \implies (x, y) \neq (y, x)$$

(whereas, as we said above, $\{x, y\} = \{y, x\}$ even if $x \neq y$).

Exercises A.2.3. (1) Prove that $(x, y) \neq \{x, y\}$ no matter what x and y are.

(2) If $x \in X$ and $y \in Y$, show that $(x, y) \subset \mathcal{P}(X \cup Y)$.

(3) Given that $0 = \varnothing$, $1 = \{0\}$, and $2 = \{0, 1\}$ (see page A67), determine whether the doubleton $\{1, 2\}$ is an ordered pair.

(4) Give a definition for the notion of an *ordered triple* (x, y, z). Prove the analog of Proposition A.2.2 for ordered triples.

Definition A.2.4. Let X and Y be sets. The **product** of X and Y is the set

$$X \times Y = \{ (x, y) : x \in X, y \in Y \}$$

consisting of all ordered pairs whose first coordinate belongs to X and second coordinate belongs to Y.

To distinguish such a product from a product such as $2 \cdot 3$, we may call $X \times Y$ the **cartesian product** of X and Y (after René Descartes, who used ordered pairs of numbers to describe geometry in the plane algebraically). The symbol $\times$ may be read as "times" or "cross."

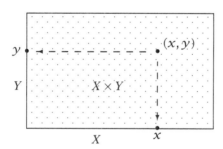

Figure A.5: Product of two sets.

The product of sets X and Y may be visualized by a diagram as follows. Represent each of X and Y as adjacent horizontal and vertical sides of a rectangle—see Figure A.5. Then $X \times Y$ is represented as all points on and inside that rectangle. The first and second coordinates x and y of an element $(x, y) \in X \times Y$ are obtained by projecting the corresponding point onto the sides X and Y. Such a diagram faithfully represents the product when X and Y are themselves closed intervals in $\mathbb{R}$. When drawing such a diagram, we no more mean to imply that X and Y are lines or subsets of lines than, when drawing a Venn diagram, we mean to suggest that the sets are contained in the plane.

Exercises A.2.5. (1) What is $X \times Y$ if X or Y is empty?

(2) For which sets X and Y, if any, is $X \times Y = Y \times X$?

(3) Prove that if $A \subset X$ and $B \subset Y$, then $A \times B \subset X \times Y$. Is the converse true?

(4) If A, B, C are sets, show that $A \times (B \cup C) = (A \times B) \cup (A \times C)$ and $A \times (B \cap C) = (A \times B) \cap (A \times C)$. Draw diagrams representing these equalities.

(5) If A, B, C, and D are sets, can the intersection $(A \times B) \cap (C \times D)$ of products be written as a product of intersections? Can the union of products be written as a product of unions?

(6) Suppose X is the circle $\{x^2 + y^2 = 1 : x, y \in \mathbb{R}\}$ and Y is the closed interval $[0, 1]$. Explain how the surface of a cylinder may be used to represent the product $X \times Y$ visually.

(7) Show that $X \times Y \in \mathcal{P}\left(\mathcal{P}(\mathcal{P}(X \cup Y))\right)$. [*Hint:* See Exercise A.2.3 (2).]

(*Note:* In an axiomatic approach to set theory, this would be a way of establishing that the product of two sets is itself a set.)

A.3 Relations

In many situations we are interested in associating, connecting, comparing, or *relating* elements of one set to elements of another or the same set. For example:

- The integer a is related to the integer b if $a - b$ is even.

- The integer m is related to the integer n if $n = km$ for some integer k.

- The real number θ is related to $y \in [-1, 1]$ if $\sin \theta = y$.

- The real number x is related to the real number y if $x^2 < y^3$.

- The point p in the plane is related to the real number x if x is the first coordinate of p.

- The real number x is related to the point p in the plane if x is the first coordinate of p.

- The function f is related to the function g if $f = g'$.

To indicate which elements of a set X are related in a specified way to which elements of a set Y (perhaps $Y = X$), we can form the set of ordered pairs (x, y) for which $x \in X$, $y \in Y$, and x is so related to y. Such a set is just a subset of the product set $X \times Y$. Hence we make the following definition.

Definition A.3.1. A **relation from** a set X **to** a set Y is a subset of $X \times Y$. When $Y = X$, we speak of a relation **in** the set X. In general, if R is a relation from X to Y, then we often write $x\,R\,y$ to mean $(x, y) \in R$ and, of course, $x\,\not\!R\,y$ to mean $(x, y) \notin R$.

Thus a relation is not an assertion but rather a thing—a set. Yes, each example above could have been phrased in the form "$x \in X$ is related to $y \in Y$ if $P(x, y)$ is true," where $P(x, y)$ is a certain predicate in two variables x and y. From such a predicate we can form at once the *set*

$$\{\, (x, y) \in X \times Y : P(x, y) \,\},$$

which is a relation in the sense of the preceding definition. However, we need not specify a relation in this way. For example,

$$R = \{(0, 0), (0, 1), (2, 1), (3, 2)\} \tag{$*$}$$

is a relation—although not an especially interesting one—from $\{0, 2, 3\}$ to $\{0, 1, 2\}$; it is also a relation from $\{0, 1, 2, 3\}$ to the same set, that is, a relation in $\{0, 1, 2, 3\}$. For any sets X and Y, the empty set is a relation from X to Y (it relates no element of X to any element of Y whatsoever!).

Since a relation is a subset of a product, it may often be visualized by a region drawn inside a rectangular region such as in Figure A.5.

The **opposite** or **reverse** of a relation R from a set X to a set Y is the relation from Y to X denoted R^{-1} and defined by

$$R^{-1} = \{\, (y, x) : (x, y) \in R \,\}.$$

In other words, $x\,R^{-1}\,y \iff y\,R\,x$. For example, the relation R in ($*$) has the opposite

$$R^{-1} = \{(0, 0), (1, 0), (1, 2), (2, 3)\}.$$

Exercises A.3.2.   (1) Find all relations from $\{1, 2\}$ to $\{1, 2, 3\}$.

(2) Find all relations in $\{1, 2\}$.

(3) For any set X, the set $\Delta_X = \{\, (x, x) : x \in X \,\}$ is a relation in X called the *identity relation* of X. Draw a diagram representing Δ_X.

(4) If R is a relation in X, what does it mean intuitively, in terms of relating elements of X to elements of X, to say that $\Delta_X \subset R$? What is the appropriate diagram representing this situation?

(5) In the closed unit interval $X = [0, 1]$ in $\mathbb{R}$, let $R = \{\, (x, y) \in X \times X : x < y \,\}$. Draw a diagram showing $X \times X$, Δ_X, R, and $\Delta_X \cup R$.

(6) For what relation or relations R in a set X is it the case that $x \, R \, y$ for all $x, y \in X$?

(7) If R is a relation in a set X, what does it mean intuitively, in terms of relating elements of X to elements of X, to say that $R = R^{-1}$? What is the appropriate diagram representing this situation?

(8) Describe and draw the relation $\{ (x, y) \in \mathbb{R} \times \mathbb{R} : \cos x < \sin y \}$.

Appendices B and C examine in detail two particularly important types of relations that pervade mathematics—*functions* and *equivalence relations*. Here we look briefly at another type—*order relations*.

Order relations are used to describe how elements of a given set *compare* with respect to such aspects as size, position, or priority. For example:

- In the set $\mathbb{Z}$ of all integers, let $m \, R \, n$ mean $m < n$.

- In the set $\mathbb{Z}$ of all integers, let $m \, R \, n$ mean $m \geq n$.

- For any set X, in its power set $\mathcal{P}(X)$, let $A \, R \, B$ mean $A \subset B$.

In describing various kinds of order relations, some terminology is useful. A relation R in a set X is said to be

- **reflexive on** X if $x \, R \, x$ for all $x \in X$.

- **transitive** if $x \, R \, y \, \& \, y \, R \, z \implies x \, R \, z$ for all x, y, z.

- **antisymmetric** if $x \, R \, y \, \& \, y \, R \, x \implies x = y$ for all x, y.

A **partial ordering of** a set X is a relation R in X that is reflexive on X, antisymmetric, and transitive; such a relation is said to **partially order** the set X. The principal example is the usual ordering relation $\leq$ in $\mathbb{R}$. For this reason, arbitrary partial orderings are typically denoted the same way. In this notation, a partial ordering of a set X is a thus a relation $\leq$ in X such that

$$x \leq x \quad (x \in X),$$
$$x \leq y \, \& \, y \leq x \implies x = y \quad (x, y \in X),$$
$$x \leq y \, \& \, y \leq z \implies x \leq z \quad (x, y, z \in X).$$

When $\leq$ is a partial ordering of X, then the opposite relation $\geq$, given by

$$x \geq y \iff y \leq x \quad (x, y \in X),$$

is also a partial ordering of X.

Given an arbitrary relation R in a set X, there is an associated **strict relation** S defined by

$$x \, S \, y \text{ if and only if } x \, R \, y \, \& \, x \neq y.$$

When a partial ordering R is denoted $\leq$, then the associated strict relation is ordinarily denoted $<$ (and when R is instead denoted $\geq$, then the strict relation is denoted $>$). The strict relation associated with a partial ordering has the properties

$$x \not< x \qquad (x \in X),$$
$$x < y \implies y \not< x \qquad (x, y \in X),$$
$$x < y \, \& \, y < z \implies x < z \qquad (x, y, z \in X).$$

The relation R in $\mathcal{P}(X)$ for a given set X defined by

$$A \, R \, B \iff A \subset B$$

is a partial ordering of $\mathcal{P}(X)$ (why?). As soon as X has at least three elements, it is possible to have subsets A and B with neither $A \, R \, B$ nor $B \, R \, A$. This is unlike the situation with the usual partial ordering $\leq$ of $\mathbb{R}$, where, for any two numbers x and y, either $x \leq y$ or $y \leq x$.

A partial ordering $\leq$ of X is called a **total ordering of** X when any two members x and y of X are *comparable* in the sense that $x \leq y$ or $y \leq x$.

Exercises A.3.3. (1) Find all partial orderings of $\{1, 2, 3\}$. Which of these are total orderings? (Do not confuse a partial ordering of your own devising with the usual one in which $1 < 2 < 3$.)

(2) Which of the following relations are partial orderings? Which are total orderings?

(a) In the set $\mathbb{R}$ of all real numbers, let $x \, R \, y$ mean $|x| \leq |y|$.

(b) In the power set of $\{1, 2, 3, \ldots, 100\}$, let $A \, R \, B$ mean that A has at least as many elements as B.

(c) In any set X, let $x \, R \, y$ mean $x = y$. In other words, R is the identity relation Δ_X of X—see Exercise A.3.2 (3).

(d) In the set X of all *nonzero* integers, let $m \, R \, n$ mean $m = kn$ for some integer $k \neq 0$.

(e) In the plane $\mathbb{R} \times \mathbb{R}$, let $(x, y) \, R \, (s, t)$ mean

$$((x, y) = (s, t)) \quad \text{or} \quad (x < s) \quad \text{or} \quad (x = s \, \& \, y < t).$$

(Draw a diagram to see what this means geometrically.)

(f) In the set $\mathcal{F}$ of all functions from $\mathbb{R}$ to $\mathbb{R}$, let $f \leq g$ mean $f(t) \leq g(t)$ for all $t \in \mathbb{R}$.

(3) Let $\leq$ be a total ordering of X and let $x, y \in X$. Prove that $x \leq y$ if and only if $z < x \implies z < y$ for every $z \in X$.

(4) For a relation S in a set X, the associated *weak relation* R is defined by $x\, R\, y$ if and only if $x\, S\, y$ or $x = y$. For example, the weak relation associated with the usual strict ordering $<$ of $\mathbb{R}$ is the usual ordering $\leq$.

Suppose S is transitive; *irreflexive* in the sense that $x \not{S} x$ for each $x \in X$; and *asymmetric* in the sense that $x\, S\, y \implies y \not{S} x$ for all $x, y \in X$. Show that R is then a partial ordering of X and S is the strict relation associated with R.

Appendix B

Functions

This appendix defines the concept of function, presents ways of modifying and combining functions to get new ones, and introduces the special types of functions said to be *injective* and *surjective*. It also treats sequences and more general "families" as functions and, in particular, outlines a proof of the Ordinary Recursion Principle.

B.1 Functions and Maps

In calculus you work with functions whose inputs and outputs are not just real numbers but even pairs or triples of numbers, that is, points in the plane and three-dimensional space. Among such functions are those defined by the rules

$$f(x) = x^2,$$

$$g(\theta) = \log \sin \sqrt{\frac{\tan \theta}{\theta^2 + 16}},$$

$$h(x) = \begin{cases} x & \text{if } x \geq 0, \\ -x & \text{if } x < 0, \end{cases}$$

$$\varphi(x) = \text{the 99th digit in the decimal expansion of } x,$$

$$y(t) = (\cos t, \sqrt{t}),$$

$$L(v) = 2v.$$

These examples demonstrate several points. First, the rule defining the function need not be a single formula—indeed, need not be a formula in the conventional sense at all [as in the case of $\varphi(x)$]. Second, the formula alone may not reveal at once what the legitimate inputs x to the rule are. For example, at which real numbers t may the value $g(t)$ be computed? Are the inputs to $L(v) = 2v$ real numbers or are they, instead, two- or three-dimensional vectors, with the output $L(v)$ being a vector with the same direction and double the length of the input v? (Writing $\vec{v}$ would indicate the variable v is supposed to be a vector, but would it then be two-dimensional, three-dimensional, or higher?) Third, for a function such as g, the formula alone may not immediately make apparent what the actual outputs are. The outputs $g(t)$ are certainly real numbers, but precisely which real numbers?

The concept of a function as being simply a rule f for computing outputs from inputs is just not adequate: we need to specify, in addition, both the set of inputs and a set where the outputs lie (which may be larger than the set of actual outputs). Here is a definition.[1]

Definition B.1.1 (Function—Tentative Formulation). A function

$$f: X \to Y$$

consists of sets X and Y along with a rule f that assigns to each $x \in X$ one and only one element of Y denoted by $f(x)$ and called the **value of f at x** or the **image of x under f**. The set X is called the **domain**, the set Y the **codomain**, and the rule f the **graph** of the function. We say that the function is **from X to Y**.

The terms **map** and **mapping** are synonyms for "function"; they are used, especially, when inputs or outputs are not numbers. (In some contexts—for example, when the inputs are themselves functions—the terms **transformation** and **operator** are also used as synonyms.)

The function y from page A19 may now more properly be referred to as "the function $y: \mathbb{R} \to \mathbb{R} \times \mathbb{R}$ given by $y(t) = (\cos t, \sqrt{t})$" [where $\mathbb{R} \times \mathbb{R}$ is, as usual, the set of pairs (x, y) of real numbers]. It may also be described in displays as the function

$$y: \mathbb{R} \to \mathbb{R} \times \mathbb{R},$$

where

$$y(t) = (\cos t, \sqrt{t}) \qquad (t \in \mathbb{R}).$$

A shorter, more symbolic way of saying the same thing is

$$y: \mathbb{R} \to \mathbb{R} \times \mathbb{R}$$
$$t \mapsto (\cos t, \sqrt{t}).$$

---

[1]Because the meaning of the term "rule" is left vague for now, the definition is really not precise. The vagueness is removed in Definitions B.1.10 and B.1.11.

In general, to indicate the rule for a function we may use a *barred*
arrow, as in $x \mapsto f(x)$, and say that f **maps** or **sends** x to $f(x)$. Thus
we write

$$f: X \to Y$$
$$x \mapsto f(x).$$

In this notation, the part $f: X \to Y$ with its ordinary arrow indicates
where the rule operates—the domain and codomain—whereas the part
$x \mapsto f(x)$ indicates *what* the rule is. In this context, the letter x is a
dummy variable that could be replaced by any other name not already
in use [for example, $t \mapsto f(t)$]; this dummy variable is also called the
argument of the expression $f(x)$.

It is important not to confuse f with $f(x)$. The first denotes the
(graph of) the function, whereas the second denotes the value of the
function at a generic x—perhaps given by a formula or expression involving x—or even the value of the function at some particular x. We
really *ought not* say, for example, "the function $f(x) = x^2$" or, even
worse, "the function x^2." Countless calculus students have been per-
plexed by perpetuation of the confusion.

Sometimes, especially when combining several functions, it is convenient to put the name f of the rule along the arrow, as in

$$X \xrightarrow{f} Y.$$

In Definition B.1.1, we said, "A function $f: X \to Y$ consists of sets X
and Y along with a rule f ...." The words "along with" are vague but
meant to suggest that all three things—X, Y, and f—are constituents
of the map. To improve the definition, we should say, more precisely,
that the function is the ordered triple (X, f, Y).

Despite the actual definition of function involving a domain and
codomain along with a rule, at times everybody lapses into a laconic
mode and speaks, for example, of "a function f" to mean "a function
with graph f and a certain domain and codomain." Accordingly, the
domain and codomain of a function $f: X \to Y$ may be denoted by dom f
and codom f, respectively.

As ordered triples (X, f, Y) and (A, g, B), two maps $f: X \to Y$ and
$g: A \to B$ are equal precisely when

$$X = Y, \quad A = B, \quad f(x) = g(x) \text{ for all } x \in X.$$

Then the functions

$$f: \mathbb{R} \to \mathbb{R} \qquad \text{and} \qquad g: \{x \in \mathbb{R} : x \geq 0\} \to \mathbb{R}$$
$$x \mapsto x^2 \qquad\qquad\qquad\qquad x \mapsto x^2$$

are *different*, even though they seem to have the same rule; they are different because the rule is applied to different sets of inputs—all real numbers for f but only nonnegative real numbers for g. Likewise, the functions

$$f: \mathbb{R} \to \mathbb{R} \qquad \text{and} \qquad h: \mathbb{R} \to \{y \in \mathbb{R} : y \geq 0\}$$
$$x \mapsto x^2 \qquad\qquad\qquad x \mapsto x^2$$

are *different*, even though they have the same rule and the same inputs; they are different because their codomains are different.

Although the functions f and h above have different codomains, they have the same set of outputs—their *range*s are the same.

Definition B.1.2. The **range** of a map $f: X \to Y$ is the set range f defined by

$$\text{range} f = \{f(x) : x \in X\},$$

that is, the set of all values of f at elements of its domain.

Here are some standard examples of functions.

Examples B.1.3.   (1) Let X be any set. Then the **identity map** of X is the map

$$i_X: X \to X$$
$$x \mapsto x.$$

Thus $i_X(x) = x$ for every $x \in X$. You may think it silly to define a map that "does nothing" to each of its inputs, but as we shall see later, such a map comes in handy.

(2) At the opposite extreme from an identity map, which leaves each input alone, is a **constant map** $f: X \to Y$ whose range is a singleton $\{c\}$ for an element c of Y, that is, $f(x) = c$ for all $x \in X$.

(3) Let X be a set and A be a subset of X. The **characteristic function of A in X** is the function $c_A: X \to \{0, 1\}$ defined for all $x \in X$ by

$$c_A(x) = \begin{cases} 1 & \text{if } x \in A, \\ 0 & \text{if } x \in X \setminus A. \end{cases}$$

Suppose, for example, $X = \{1, 2, 3, 4, 5\}$ and $A = \{3, 4\}$. Then $c_A(4) = 1$ but $c_A(5) = 0$. Some basic properties of characteristic functions are indicated in Exercise 4.1.37.

(4) Let X and Y be any sets. Form their product $X \times Y = \{(x,y) : x \in X, y \in Y\}$. Then the maps

$$p: X \times Y \to X \qquad \text{and} \qquad q: X \times Y \to Y$$
$$(x,y) \mapsto x \qquad\qquad\qquad (x,y) \mapsto y$$

are called **projections**—the **first projection** and **second projection**, respectively.[2]

When $X = Y = \mathbb{R}$, then these are the usual projections to the x-axis and y-axis, respectively, as depicted in Figure B.1. In general, $z = (p(z), q(z))$ for all $z \in X \times Y$.

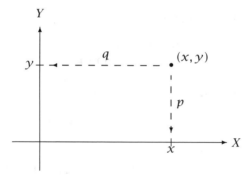

Figure B.1: First and second projection maps.

There are various ways to form new functions from old. One way is to cut down the domain of a function but otherwise leave the rule for forming values the same.

Definition B.1.4. Let $f: X \to Y$ be a map and let $A \subset X$. Then the **restriction of f to A** is the map

$$f|_A : A \to Y$$
$$x \mapsto f(x).$$

For example, let X be any set and let $i_X: X \to X$ be its identity map. Then for any subset A of X, the restriction

$$(i_X)|_A : A \to X$$
$$x \mapsto x$$

is called the **inclusion map of A into X**.

---

[2]We ought to use a pair of nested parentheses and write $p((x,y))$ for the value of p at an element (x,y) of its domain. The inner parentheses are part of the notation for the ordered pair (x,y), and the outer parentheses enclose the input (x,y) to the function. Typically, only one pair of parentheses is used in such situations.

If $f\colon X \to Y$ is a map and $B \subset \operatorname{range} f$, then the map

$$g\colon X \to B$$
$$x \mapsto f(x)$$

may be called a **codomain restriction** of $f\colon X \to Y$. (Actually, the graph g of the latter is the same as the graph f of the original map.)

A map f is called an **extension** of a map g when g is a restriction of f. For example, the function $f\colon \mathbb{R} \to \mathbb{R}$ defined by

$$f(x) = \begin{cases} (\sin x)/x & \text{if } x \neq 0, \\ 1 & \text{if } x = 0 \end{cases}$$

is an extension of the function

$$g\colon \mathbb{R} \setminus \{0\} \to \mathbb{R}$$
$$x \qquad \mapsto (\sin x)/x,$$

and so is the function $h\colon \mathbb{R} \setminus \{0\} \to \mathbb{R}$ defined by

$$h(x) = \begin{cases} (\sin x)/x & \text{if } x \neq 0, \\ 300 & \text{if } x = 0 \end{cases}$$

(For some purposes—such as making the extension continuous—f is a nicer extension of g than is h).

Some things you may not be accustomed to thinking of as maps actually are maps, as in the following examples.

Examples B.1.5.   (1) Let $C[0, 2\pi]$ be the set of all continuous functions from the closed interval $[0, 2\pi]$ to $\mathbb{R}$. Then forming definite integrals of such functions may be regarded as a map

$$J\colon C[0, 2\pi] \to \mathbb{R}$$
$$f \mapsto \int_0^{2\pi} f(t)\, dt.$$

(2) Let C be the set of all continuous functions from $\mathbb{R}$ to $\mathbb{R}$. Then forming definite integrals of such functions over closed intervals may be regarded as a map

$$I\colon C \times \mathbb{R} \times \mathbb{R} \to \mathbb{R}$$
$$(f, a, b) \mapsto \int_a^b f(t)\, dt.$$

For maps in general, the value of a map f at an element x of its domain is denoted $f(x)$—the function name is a *prefix*, and the argument x is enclosed in parentheses. For particular maps, there are many variations. The parentheses may simply be omitted, as in $\cos x$. The function name may be a suffix, as with the factorial function ($n!$); or a raised suffix, as in transposition to interchange rows and columns of a matrix (M' or M^T). It may be placed above the argument, as in conjugation of complex numbers ($\bar{z}$); or may wrap around from the left to above the argument, as for square root ($\sqrt{x}$). Sometimes, even, the function name may be an infix between parts of the argument, as for addition ($x + y$). Yes, addition is really a function—it just happens to take a pair of numbers as argument; instead of writing $+((x, y))$ we normally write $x + y$.

Definition B.1.6. If X is a set, then a map from $X \times X$ to X is called an **operation on** X.

When expressed in the usual, prefix notation for functions, the commutative law of addition (for real numbers, say) is $+(x, y) = +(y, x)$. Such expressions do look peculiar, but they are correct.

Exercises B.1.7. (1) Let $X = \{1, 2, 3\}$ and $Y = \{-1, -2\}$. Construct all maps from X to Y and all maps from Y to X. Find the range of each.

(2) Determine the range of $f: \mathbb{R} \to \mathbb{R}$ if:

    (a) $f(x) = 2x + x^2$.

    (b) $f(x) = |x|$.

(3) Let $f: X \to Y$ be a map and let $A \subset B \subset X$. Show that $(f|_B)|_A = f|_A$.

(4) Use prefix notation for functions to express the associative law $(x + y) + z = x + (y + z)$ and distributive law $x(y + z) = xy + xz$ about the operations of addition and multiplication of natural (or real) numbers.

(5) Let I be any set and let Δ be the *diagonal* of $I \times I$, that is, $\Delta = \{(i, i) : i \in I\}$. What is the characteristic function c_Δ of Δ in $I \times I$?

(6) Let X be a set.

    (a) Show that two subsets S and E of X are equal if and only if their characteristic functions c_S and c_E are equal.

    (b) For subsets A and B of X, express the characteristic function $c_{A \triangle B}$ in terms of c_A and c_B. Here $A \triangle B$ is the symmetric difference of A and B—see Exercise A.1.5 (5). See also Exercise 4.1.37.

(c) Prove that symmetric difference is associative, that is, $A \triangle (B \triangle C) = (A \triangle B) \triangle C$ for all subsets A, B, and C of X, by proving and using equality of corresponding characteristic functions. [*Note:* This is a considerably easier way to prove associativity of symmetric difference than doing it directly as in Exercise A.1.5 (5).]

(7) Express taking derivatives as a map.

(8) For each continuous function f and each real number x, we may form the integral $\int_0^x f(t)\, dt$. Express forming such integrals as a map.

Instead of forming the image under a map of a single element of its domain, we may form the set of all images of elements of a subset of the domain. Also, for a subset of the codomain, we may form the set of all elements of the domain whose images are in the given subset of the codomain.

Definition B.1.8. Let $f : X \to Y$ be a map. For $A \subset X$, its **image under** f is the subset $f(A)$ of Y defined by

$$f(A) = \{ f(x) : x \in A \}.$$

For $D \subset Y$, its **preimage under** f is the subset $f^{-1}(D)$ of X defined by

$$f^{-1}(D) = \{ x \in X : f(x) \in D \}.$$

For example, if $X = \{ x \in \mathbb{R} : x \geq 0 \}$ and $f : X \to \mathbb{R}$ is given by $f(x) = \sqrt{x}$, then $f([9, 25]) = [3, 5]$ and $f^{-1}([3, 5]) = [9, 25]$. If $g : \mathbb{R} \to \mathbb{R}$ is given by $g(x) = x^2$, then $g([3, 5]) = [9, 25]$ but $g^{-1}([9, 25]) = [-5, -3] \cup [3, 5]$.

For any map $f : X \to Y$, we have

$$\text{range } f = f(X) = f(\text{dom } f).$$

Exercises B.1.9. (1) Determine $f^{-1}(\{0\})$, $f^{-1}(\{1\})$, and $f^{-1}([0, 1])$ for the function $f : \mathbb{R} \to \mathbb{R}$ given by $f(x) = \sin 2\pi x$.

(2) Let $f : X \to Y$ be a map.

(a) What set is $f(\{x\})$ for $x \in X$?

(b) Complete the following proof that $f(A \cap B) \subset f(A) \cap f(B)$ for all subsets A and B of X: Let $A, B \subset X$. Let $y \in f(A \cap B)$. Then $y = f(x)$ for some $x \in A \cap B$. Since $x \in A$, we have $f(x) \in f(A)$.

(c) Give an example of a map $f\colon \mathbb{R} \to \mathbb{R}$ and subsets A and B of $\mathbb{R}$ such that $f(A \cap B) \neq f(A) \cap f(B)$.

(d) Prove that $f(A \cup B) \subset f(A) \cup f(B)$ for all subsets A and B of X. Must equality hold?

(e) Describe in symbols and words what $f^{-1}(\{y\})$ is for $y \in Y$. When is $f^{-1}(\{y\})$ nonempty?

(f) Complete the following proof that $f^{-1}(D \cup E) = f^{-1}(D) \cup f^{-1}(E)$ for all $D, E \subset Y$: Let $D, E \subset Y$. Assume $x \in f^{-1}(D \cup E)$. Then $f(x) \in D \cup E$, and so $f(x) \in D$ or $f(x) \in E$. If $f(x) \in D$, then $\ldots$; if $f(x) \in E$, then $\ldots$. It follows that $x \in f^{-1}(D) \cup f^{-1}(E)$. Conversely, assume $x \in f^{-1}(D) \cup f^{-1}(E)$. Then $\ldots$.

(g) If $A \subset X$, show that $A \subset f^{-1}(f(A))$. Must equality hold?

(h) If $D \subset Y$, show that $f(f^{-1}(D)) \subset D$. Must equality hold?

(i) For a subset D of Y, what is the relationship between $f^{-1}(Y \setminus D)$ and $X \setminus f^{-1}(D)$?

(j) Prove that $f^{-1}(Y \setminus f(X \setminus A)) \subset A$ for all $A \subset X$.

(3) For sets X and Y, let $p\colon X \times Y \to X$ and $q\colon X \times Y \to Y$ be the projections. Must $E = p(E) \times q(E)$ for every nonempty $E \subset X \times Y$?

(4) Let $f\colon X \to Y$ be a given map. Construct a map whose graph f^* satisfies $f^*(D) = f^{-1}(D)$ for every $D \subset Y$.

For everyday work with functions, the tentative formulation of Definition B.1.1 is ordinarily sufficient. Sometimes, though, a more precise definition is required.[3] After all, what do we mean—really—when we say that the graph of a function is a rule? The answer can be precisely expressed in terms of relations (see Section A.3).

The rule f of a function $f\colon X \to Y$ is just a *relation f in X to Y* with the following two properties:

- For each $x \in X$, there is at least one $y \in Y$ such that x is related to y by f.

- For each $x \in X$ and for all $y_1, y_2 \in Y$, if x is related to both y_1 and y_2 by f, then $y_1 = y_2$.

Now an arbitrary relation in X to Y is a subset of the product $X \times Y$, that is, a set of pairs (x, y) with $x \in X$ and $y \in Y$. Then the relation f constituting the rule of a function $f\colon X \to Y$ is the kind named in the next definition.

---

[3]This is the case, for example, in the proof of the Ordinary Recursion Principle (see Exercise B.4.2).

Definition B.1.10. Let X and Y be sets. A relation f from X to Y is said to be **functional** when:

- For each $x \in X$, there exists at least one $y \in Y$ such that $(x, y) \in f$.

- For each $x \in X$ and for all $y_1, y_2 \in Y$, if $(x, y_1) \in f$ and $(x, y_2) \in f$, then $y_1 = y_2$.

In other words, f is functional when, for each $x \in X$, there is exactly one $y \in Y$ for which $(x, y) \in f$.

At last we can make completely precise our earlier, tentative definition of function.

Definition B.1.11 (Function—Precise Formulation). A **function**, also called a **map**, is an ordered triple (X, f, Y) consisting of a set X called the **domain** of the function, a functional relation f from X to Y called the **graph** of the function, and a set Y called the **codomain** of the function. Such an ordered triple is also denoted by $f: X \to Y$ and is said to be a function **from X to Y.**

Thus the graph of a function $f: X \to Y$ *is* the subset

$$\{(x, f(x)) : x \in X\}$$

of $X \times Y$. For each $x \in X$, the value $f(x)$ of f at x is the *unique* $y \in Y$ for which $(x, y) \in f$.

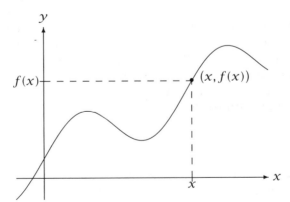

Figure B.2: Graph of a function.

Regarding the rule of a function this way as a certain kind of relation is what you are accustomed to doing when you plot the graph of a function in the plane, as in Figure B.2. (That is why we used the term "graph" in the first place.) Visualizing the function's graph as a subset of the plane can be useful even when the function's domain or codomain is not actually a subset of $\mathbb{R}$.

A functional relation f whose domain is only a subset of a set X and whose range is a subset of a set Y is called a **partial function from** X **to** Y. For example, tan is a partial function in $\mathbb{R}$ to $\mathbb{R}$ (the odd multiples of $\pi/2$ do not belong to its domain). Subtraction of natural numbers is a partial function from $\mathbb{N} \times \mathbb{N}$ to $\mathbb{N}$—an operation *in* rather than *on* $\mathbb{N}$— because the difference $m - n$ of natural numbers is a natural number only when $m \geq n$.

Exercises B.1.12. (1) Suppose $f: X \to Y$ is a map and $A \subset X$. As a relation, what is the graph of the restriction $f|_A: A \to Y$ of f to A?

(2) Which of the following are functional relations from $\mathbb{R}$ to $\mathbb{R}$?

  (a) $\{ (x, y) \in \mathbb{R} \times \mathbb{R} : x^2 + y^2 = 1 \text{ and } y \geq 0 \}$.

  (b) $\{ (x, y) \in \mathbb{R} \times \mathbb{R} : x^2 + y^2 = 1 \text{ and } x > 0 \}$.

  (c) $\{ (x, y) \in \mathbb{R} \times \mathbb{R} : y^3 = x^2 \}$.

(3) If Y is a set, what maps $\varnothing \to Y$ are there? What is the graph of each (if there are any)?

(4) Let C be the set of all continuous functions from $\mathbb{R}$ to $\mathbb{R}$ and let $\mathcal{D}$ be the set of all differentiable functions from $\mathbb{R}$ to $\mathbb{R}$. Is the relation

$$\{ (f, g) \in C \times \mathcal{D} : g' = f \}$$

a functional relation from C to $\mathcal{D}$? (The prime denotes taking the derivative.)

B.2 Composition of Functions

The outputs of one function can serve as the inputs to another. For example, a typical output $\sin x$ of the sine function can serve as the input to the natural logarithm function log, giving the combined value $\log(\sin x)$—provided, of course, that the output $\sin x$ of the first function is a legitimate input of the second. Now $\log y$ is not defined (at least in terms of real numbers) unless $y > 0$, and so to form $\log(\sin x)$, we must take care that $\sin x > 0$. So we should restrict the domain of sin to, say, the open interval $X = (0, \pi)$. For $x \in X$, then, $\sin x \in Y$, where $Y = \{ y \in \mathbb{R} : y > 0 \}$, that is, Y is the open ray $(0, \infty)$. True, not every number in Y is $\sin x$ for some suitable $x \in X$ (why?), but definitely each number in Y *is* a legitimate input to log. Then for $x \in X$, the combined value $\log(\sin x) \in Z$, where $Z = \mathbb{R}$. Thus, to form a function with domain X a subset of $\mathbb{R}$ and taking at an $x \in X$ the value $\log(\sin x)$, we should consider the two functions

$$f: x \to Y \qquad \text{and} \qquad f: Y \to Z$$
$$x \mapsto \sin x \qquad\qquad\qquad y \mapsto \log y,$$

where

$$X = (0, \pi), \quad Y = (0, \infty), \quad Z = \mathbb{R},$$

for which the codomain of the first function f is the same as the domain of the second function g.

The same considerations apply whenever we want to combine two functions by applying one and then the other.

Definition B.2.1. Let $f: X \to Y$ and $g: Y \to Z$ be two maps with the codomain of f the same as the domain of g. Then their **composite** is the map $g \circ f: X \to Z$ whose domain is the domain of f, whose codomain is the codomain of g, and which is defined by

$$(g \circ f)(x) = g(f(x)) \qquad (x \in X).$$

Simply put, the composite $g \circ f$ assigns to each $x \in \operatorname{dom} f$ the value of g at the element $f(x) \in \operatorname{dom} g$.

Example B.2.2. Let $\mathbb{R}^+ = \{\, y \in \mathbb{R} : y \geq 0 \,\}$, that is, $\mathbb{R}^+$ is the closed ray $[0, \infty)$. Then the composite of

$$f: \mathbb{R} \to \mathbb{R}^+ \qquad \text{and} \qquad f: \mathbb{R}^+ \to \mathbb{R}$$
$$x \mapsto e^x \qquad\qquad\qquad y \mapsto 2y$$

is

$$g \circ f: \mathbb{R} \to \mathbb{R}$$
$$x \mapsto 2e^x.$$

Their composite in the opposite order is

$$f \circ g: \mathbb{R}^+ \to \mathbb{R}^+$$
$$y \mapsto e^{2y}.$$

Evidently here $g \circ f \neq f \circ g$. Thus:

Composition of functions is not in general commutative!

Exercises B.2.3.   (1) Let $f, g: \mathbb{R} \to \mathbb{R}$ be the functions given by $f(x) = 2x + x^2$ and $g(x) = |x|$. Determine $\operatorname{range}(g \circ f)$ and $\operatorname{range}(f \circ g)$.

(2) Find a set X and functions $f: X \to X$ and $g: X \to X$ such that $g \circ f \neq f \circ g$. (In the preceding example, the two composites could not possibly have been the same since their domains were different and their codomains were different.)

Although composition of maps is not commutative, it is associative, according to part of the following proposition.

Proposition B.2.4.   *1. If $f: X \to Y$ is any map, then*

$$f \circ i_X = f, \qquad i_Y \circ f = f$$

(where i_X and i_Y are the identity maps of X and Y, respectively).

2. If $f: X \to Y$, $g: Y \to Z$, and $h: Z \to W$ are maps, then

$$(h \circ g) \circ f = h \circ (g \circ f).$$

Proof.   1. Exercise.

2. Let f, g, and h be maps with the stated domains and codomains. To show that $(h \circ g) \circ f$ and $h \circ (g \circ f)$ are the same, we must show they have the same domain, have the same codomain, and take the same value at each element of their domain.

That the domains are the same and the codomains are the same is probably obvious, but let us spell out the reasons nonetheless. First, $\mathrm{dom}(h \circ g) = \mathrm{dom}\, g = Y = \mathrm{codom}\, f$. Hence $\mathrm{dom}((h \circ g) \circ f) = \mathrm{dom}\, f = X$. Similarly, $\mathrm{dom}(h \circ (g \circ f)) = X$. Second, $\mathrm{codom}(h \circ g) = \mathrm{codom}\, h = W$, and so $\mathrm{codom}((h \circ g) \circ f) = W$, too. Similarly, $\mathrm{codom}(h \circ (g \circ f)) = W$.

Let $x \in X$. Then

$$\begin{aligned}
((h \circ g) \circ f)(x) &= (h \circ g)(f(x)) \\
&= h(g(f(x))) \\
&= h((g \circ f)(x)) \\
&= (h \circ (g \circ f))(x). \quad \square
\end{aligned}$$

The relationship of two maps $f: X \to Y$ and $g: Y \to Z$ to their composite $h = g \circ f: X \to Z$ may be displayed in the triangular diagram:

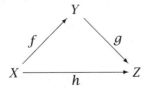

In general, such a diagram shows where the maps are from and where they are to. The diagram is said to be **commutative** when

$$g \circ f = h,$$

in other words, for arbitrary $x \in X$, the result $g(f(x))$ of applying first f and then g on the path from X to Z through Y is the same as the result $h(x)$ of applying h to x on the direct path from X to Y.

Similarly, four maps $f: X \to Y$, $g: Z \to W$, $h: X \to Z$, and $k: Y \to W$ may be displayed in the square diagram:

Such a diagram is said to be **commutative** when

$$g \circ h = k \circ f,$$

in other words, for arbitrary $x \in X$, the result $g(h(x))$ of applying first h and then g on the path from X to W through Z is the same as the result $k(f(x))$ of applying first f and then k on the path from X to W through Y. More complicated diagrams of maps are said to be commutative when their component triangles and squares are commutative.

Examples B.2.5.   (1) Let $f: X \to Y$ be a map, A be a subset of X, and $j: A \to X$ be the inclusion map. Then the triangular diagram

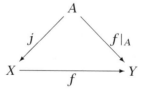

is commutative, that is, $f|_A = f \circ j$ (why?).

(2) Let $f: X \to Y$ be a map. Define a map $h: X \to \mathcal{P}(X)$ [where $\mathcal{P}(X)$ is the power set of X—the set of all subsets of X] by $h(x) = \{x\}$ for each $x \in X$. Similarly define $k: Y \to \mathcal{P}(Y)$. Let $f_*: \mathcal{P}(X) \to \mathcal{P}(Y)$ be the map defined by

$$f_*(A) = f(A) = \text{the image of } A \text{ under } f \qquad (A \in \mathcal{P}(X)).$$

Then the square diagram

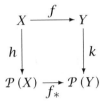

is commutative. In fact, for each $x \in X$,

$$(f_* \circ h)(x) = f_*(h(x)) = f_*(\{x\}) = f(\{x\}) = \{f(x)\}$$
$$= k(f(x)) = (k \circ f)(x).$$

Exercises B.2.6. (1) Let $f: X \to Y$ and $g: Y \to Z$ be maps and $A \subset X$. Draw a commutative diagram (which may consist of more than one triangle or square) illustrating the fact that $(g \circ f)|_A = g \circ (f|_A)$ and prove that this equality of maps is correct. (*Hint:* Express the restriction $f|_A$ in terms of the inclusion map $j: A \to X$.)

(2) Let $p: X \times Y$ and $q: X \times Y$ be the projections for the product of two sets X and Y. Suppose $f: Z \to X$ and $g: Z \to Y$ are maps. Show that there exists a unique map $h: Z \to X \times Y$ for which the following diagram is commutative:

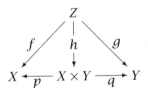

B.3   Injections and Surjections

To each element of the domain X of a map $f: X \to Y$, some element of the codomain Y is related. However, it need not be the case that each element of the codomain is related by f to some element of the domain. In other words, each element of the domain *has* an image, but it need not be the case that each element of the codomain *is* an image. For example, the squaring function $f: \mathbb{R} \to \mathbb{R}$ defined by $f(x) = x^2$ takes only nonnegative values, and so no negative number in the codomain $\mathbb{R}$ of this map is an image of any element of the domain.

   When each element of the codomain *is* an image, we have a special type of map.

Definition B.3.1. A map $f: X \to Y$ is said to be **surjective** and is called a **surjection**, when:

For each $y \in Y$, there is at least one $x \in X$ such that $y = f(x)$.

In this case, we also say that f maps X **onto** Y.

The word "onto" is a preposition, not an adjective, so please do not commit the grammatical atrocity of saying that this or that map "is onto".

Another way of saying that a map is surjective is that its range equals its codomain. Then it is easy to manufacture surjections: take any map whatsoever and replace its codomain with its range $f(X)$. For example, the squaring map $f: \mathbb{R} \to \mathbb{R}$ is not surjective, but its codomain restriction $f: \mathbb{R} \to \{y \in \mathbb{R} : y \geq 0\}$ is surjective.

Given any map $f: X \to Y$ whatsoever, we can, by that process, manufacture a new map $f: X \to f(X)$ that is surjective and is given by the same rule—that is, functional relation—f as the given map. Then why bother with any codomain other than the range in the first place? The answer was suggested earlier: For a given set of inputs—a domain—it may be quite easy to identify a set to which all the outputs of a given functional relation belong—a codomain—but it may not be possible to decide without further analysis what precisely the set of all the outputs is.

By definition, a map $f: X \to Y$ relates only one $y \in Y$ to any given $x \in X$; that is, distinct elements of Y cannot be related to the same element of X. However, f can relate the same element of Y to distinct elements of X. In other words, f assigns only a single value to any given element of its domain, but f may assign the same value to different elements of its domain. For example, the squaring map $f: \mathbb{R} \to \mathbb{R}$ has $f(-2) = 4 = f(2)$.

When distinct elements of the domain are never mapped to the same element of the codomain, then again we have a special type of map.

Definition B.3.2. A map $f: X \to Y$ is said to be **injective** and is called an **injection** when:

For all $x_1, x_2 \in X$, if $x_1 \neq x_2$, then $f(x_1) \neq f(x_2)$.

In this case, we also say that f is **one-to-one** (abbreviated 1-1).

The condition above that a map be injective may be written equivalently as:

For $x_1, x_2 \in X$, if $f(x_1) = f(x_2)$, then $x_1 = x_2$.

When working with injections, sometimes you will find the original definition more convenient, and sometimes the equivalent form.

Examples B.3.3.   (1) The squaring function

$$f: \mathbb{R} \to \mathbb{R}$$
$$x \mapsto x^2$$

is neither injective nor surjective.

(2) The squaring function

$$f: \mathbb{R}^+ \to \mathbb{R}$$
$$x \mapsto x^2$$

with domain the set $\mathbb{R}^+ = \{ x \in \mathbb{R} : x \geq 0 \}$ of nonnegative real numbers is injective but not surjective.

(3) The squaring function

$$f: \mathbb{R} \to \mathbb{R}^+$$
$$x \mapsto x^2$$

is surjective but not injective.

(4) The squaring function

$$f: \mathbb{R}^+ \to \mathbb{R}^+$$
$$x \mapsto x^2$$

is both injective and surjective.

(5) The cubing function

$$f: \mathbb{R} \to \mathbb{R}$$
$$x \mapsto x^3$$

is both injective and surjective.

(6) For any set X, its identity function

$$i_X: X \to X$$
$$x \mapsto x$$

is both injective and surjective.

Exercises B.3.4. Is the map injective? Is it surjective?

(1) The successor function $\sigma: \mathbb{N} \to \mathbb{N}$ given by $\sigma(n) = n + 1$.

(2) The function $\tau: \mathbb{Z} \to \mathbb{Z}$ given by $\tau(n) = n + 1$.

(3) For a set X and a subset A of X, the inclusion map $j: A \to X$ defined by $j(x) = x$.

(4) For any two sets X and Y, the first projection

$$p: X \times Y \to X$$
$$(x, y) \mapsto x.$$

Likewise the second projection $(x, y) \mapsto y$ of $X \times Y \to Y$.

(5) For an integer $m > 1$, the map

$$f: \mathbb{Z} \to \mathbb{Z}_m$$
$$n \mapsto [n],$$

where, for each integer n, $[n]$ denotes its congruence class modulo m.

(6) The same as the map in (5) except that $m = 1$.

Maps that are both injective and surjective deserve a special name.

Definition B.3.5. A map is said to be **bijective** and is called a **bijection** when it is both injective and surjective.

Proposition B.3.6. *Let $f: X \to Y$ and $g: Y \to Z$ be maps.*

1. If f and g are both injective, then so is their composite $g \circ f$.

2. If f and g are both surjective, then so is their composite $g \circ f$.

3. If f and g are both bijective, then so is their composite $g \circ f$.

Proof. 1. Assume f and g are injective. Let $x_1, x_2 \in X$. Suppose $(g \circ f)(x_1) = (g \circ f)(x_2)$. We must show that $x_1 = x_2$. (Finish the proof.)
    2 and 3. Exercise.   □

Exercise B.3.7. Let $f: X \to Y$ and $g: Y \to Z$ be maps whose composite $g \circ f: X \to Z$ is bijective. Prove that then f must be injective and g must be surjective.

A map $f: X \to Y$ is bijective if and only if both of the following properties hold:

- The map f is injective; that is, for each $y \in Y$, there is at most one $x \in X$ such that $f(x) = y$.

- The map f is surjective; that is, for each $y \in Y$, there is at least one $x \in X$ such that $f(x) = y$.

In other words, f is bijective if and only if:

> For each $y \in Y$, there is exactly one $x \in X$ such that $f(x) = y$.   (*)

Thus, a bijection matches up—pairs off—elements of its domain with elements of its codomain. For this reason, a bijection from X to Y is also called a **one-to-one correspondence** between X and Y.

Characterization (*) of bijections justifies the following definition.

Definition B.3.8. Let $f : X \to Y$ be a bijection. Then the map $g : Y \to X$ defined by

$$g(y) = \text{the } x \in X \text{ such that } f(x) = y \quad (y \in Y)$$

is called the **inverse** of f and is denoted by f^{-1}.

For example, the inverse of the squaring function $f : \mathbb{R}^+ \to \mathbb{R}^+, x \mapsto x^2$ is the square root function $f^{-1} : \mathbb{R}^+ \to \mathbb{R}^+, y \mapsto \sqrt{y}$.

For a bijection $f : X \to Y$, its inverse $f^{-1} : Y \to X$ satisfies

> $$f^{-1}(y) = x \iff f(x) = y \quad (x \in X, y \in Y).$$

In this sense, the inverse f^{-1} of a bijection $f : X \to Y$ undoes whatever f does. Indeed,

$$f^{-1}(f(x)) = x \quad (x \in X),$$
$$f(f^{-1}(y)) = y \quad (y \in Y).$$

In terms of composites, we can say that more concisely (without referring explicitly to elements of the domain or codomain):

> $$f^{-1} \circ f = i_X, \quad f \circ f^{-1} = i_Y.$$

Does the inverse of a bijection itself have an inverse? Think about the intuitive meaning of the terms before studying the following proposition.

Proposition B.3.9.   *1. If $f : X \to Y$ is a bijection, then so is its inverse $f^{-1} : Y \to X$, and $(f^{-1})^{-1} = f$.*

2. If $f : X \to Y$ and $g : Y \to Z$ are bijections, then so is their composite $g \circ f : X \to Z$, and $(g \circ f)^{-1} = f^{-1} \circ g^{-1}$.

Proof. Exercise.   □

Notice that $(g \circ f)^{-1} = g^{-1} \circ f^{-1}$ is ordinarily *not* true for bijections f and g.

Exercise B.3.10. For which real values of α and β is the affine linear map $f\colon \mathbb{R} \to \mathbb{R}$ given by $f(x) = \alpha x + \beta$

(a) constant?

(b) injective?

(c) surjective?

(d) bijective? In this case, give a formula for $f^{-1}(y)$ in terms of α and β.

Using f^{-1} to denote the inverse of a bijection and in the definition of preimages can be a source of confusion. If $f\colon X \to Y$ is a bijection, then for $y \in Y$, the image $f^{-1}(y)$ of y under the inverse $f^{-1}\colon Y \to X$ is an element of X, namely, the unique $x \in X$ with $f(x) = y$. By way of contrast, the preimage $f^{-1}(\{y\})$ of the one-element subset $\{y\}$ of Y is, in this case of a bijection, the one-element subset $\{x\}$ of X.

Exercise B.3.11. Let $f\colon X \to Y$ be an injection. If $g_1, g_2\colon Z \to X$ are maps such that $f \circ g_1 = f \circ g_2$, show that then $g_1 = g_2$.

Sometimes, the easiest way to check that a given map from a set X to a set Y is bijective is to find a map that "goes backward" from Y to X and "undoes" f in precisely the way that f^{-1} would.

Proposition B.3.12. *Let $f\colon X \to Y$ and $g\colon Y \to X$ be maps such that*

$$g \circ f = i_X \quad and \quad f \circ g = i_Y.$$

Then f is bijective, and $g = f^{-1}$.

Proof. Exercise. $\square$

Exercises B.3.13.   (1) Let $f\colon X \to Y$ be a map. Suppose there exists some map $s\colon Y \to X$ such that $f \circ s = i_Y$. Show that then f must be surjective.

(2) Let $f\colon X \to Y$ and $g\colon Y \to Z$ be maps. Prove: If $g \circ f$ is injective, then f is injective. Is the converse true?

(3) Let $f\colon X \to Y$ be a map. Prove: If there exists a map $r\colon Y \to X$ such that $r \circ f = i_X$, then f is injective. Is the converse true?

(4) Let $f\colon (2,4) \to (2,4)$ be a bijection and $g = f^{-1}$. Suppose $f(x) \cdot g(x) = x^2$ for every $x \in (2,4)$. Prove that $f(3) = g(3) = 3$.

B.4 Sequences and Families

The idea of a *sequence* is that it is a list with an initial entry s_0, a next entry s_1, a next entry s_2 after that, and so forth, that is, a list that goes on forever. But what, exactly, is meant by "a list that goes on forever"? One answer is that a list is something of the form

$$s_0, s_1, s_2, \ldots . \qquad (*)$$

But what *is* something of the form (*)—really?

The subscripts in the list (*) indicate that the successive entries can be numbered 0, 1, 2, etc.[4] The numbering accomplished by these subscripts assigns to each natural number n an object s_n like this:

$$
\begin{array}{ccccccc}
0 & 1 & 2 & \ldots & n & \ldots \\
\downarrow & \downarrow & \downarrow & & \downarrow & \\
s_0 & s_1 & s_2 & \ldots & s_n & \ldots
\end{array}
$$

In other words, such a numbering may be represented as a *function* $n \mapsto s_n$ with domain the set $\mathbb{N}$ of natural numbers.

Definition B.4.1. Let X be a set. A **sequence** in X is a function $s \colon \mathbb{N} \to X$. For each $n \in \mathbb{N}$, the value $s(n)$ of the sequence at n is called the nth **entry** in the sequence and is also denoted with a subscript as s_n. The sequence s itself is variously denoted by $(s_n)_{n \in \mathbb{N}}$ or $(s_n)_{n=0,1,2,\ldots}$ or even $(s_0, s_1, s_2, \ldots)$.

For sequences in which the initial entry is numbered 1 instead of 0, the notation would be $(s_n)_{n \in \mathbb{N}^*}$ or $(s_n)_{n=1,2,\ldots}$ or $(s_1, s_2, s_3, \ldots)$. Numbering origins other than 0 and 1 are possible, too.

Nothing in the definition above precludes that the values s_n may be the same for different subscripts n. If $s_i \neq s_j$ whenever $i \neq j$, then we shall say that the sequence $s = (s_n)_{n \in \mathbb{N}}$ has **distinct** entries. In other words, the sequence has distinct entries when the function $s \colon \mathbb{N} \to X$ is injective.

The individual entries in a sequence need not be numbers but may be sets of numbers or other objects. We could have, for example, a sequence $([a_n, b_n])_{n \in \mathbb{N}}$ of closed intervals in $\mathbb{R}$—say $[a_n, b_n] = [1/(n+1), n/(n+1)]$ for each n. Or we could have a sequence $(f_n)_{n \in \mathbb{N}}$ of functions—say $f_n(t) = t^n$ (for every $t \in \mathbb{R}$) for each n.

To define a specific sequence, one may specify the rule $n \mapsto s_n$ explicitly, as for any function. For example, $s_n = n/(n+1)$. Or, one may

---

[4]You may wonder, "Why not give the number 1 to the first entry in a list?" In many contexts it is convenient to begin with 0 instead. Suppose, for example, the entries in the list represent the distances, at successive seconds of time, of a moving object from some reference point; then it is natural to measure the elapsed time, starting with initial time 0, and accordingly denote the distance at the start by s_0.

specify the rule implicitly, with an initial condition and a recurrence relation. For example,

$$\begin{cases} s_0 & = 1, \\ s_{n+1} = 3s_n^2 - 4s_n + 5 & (n \geq 0). \end{cases}$$

This seems to define the sequence "in terms of itself." That a sequence meeting such an implicit specification actually exists must be justified. The justification is furnished by the Ordinary Recursion Principle— Theorem 1.2.9.

Recall the statement of that theorem:

> Let X be a set, let $z \in X$ be a given element, and let $G\colon X \to X$ be a given function. Then there exists one and only one sequence $(s_n)_{n \in \mathbb{N}}$ in X such that
>
> $$\begin{cases} s_0 & = z, \\ s_{n+1} = G(s_n) & (n \in \mathbb{N}). \end{cases}$$

The following exercise outlines a proof of the Ordinary Recursion Principle. To facilitate the proof, it is desirable to reformulate the theorem's conclusion explicitly in terms of functions: There exists one and only one function $s\colon \mathbb{N} \to X$ such that

$$\begin{cases} s(0) = z, \\ s(n + 1) = G(s(n)) & (n \in \mathbb{N}). \end{cases}$$

To complete the proof of the Ordinary Recursion Principle outlined in the following exercise, you will need the precise definition of a map (Definition B.1.11) in terms of functional relations.

Exercise B.4.2. Given a set X, an element $z \in X$, and a function $G\colon X \to X$, prove that there exists one and only one function $s\colon \mathbb{N} \to X$ as above by showing the following.

(a) Let T be the set of all functions t with domain $\{0, 1, \ldots, n\}$ for some $n \in \mathbb{N}$ and codomain X that satisfy $t(0) = z$ and $t(k + 1) = G(t(k))$ for all $k = 0, 1, \ldots, n$. If $t, u \in T$ with $\mathrm{dom}\, t \subset \mathrm{dom}\, u$, then u is an extension of t.

(b) Regarding each $t \in T$ as a relation from $\mathbb{N}$ to X, that is, as a subset of $\mathbb{N} \times X$, let $s = \bigcup_{t \in T} t$; in other words, let s be the subset of $\mathbb{N} \times X$ consisting of the union of the graphs of all the functions t belonging to T. Then the relation s from $\mathbb{N}$ to X is a functional relation.

(c) The function s has domain $\mathbb{N}$.

(d) The function $s: \mathbb{N} \to X$ satisfies $s(0) = z$ and $s(n + 1) = G(s(n))$ for all $n \in \mathbb{N}$.

This establishes the existence of the desired s.

(e) Such an s must be unique.

A sequence, as formally defined above, represents an infinite list—one whose entries "go on forever." What about a finite list—one in which the entries eventually stop? We are talking about something of the form

$$x_1, x_2, \ldots, x_n \tag{*}$$

for a fixed positive integer n. The formal representation of such a finite list is similar to that for a sequence.

Definition B.4.3. Let X be a set and $n \in \mathbb{N}^*$. An **n-tuple in** X is a function $x: \{1, 2, \ldots, n\} \to X$. For each $j \in \{1, 2, \ldots, n\}$, the value $x(j)$ of the n-tuple at j is called the **jth entry** in the n-tuple and is also denoted by a subscript as x_j. The n-tuple itself is variously denoted by $(x_j)_{j \in \{1,2,\ldots,n\}}$, $(x_j)_{j=1,2,\ldots,n}$, or $(x_j)_{1 \le j \le n}$. The set of all n-tuples in a set X is denoted by X^n.

It follows from the meaning of equality of functions that, for a given positive integer n and two n-tuples $(x_j)_{j=1,2,\ldots,n}$ and $(y_j)_{j=1,2,\ldots,n}$ in a set X,

$$(x_j)_{j=1,2,\ldots,n} = (y_j)_{j=1,2,\ldots,n} \iff (\forall j \in \{1, 2, \ldots, n\})\, (x_j = y_j).$$

Let us examine, in particular, the case $n = 2$. Consider two 2-tuples—can you say that quickly 10 times?—$(x_j)_{j=1,2}$ and $(y_j)_{j=1,2}$ in a set X. Then

$$(x_j)_{j=1,2} = (y_j)_{j=1,2} \iff x_1 = y_1 \, \& \, x_2 = y_2.$$

This is suspiciously like the fundamental property

$$(x_1, x_2) = (y_1, y_2) \iff x_1 = y_1 \, \& \, x_2 = y_2$$

of *ordered pairs* of elements in a set—see Proposition A.2.2. Now an ordered pair (x_1, x_2) is, by definition, the set $\{\{x_1\}, \{x_1, x_2\}\}$, whereas the 2-tuple $(x_j)_{j=1,2}$ is a map with domain $\{1, 2\}$. These are two different representations of the same concept—two objects in an order, first x_1 and then x_2. The two representations ought to be essentially the same. In fact, they are. There is a natural one-to-one correspondence between the set X^2 of all 2-tuples in X and the set $X \times X$ of all ordered pairs of elements of X.

Proposition B.4.4. *Let X be a set. Then the map*

$$\varphi: X^2 \to X \times X$$
$$x \mapsto (x_1, x_2)$$

is a bijection.

Proof. For each ordered pair $(x, y) \in X \times X$, let $c_{x,y}$ denote the map $\{1, 2\} \to X$ defined by

$$c_{x,y}(1) = x, \qquad c_{x,y}(2) = y.$$

Then we have a map

$$\psi: X \times X \to X^2$$
$$(x, y) \mapsto c_{x,y}$$

going backward. Check that the composites $\psi \circ \varphi$ and $\varphi \circ \psi$ are the identity maps of X^2 and $X \times X$, respectively. It follows (why?) that φ is a bijection (and ψ is its inverse). $\square$

In view of the preceding proposition, we often disregard the difference between X^2 and $X \times X$ and treat them as if they were actually the same. Indeed, in Exercise 1.2.14 (4) we already denoted $X \times X$ by X^2. Thus, we may speak interchangeably of the plane as $\mathbb{R}^2$ or $\mathbb{R} \times \mathbb{R}$.

Exercise B.4.5.   (a) Construct a bijection between X^1 and X.

(b) Show that ordered triples as treated in Exercise A.2.3 (4) are essentially the same as 3-tuples.

(c) Generalize Proposition B.4.4 to the case of arbitrary $n \in \mathbb{N}^*$. [The generalization of ordered pairs and ordered triples is indicated in Exercise 1.2.14 (4) as involving a recursive definition.]

In view of the preceding proposition and exercise, there is no great harm in using (x_1, x_2), (x_1, x_2, x_3), and, in general, $(x_1, x_2, \ldots, x_n)$ as alternate notations for $(x_j)_{j=1,2}$, $(x_j)_{j=1,2,3}$, and $(x_j)_{j=1,2,\ldots,n}$, respectively.

Sequences and n-tuples have a common generalization—the notion of a *family*.

Definition B.4.6. Let X be a set. A map $x: I \to X$ to X is called a **family in X indexed by I**. The domain I of such a map is called the **index set**, and each element i of I is called an **index** of the family. For an index i, the value $x(i)$ of x at i is called the ith **entry** or ith **coordinate** of the family and is typically denoted with a subscript as x_i. Then the family x itself is denoted by $(x_i)_{i \in I}$.

Thus, a sequence in a set X is just a family in X with index set $\mathbb{N}$ (or $\mathbb{N}^*$); an n-tuple in X is just a family in X with index set $\{1, 2, \ldots, n\}$. Families with other index sets are common if you just look for them. For example, the differential equation $dx/dt = t$ has, for each real constant c, the particular solution x_c defined by $x_c(t) = t^2/2 + c$, which satisfies the initial condition $x_c(0) = c$. All these particular solutions combine to form the family $(x_c)_{c \in \mathbb{R}}$ of functions from $\mathbb{R}$ to $\mathbb{R}$. Such a family with index set $\mathbb{R}$ is called a *one-parameter family,* and the index c is then called the *parameter.*

The letter i used in the notation for a family is, of course, a dummy variable that may be replaced by any other letter not already in use for something else. You could write $(x_j)_{j \in I}$ or $(x_k)_{k \in I}$, etc.

As for any function, so for a family you must carefully distinguish between the set of values of the family and the family itself. Let $x = (x_i)_{i \in I}$ be a family in a set X. Then the set of values of this family—the range of the map x—is the set

$$\{x_i : i \in I\},$$

which may also be denoted by $\{x_i\}_{i \in I}$. The family itself, by way of contrast, is the map

$$x : I \to X$$
$$i \mapsto x_i.$$

This distinction applies, in particular, when the index set $I = \mathbb{N}$, that is, for sequences. Despite the regrettable confusion in notation you may see elsewhere, a sequence $(x_n)_{n=0,1,2,\ldots}$ is simply *not* in general the same thing as the set $\{x_n\}_{n=0,1,2,\ldots}$ of its values. Confusing the two is unforgivable—unless the sequence has distinct elements.

The general notion of family facilitates treating unions and intersections of both sequences of sets and n-tuples of sets in a unified way. When we speak of a *family of subsets* of a set X, we mean a family in the power set $\mathcal{P}(X)$.

Definition B.4.7. Let $(A_i)_{i \in I}$ be a family of subsets of a set X. Then the **intersection** and **union** of this family are the subsets

$$\bigcap_{i \in I} A_i = \{x \in X : x \in A_i \text{ for all } i \in I\}$$
$$\bigcup_{i \in I} A_i = \{x : x \in A_i \text{ for some } i \in I\}$$

of X, respectively.

Recall that trying to form the intersection of the empty collection of sets causes big trouble (see page A11). In the preceding definition, by contrast, nothing prohibits the index set I from being empty: $\bigcap_{i \in \varnothing} A_i = X$ (why?). Remember that a family of sets, being a map, has a codomain, namely, the power set $\mathcal{P}(X)$ for some particular set X.

When for some $n \in \mathbb{N}^*$, the index set I of a family of sets is the set $\{1, 2, \ldots, n\}$, then the notation $\bigcap_{j=1}^{n} A_j$ is also used for the union of the family. When $I = \mathbb{N}$, then the notation $\bigcup_{n=0}^{\infty} A_n$ is also used. Similarly for intersection.

Examples B.4.8.   (1) Let $(A_i)_{i=1,2}$ be a family of two subsets of a set X. Then

$$\bigcap_{i=1}^{2} A_i = A_1 \cap A_2, \qquad \bigcup_{i=1}^{2} A_i = A_1 \cup A_2.$$

Thus the intersection and union of a family of sets generalize the intersection and union, respectively, of two sets as defined in Appendix A.

(2) Let

$$A_j = \left(-\frac{1}{n}, \frac{n}{n+1} \right) \qquad (j = 1, 2, \ldots).$$

Then

$$\bigcap_{j=1}^{\infty} A_j = [0, 1/2), \qquad \bigcup_{j=1}^{\infty} A_j = (-1, 1),$$

and

$$\bigcap_{j=1}^{3} A_j = (-1/3, 1/2), \qquad \bigcup_{j=1}^{3} A_j = (-1, 3/4).$$

Exercises B.4.9.   (1) Let $A_1 = \{0, 1\}$, $A_2 = \{1, 2\}$, $A_3 = \{2, 3\}$, and $A_n = \{3\}$ for each integer $n \geq 4$. Determine $\bigcap_{j=1}^{3} A_j$, $\bigcup_{j=1}^{3} A_j$, $\bigcap_{j=1}^{\infty} A_j$, and $\bigcup_{j=1}^{\infty} A_j$.

(2) Call a subset K of the plane $\mathbb{R}^2$ *convex* when, for any two points a and b of K, the entire line segment joining a to b is contained in K. (For example, the unit square $[0, 1] \times [0, 1]$ of $\mathbb{R}^2$ is convex, whereas the "punctured plane" $\mathbb{R}^2 \setminus \{(0, 0)\}$ is not.) Suppose $(K_i)_{i \in I}$ is a family of convex subsets of $\mathbb{R}^2$ indexed by a nonempty set I. Must $\bigcap_{i \in I} K_i$ and $\bigcup_{i \in I} K_i$ also be convex?

(3) Let $(A_i)_{i \in I}$ be a family of subsets of a set X indexed by a nonempty set I. Prove the *generalized De Morgan law*

$$X \setminus \bigcup_{i \in I} A_i = \bigcap_{i \in I} (X \setminus A_i)$$

as well as the corresponding generalized De Morgan law where union and intersection are interchanged.

(4) Let $(A_n)_{n \in \mathbb{N}}$ be a sequence of subsets of a set X. Let

$$L = \bigcup_{i=0}^{\infty} \bigcap_{j=i}^{\infty} A_j, \qquad U = \bigcap_{i=0}^{\infty} \bigcup_{j=i}^{\infty} A_j.$$

(a) Prove that $L \subset U$.

(b) What are L and U in the case where $A_0 \subset A_1 \subset \cdots \subset A_n \subset A_{n+1} \cdots$?

(c) Repeat (b) for the inclusions going the opposite way.

Any collection of sets (indeed, any set whatsoever) can be indexed by itself! For a collection $\mathcal{A}$ of subsets of a set X, we may form the family $(A)_{A \in \mathcal{A}}$. [As a map, this family is just the inclusion map of $\mathcal{A}$ into $\mathcal{P}(X)$.] Then

$$\bigcup \mathcal{A} = \bigcup_{A \in \mathcal{A}} A, \qquad \bigcap \mathcal{A} = \bigcap_{A \in \mathcal{A}} A.$$

(What happens if the collection $\mathcal{A}$ is empty here?) Thus the union and intersection of a collection of sets, as defined in Appendix A, may be viewed as the union and intersection, respectively, of a family of sets.

The two-dimensional analog of a (finite-length) list is a table—or *matrix*—which consists of some number m of rows that stretch horizontally and some number n of columns that stretch vertically. If A is such a matrix, then the object in the ith row and jth column may be denoted $A_{(i,j)}$, $A_{i,j}$, or just A_{ij} (the juxtaposition of i and j in the subscript is *not* supposed to indicate multiplication). Then the entire matrix may be regarded as the family

$$(A_{ij})_{(i,j) \in \{1,2,\ldots,m\} \times \{1,2,\ldots,n\}}$$

whose index set consists of all pairs (i, j) with $1 \le i \le m$ and $1 \le j \le n$. But the matrix may also be regarded as the m-tuple $(R_i)_{i=1,2,\ldots,m}$ where, for each $i = 1, 2, \ldots, m$, the ith entry R_i is in turn the n-tuple $(A_{ij})_{j=1,2,\ldots,n}$. Indeed, this is exactly the way a matrix is treated by some programming languages, such as that used in MATHEMATICA.

Exercise B.4.10. What bijection is suggested by the preceding discussion of matrices as families? What bijection is suggested if a matrix is regarded, instead, as an n-tuple of columns?

Appendix C

Equivalence Relations

Classifying things as being alike or unalike is a basic human mental activity. When we consider things as being alike, we do not require that they be identical, only that they have in common some property of interest.

Classification may be represented mathematically in either of two ways—equivalence relations and partitions. We look in turn at each of these two mathematical notions and show that they are essentially the same.

This appendix should be read in conjunction with one of the extended discussions of a specific equivalence relation, such as in Sections 2.4 and 2.5. (Additional examples of equivalence relations appear in Exercise 3.2.21, Example 3.4.10, Exercises 3.4.12 and 3.5.46, and Section 4.1.)

C.1   Equivalence Relations

When we say that some things are alike while others are unalike with respect to some property or other, we ordinarily expect the following to be true:

- Each thing is like itself.

- If one thing is like a second, then the second is like the first.

- If one thing is like a second and the second is like a third, then the first is like the third.

Then "alikeness" classification may be represented by a certain kind of relation—an *equivalence relation*—in the set of objects being classified.

Definition C.1.1. A relation $\sim$ in a set X is said to be an **equivalence relation on** X when it is

- **reflexive on** X, that is, $x \sim x$ for every $x \in X$;

- **symmetric**, that is, $x \sim y \implies y \sim x$ for all $x, y \in X$; and

- **transitive**, that is, $x \sim y \,\&\, y \sim z \implies x \sim z$ for all $x, y, z \in X$.

Thus the set X is the given set of things being classified, and the equivalence relation $\sim$ provides the rule for classifying them: $x \sim y$ is interpreted to mean that x is like y.

Examples C.1.2.　　(1) Fix a positive integer m. For elements a and b of the set $\mathbb{Z}$ of all integers, let $a \sim b$ mean that $a - b$ is an integral multiple of m, that is,

$$a \sim b \iff a - b = km \text{ for some integer } k.$$

Then $\sim$ is an equivalence relation on $\mathbb{Z}$ (the proof appears in Section 2.4). The equivalence relation $\sim$ on $\mathbb{Z}$ is also called *congruence modulo* m, and the standard notation is $a \equiv b \pmod{m}$ instead of $a \sim b$.

For one specific example, take $m = 2$. Then $a \equiv b \pmod 2$ means that a and b are both even or both odd. Thus the classification effected by this equivalence relation is that of having the same *parity*—evenness or oddness.

(2) In the set $\mathbb{R}$ of all real numbers, let $\sim$ be the relation defined by

$$x \sim y \iff y - x \in \mathbb{Z}.$$

In other words, real numbers are being classified as being alike exactly when they differ by an integer. For example, $\sqrt{2} \sim \sqrt{2} + 1995$ and $1/2 \sim 5/2$, whereas $\pi \nsim \pi + 1/2$.

The relation $\sim$ in $\mathbb{R}$ is an equivalence relation on $\mathbb{R}$. In fact, it is reflexive on $\mathbb{R}$ because, for each $x \in \mathbb{R}$, we have $x - x = 0 \in \mathbb{Z}$. It is symmetric because if $x \sim y$ for $x, y \in \mathbb{R}$, then $y - x \in \mathbb{Z}$; then $x - y = -(y - x) \in \mathbb{Z}$, also, and that means $y \sim x$. Finally, the relation is transitive because, if $x \sim y$ and $y \sim z$ for $x, y, z \in \mathbb{R}$, then $y - x \in \mathbb{Z}$ and $z - y \in \mathbb{Z}$ whence $z - x = (z - y) - (y - x) \in \mathbb{Z}$, too, so that $x \sim z$.

(3) For two triangles T and S in the plane, define $T \sim S$ to mean that T is congruent to S in the usual geometric sense, that is, there

is some rigid motion[1] of the plane mapping T to S. Then the relation $\sim$ is an equivalence relation on the set of all triangles in the plane.

(4) In any set X whatsoever, define $\sim$ to be the relation that classifies two elements as alike if and only if they are exactly the same, that is,

$$x \sim y \iff x = y.$$

Then $\sim$ is an equivalence relation on X.

Equality is the "finest" possible equivalence relation on a set X: for an arbitrary equivalence relation $\sim$ on X, for all $x, y \in X$, we have $x = y \implies x \sim y$. Thus the notion of equivalence relation on a set X generalizes the notion of equality in that set.

(5) Let $\mathcal{A}$ be a given collection of sets. (For example, $\mathcal{A}$ could consist of all the subsets of $\mathbb{R}^n$ for all the various positive integers n.) For $A, B \in \mathcal{A}$, let $A \approx B$ mean that there is some bijection from A to B, that is, a one-to-one correspondence between A and B. That this relation is an equivalence relation on $\mathcal{A}$ follows from basic properties of composition of bijections—see Section B.3 and, in particular, Proposition B.3.6. (The equivalence relation $\approx$ is examined in depth in Chapter 4; see, especially, Proposition 4.1.3.)

Exercise C.1.3. Which of the following relations are equivalence relations on the specified sets? For those that are not, see if you can make them into equivalence relations either by restricting attention to a subset of the given set or by altering the given relation in some minor way.

(a) In the set $\mathbb{R}$, let $x \sim y$ mean $y - x \in \mathbb{Q}$.

(b) In the set $\mathbb{Z}$, let $m \sim n$ mean there exists some nonzero integer k with $n = km$.

(c) In the set $\mathbb{R}$, let $x \sim y$ mean that $|x| = |y|$.

(d) For any set X whatsoever, let $x \sim y$ hold for all $x, y \in X$. (In the technical meaning of "relation," we have $\sim = X \times X$.)

(e) In the plane $\mathbb{R}^2$, let $p \sim q$ for points $p, q \in \mathbb{R}^2$ mean that p and q lie on the same line through the origin.

---

[1] A rigid motion of the plane is a one-to-one correspondence between the plane and itself that preserves distances between points—the distance between any two points is the same as the distance between their images under the correspondence. Any such rigid motion is a composite of translations, rotations, and reflections.

(f) In the plane $\mathbb{R}^2$, let $p \sim q$ for points $p, q \in \mathbb{R}^2$ mean that p and q lie on the same vertical line.

(g) In the plane $\mathbb{R}^2$, let $p \sim q$ for points $p, q \in \mathbb{R}^2$ mean that p and q lie on the same line.

(h) In the set of all triangles in the plane, let $T \sim S$ mean that the triangle T is similar to the triangle S in the usual geometric sense. (What does "similar" mean here?)

(i) In the set $\mathbb{R} \times \mathbb{R}$, let $x \sim y$ mean $x^2 = y^2$.

(j) In the set $\mathbb{Z}$, let $m \sim n$ mean $m \le n$.

(k) In the set $\mathcal{P}(\mathbb{R})$ of all subsets of $\mathbb{R}$, let $A \sim B$ mean $A \cap B \ne \varnothing$.

(l) In the set $\mathcal{F}$ of all functions from $\mathbb{R}$ to $\mathbb{R}$, let $f \sim g$ mean that g differs from f by a constant, that is, that there is some $c \in \mathbb{R}$ with $g(t) = f(t) + c$ for all $t \in \mathbb{R}$.

(m) In the set $\mathcal{F}$ of all functions from $\mathbb{R}$ to $\mathbb{R}$, let $f \sim g$ mean that g is a constant multiple of f, that is, that there is some $k \in \mathbb{R}$ with $g(t) = kf(t)$ for all $t \in \mathbb{R}$.

(n) In the set $\mathcal{D}$ of all differentiable functions from $\mathbb{R}$ to $\mathbb{R}$, let $f \sim g$ for functions $f, g \in \mathcal{D}$ mean that $f' = g'$, where the prime denotes the derivative.

Exercises C.1.4.   (1) In a set X, let $\sim$ be a relation that is symmetric and transitive. Suppose the domain of $\sim$ is X, that is, for each $x \in X$, there is some $y \in X$ such that $x \sim y$. Prove that $\sim$ is necessarily reflexive on X, hence is an equivalence relation on X.

(2) Consider a relation in $\mathbb{R}$ in its technical sense as a subset of the plane $\mathbb{R} \times \mathbb{R}$. What does it mean, in geometric terms, to say that a relation $\sim$ in $\mathbb{R}$ is reflexive on $\mathbb{R}$? Symmetric?

(3) Let R be a relation in a set X that is reflexive on X and transitive. Define $\sim$ to be the relation in X given by $x \sim y$ if and only if $x \, R \, y$ & $y \, R \, x$. Show that $\sim$ is an equivalence relation on X. Also, give an example of such an R that is not already an equivalence relation, and the resulting equivalence relation $\sim$.

For an element in the domain of a given equivalence relation, we can group together all the elements that are equivalent to it.

Definition C.1.5. Let $\sim$ be an equivalence relation on a set X. For each $x \in X$, the **equivalence class of x under** $\sim$ is the subset of X denoted by $[x]$ and defined by

$$[x] = \{\, y \in X : x \sim y \,\}.$$

For $x \in X$, each $y \in [x]$ is called a **representative** of this equivalence class.

The **quotient of X under** $\sim$, denoted by $X/\sim$, is the collection of all equivalence classes of elements of X under $\sim$, that is,

$$X/\sim \; = \{\, [x] : x \in X \,\}.$$

The map

$$q : X \to X/\sim$$
$$x \mapsto [x]$$

which sends each element of X to its equivalence class under $\sim$ is called the **quotient map induced by** $\sim$.

Examples C.1.6.   (1) Let $\sim$ be the equivalence relation on $\mathbb{Z}$ of congruence modulo 2—see Example C.1.2(1). Then for $n \in \mathbb{Z}$,

$$[n] = \begin{cases} E & \text{if } n \text{ is even,} \\ O & \text{if } n \text{ is odd,} \end{cases}$$

where $E = \{\ldots, -4, -2, 0, 2, 4, \ldots\}$ is the set of all even integers and $O = \{\ldots, -3, -1, 1, 3, \ldots\}$ is the set of all odd integers. Thus the quotient set $\mathbb{Z}/\sim \; = \{E, O\}$, a set of two elements. For an integer n, the image $q(n)$ under the quotient map $q : \mathbb{Z} \to \mathbb{Z}/\sim$ is E if n is even, and O if n is odd.

More generally, for an integer $m > 1$, the equivalence class $[a]$ of an $a \in \mathbb{Z}$ under the relation of congruence modulo m is also called the **congruence class of a modulo** m; to make clear its dependence on the modulus m, the congruence class may also be denoted by $[a]_m$. Then,

$$[a]_m = \{\, b \in \mathbb{Z} : a - b \text{ is a multiple of } m \,\}$$
$$= \{\, a + jm : j \in \mathbb{Z} \,\}$$

(explain!). The quotient of $\mathbb{Z}$ under the relation of congruence modulo m is commonly denoted by $\mathbb{Z}_m$. This quotient set consists of exactly m different equivalence classes (why?). (See Section 2.5.)

(2) Let ~ be the equivalence relation on $\mathbb{R}$ given by $x \sim y$ if and only if $y - x \in \mathbb{Z}$—see Example C.1.2 (2). For an arbitrary real number x, there exists a unique real t with $0 \le t < 1$ such that $x \sim t$. In fact, $t = x - n$ where n is the unique integer n such that $n \le x \le n + 1$. In this case, $[x] = [t]$. Thus $\mathbb{R}/\sim \; = \{\, [t] : t \in [0, 1) \,\}$, and $[t] \ne [s]$ whenever $t, s \in [0, 1)$ with $t \ne s$.

(3) Let ~ be the relation of equality in a set X. Then for each $x \in X$, the equivalence class $[x] = \{x\}$, the one-element set containing x. Hence the quotient set $X/\sim$ is the collection $\{\, \{x\} : x \in X \,\}$ of all such one-element subsets of X.

For an equivalence relation ~ on a set X,

$$x \in [x] \qquad (x \in X).$$

It is critical to realize, though, that $[x] = [y]$ does *not* necessarily mean that $x = y$. For example, for the equivalence relation of congruence of integers modulo 2, the equivalence classes of 2 and 356 are the same—each equivalence class is the set of all even integers—yet certainly $2 \ne 356$.

Lemma C.1.7. *Let ~ be an equivalence relation on a set X and let $x, y \in X$. Then:*

1. $y \in [x] \iff x \sim y$.

2. $[x] = [y] \iff x \sim y$.

3. $y \in [x] \iff [x] = [y]$.

Proof.   1. This is just the definition of the equivalence class $[x]$.

2. First, assume $[x] = [y]$. Since $y \in [y]$, then also $y \in [x]$. But this means $x \sim y$.

Conversely, assume $x \sim y$. We shall prove that $[x] \subset [y]$; the proof of the reverse inclusion $[y] \subset [x]$ is similar. Let $z \in [x]$ be arbitrary. (We want to deduce that $z \in [y]$.) Then

$$x \sim z.$$

By assumption, $x \sim y$, and so, by symmetry,

$$y \sim x.$$

By transitivity,

$$y \sim z.$$

But this means $z \in [y]$.

3. This is an immediate consequence of parts 1 and 2. ☐

From the preceding lemma it will follow that any two equivalence classes that are distinct are actually *disjoint,* that is, have no elements in common.

Proposition C.1.8. *Let ~ be an equivalence relation on a set X. If A and B are any two equivalence classes under ~, then either A = B or else A ∩ B = ∅.*

Proof. Exercise. (Compare the proof of Proposition 2.5.6.) ☐

Exercises C.1.9. (1) For each equivalence relation in Exercise C.1.3, describe in as simple a way as possible the equivalence class of each element of the domain of the relation.

(2) Verify that the specified relation ~ in X is an equivalence relation on X and determine the equivalence class of each $x \in X$.

(a) Let $X = \mathbb{Z} \times (\mathbb{Z} \setminus \{0\})$. For $(m, n) \in X$ and $(i, j) \in X$, let $(m, n) \sim (i, j)$ mean $mj = ni$. (Compare Example 3.4.10.)

(b) Let $X = \mathbb{N} \times \mathbb{N}$. For $(m, n) \in X$ and $(i, j) \in X$, let $(m, n) \sim (i, j)$ mean $m + j = n + i$. (Compare Exercise 3.4.12.)

(c) Let $X = [0, 1] \times [0, 1]$ (the unit square in the plane). For $(x, y) \in X$ and $(t, s) \in X$, let $(x, y) \sim (t, s)$ mean either $(x, y) = (t, s)$ or else $x = t$ and $y = 1 - s$.

(3) Given a function $f: X \to Y$, define the relation ~ in X by

$$x \sim y \iff f(x) = f(y). \tag{*}$$

(a) Show that ~ is an equivalence relation on X.

(b) Describe in the simplest and most direct way that you can the equivalence relation ~ in the case of the map $f: \mathbb{R} \to \mathbb{R}$ given by $f(x) = \cos 2\pi x$. Determine all the equivalence classes under ~.

(c) Repeat (b) for the case of the map $f: \mathbb{R} \to \mathbb{R}$ given by $f(x) = |x|$.

(d) Prove that an arbitrary equivalence relation ~ on a set X has the form (*) for some function f from X to some set Y.

When working with a quotient set, one often has occasion to define a map whose domain is the quotient set.

Example C.1.10. Let X be the quotient set $\mathbb{Z}_4$—see Exercise C.1.2 (1)—and let $Y = \mathbb{Z}_{12}$. We propose to define a map $F \colon X \to Y$ by the rule

$$F([a]_4) = [3a]_{12} \qquad (a \in \mathbb{Z}). \qquad\qquad (*)$$

But does this really makes sense? The problem is that an element of X is not just an integer but an entire equivalence class of integers; for an equivalence class $A \in X = \mathbb{Z}_4$, we need to know how to define the image of the element $F(A)$.

It is true here that each such A has the form $[a]_4$, but it has this form for infinitely many different values of a (for example, $[2]_4 = [6]_4 = [10]_4 = [-2]_4$, etc.). What happens if we start with two different representatives a and a' of a given $A \in X$? Certainly $3a$ and $3a'$ will be different, but what about the elements $[3a]_{12}$ and $[3a']_{12}$ of $Y = \mathbb{Z}_{12}$? Since $a \sim a'$, that is, $a \equiv a' \pmod 4$, then $a - a'$ is a multiple of 4. But then $3a - 3a' = 3(a - a')$ is a multiple of $3 \cdot 4 = 12$, that is, $3a \equiv 3a' \pmod{12}$. This means $3a \sim 3a'$, and so $[3a]_{12} = [3a']_{12}$ as elements of $Y = \mathbb{Z}_{12}$ after all. In other words, the definition of $F(A)$ as $[3a]_{12}$ *does not depend on the choice of the representative a of the equivalence class A.*

When, as in the preceding example, the proposed definition of the image of each equivalence class does not depend on the choice of representative of the equivalence class, we say that the proposed rule makes the function **well-defined**.

All such examples can be handled by an ad hoc analysis like the above, but it is nice to have a general principle to cover them more cleanly. In the example, we really started with a map $a \mapsto [3a]_{12}$ from $\mathbb{Z}$ to $Y = \mathbb{Z}_{12}$. The question was whether we could, from that, legitimately obtain the desired map $F \colon X = \mathbb{Z}/\!\sim \; \to Y$.

Proposition C.1.11 (Passing to Quotients). *Let $f \colon X \to Y$ be a map and let $\sim$ be an equivalence relation on X. Suppose f is constant on each equivalence class under $\sim$, that is,*

$$x \sim x' \;\Rightarrow\; f(x) = f(x') \qquad (x, x' \in X).$$

Then there exists a unique map $F \colon X/\!\sim \; \to Y$ such that

$$F([x]) = f(x) \qquad (x \in X).$$

Proof. Existence. Let $\mathcal{A} = X/\!\sim$. Define F to be the relation from $\mathcal{A}$ to Y that, as a subset of $\mathcal{A} \times Y$, is

$$\{\, (A, y) \in \mathcal{A} \times Y : (\exists x \in A)\, (y = f(x)) \,\}.$$

First we are going to show that F is a functional relation from $\mathcal{A}$ to Y.

Let $A \in \mathcal{A}$. There exists some $x \in A$. Then $(A, f(x)) \in F$.

Next, let $A \in \mathcal{A}$ and $y, y' \in Y$ such that both $(A, y) \in F$ and $(A, y') \in F$. There exist $x, x' \in A$ such that $y = f(x)$ and $y' = f(x')$. Since f is constant on the equivalence class A, then $f(x) = f(x')$. Then $y = y'$. This completes the proof that F is functional.

The map from $\mathcal{A}$ to Y with graph F is the desired one because the ordered pair $([x], f(x)) \in F$ for each $x \in X$.

Uniqueness. Suppose $F: X/\sim \to Y$ and $G: X/\sim \to Y$ are maps such that $F([x]) = f(x)$ and $G([x]) = f(x)$ for each $x \in X$. Let $A \in X/\sim$. There exists some $x \in A$. Then $F(A) = F([x]) = f(x) = G([x]) = G(A)$. □

In the notation of the preceding proposition, let $q: X \to X/\sim$ be the quotient map. The requirement $F([x]) = f(x)$ means $F(q(x)) = x$, and so the conclusion of the proposition is that there exists a unique map $F: X/\sim \to Y$ making the following diagram commutative:

This map F is said to be **obtained by passing to quotients under** $\sim$.

To use Proposition C.1.11 for the example that motivated it, start with the map $f: \mathbb{Z} \to \mathbb{Z}_{12}$ given by $f(a) = [3a]_{12}$ for each $a \in \mathbb{Z}$. If $a, a' \in \mathbb{Z}$ with $a \sim a'$, then (as before) $[3a]_{12} = [3a']_{12}$, that is, $f(a) = f(a')$. Without any further ado, we may assert that there is a unique map $F: \mathbb{Z}_4 = \mathbb{Z}/\sim \to \mathbb{Z}_{12}$ such that $F([a]_4) = [3a]_{12}$ for every $a \in \mathbb{Z}$.

Exercises C.1.12.   (1) Let $\sim$ be the equivalence relation on $S = \mathbb{N} \times \mathbb{N}$ from Exercise C.1.9(2)(b). Use Proposition C.1.11 to construct a map from $S/\sim$ to $\mathbb{Z}$ that is a bijection. (Compare Exercise 3.4.12.)

(2) Formulate and prove a result similar to Proposition C.1.11 for obtaining an operation on a quotient set $X/\sim$ starting with an operation on X. Apply it obtain an operation of addition on the quotient set $S/\sim$ in (1).

C.2 Partitions

One way to classify objects in a set is to say when two are to be regarded as alike. This is exactly what an equivalence relation does. Another, equally cogent way is to divide the objects into different classes— subsets of the original set. This is the idea of a *partition*.

Definition C.2.1. A **partition of** a set X is a collection $\mathcal{A}$ of subsets of X with the following properties:

- Each $A \in \mathcal{A}$ is nonempty.

- Each element of X belongs to at least one member of $\mathcal{A}$.

- No element of X belongs to more than one member of $\mathcal{A}$, that is, each two distinct members of $\mathcal{A}$ are disjoint.

The second property in this definition means that X is the *union* of all the members of $\mathcal{A}$. The third means that the collection $\mathcal{A}$ is *pairwise disjoint.*

Examples C.2.2.   (1) Let E be the set of all even integers and O be the set of all odd integers. Then the two-member collection $\{E, O\}$ is a partition of $\mathbb{Z}$.

(2) More generally, let A be a subset of a set X with $\varnothing \neq A \neq X$. Then the two-element collection $\{A, X \setminus A\}$ is a partition of X.

(3) The collection

$$\{[0, 1/4], [1/4, 1/2], [1/2, 3/4], [3/4, 1]\}$$

of subintervals of the closed interval $[0, 1]$ is *not* a partition of $[0, 1]$ (even though some calculus books call it one): the intervals in it have endpoints in common. However, the modified collection

$$\{[0, 1/4), [1/4, 1/2), [1/2, 3/4), [3/4, 1]\}$$

of subintervals is a partition of $[0, 1]$.

(4) The collection of all vertical lines is a partition of the plane.

(5) The empty set $\varnothing$ has exactly one partition, namely, the empty collection $\varnothing$ (why?). No, we did *not* just say that $\mathcal{A} = \{\varnothing\}$ is a partition of $\varnothing$; it cannot be, since its sole member $\varnothing$ would not be nonempty.

By way of contrast, a partition $\mathcal{A}$ of a nonempty set X cannot be empty (why?).

You have seen the properties of a partition before: they hold for the collection $X/\sim$ of all equivalence classes under an equivalence relation $\sim$ on a set X. The third property is just Proposition C.1.8. The first two properties are evident. In fact, if $A \in X/\sim$, then $A = [a]$ for some $a \in X$ and so $a \in A$; if $x \in X$, then $x \in [x] \in X$. This proves the following.

Proposition C.2.3. *Let $\sim$ be an equivalence relation on a set X. Then the quotient set $X/\sim$ is a partition of X.*

Thus, each equivalence relation on a set gives rise to a partition of that set. Now partitions and equivalence relations are supposed to be two different ways of looking at the same idea of classification, and so it ought to be possible to go backward—to obtain an equivalence relation from a partition. This is indeed the case, and the way to do it is quite simple: say that two elements are alike if they belong to the same member of the partition.

Definition C.2.4. Let $\mathcal{A}$ be a partition of a set X. The relation $\sim_{\mathcal{A}}$ in X defined by

$$x \sim_{\mathcal{A}} y \iff (\exists A \in \mathcal{A})(x \in A \text{ and } y \in A)$$

is said to be **induced by** $\mathcal{A}$.

Proposition C.2.5. *Let $\mathcal{A}$ be a partition of a set X. Then the relation $\sim_{\mathcal{A}}$ in X induced by $\mathcal{A}$ is an equivalence relation on X.*

Proof. Exercise. □

Exercise C.2.6. Verify that the given collection $\mathcal{A}$ is a partition of the given set X. Then describe the equivalence relation $\sim_{\mathcal{A}}$ on X induced by $\mathcal{A}$.

(a) X is any set and $\mathcal{A} = \{X\}$.

(b) $X = [0,1]$ and $\mathcal{A} = \{0,1\} \cup \{\{t\} : 0 < t < 1\}$.

(c) $X = \mathbb{R}$ and $\mathcal{A} = \{\mathbb{Z}\} \cup \{\{t\} : t \in \mathbb{R} \setminus \mathbb{Z}\}$.

We said that equivalence relations and partitions are two ways of looking at classification and that they are essentially the same. The sense in which they are essentially the same is the following. Let X be a set—the set of objects being classified. Then we have a way of constructing, from each equivalence relation on X, a partition of X, namely, the quotient set $X/\sim$. Thus we have a map

$$\sim \,\mapsto X/\sim$$

from the set of all equivalence relations on X to the set of all partitions of X. We also have a way of constructing, from each partition $\mathcal{A}$ of X, an equivalence relation on X, namely, the relation $\sim_{\mathcal{A}}$ induced by $\mathcal{A}$. Thus we also have a map

$$\mathcal{A} \mapsto \sim_{\mathcal{A}}$$

from the set of all partitions of X to the set of all equivalence relations on X. According to the next proposition, the equivalence relation induced by the quotient set of X under a given equivalence relation on X *is* that given equivalence relation; reciprocally, the partition of X that is the quotient set of X under the equivalence relation induced by a given partition of X is that given partition. Thus these two maps establish a one-to-one correspondence between the totality of all equivalence relations on X and the totality of all partitions on X.

Proposition C.2.7. *Let X be a set. Then:*

1. For each equivalence relation $\sim$ on X,

$$\sim\!(X/\!\sim) \;=\; \sim .$$

2. For each partition $\mathcal{A}$ of X,

$$(\sim_{\mathcal{A}})/\!\sim \;=\; \mathcal{A}.$$

Proof. Exercise.   □

Appendix D

The Integers

God made the integers, everything else is the work of man.
— Leopold Kronecker

This appendix presents the fundamental properties of the set ℕ of all natural numbers and the larger set ℤ of all integers that are used throughout this book. Ordinarily, these are properties you should feel free to use in proofs without further justification and even (unless your instructor decrees otherwise) without explicit citation.

In a few special—and fairly evident—circumstances, you should not feel free to use these properties. For example, in Chapter 1, after we define addition of natural numbers by recursion, you are asked to use induction to prove that $0 + n = n$ for every natural number n. [See Example 1.2.10(2) and Exercises 1.2.11.] There, it would hardly be legitimate to assume the commutative law of addition until you have proved it; in fact, establishing that always $0 + n = n$ is one of the steps needed to prove commutativity of addition! In such circumstances, you must pay attention not only to what you know but also when you are supposed to know it.

Not all the properties of the integers that you would regard as basic appear in this appendix. One such property is the possibility of representing any integer in terms of powers of 10, with decimal digits; another is the factorization of any positive integer into a product of primes. Such properties are considered in Chapter 2. Moreover, mathematical induction in its various forms, which is included here as one of the basic properties, is treated at length in Chapter 1.

There are no explicit exercises in this appendix. However, you may regard as an implicit exercise to prove each proposition, lemma, or other nonassumed statement whose proof is not provided.

D.1   Natural Numbers: Working Assumptions

The **set of natural numbers** is denoted by $\mathbb{N}$. About this set we generally assume the existence of two operations of **addition** and **multiplication,** denoted in the usual way. That addition and multiplication are operations on $\mathbb{N}$ means they are functions from the set $\mathbb{N} \times \mathbb{N}$, consisting of all ordered pairs (m, n) of elements of $\mathbb{N}$, to the set $\mathbb{N}$. Then:

- For all $m, n, m', n' \in \mathbb{N}$, if $m = m'$ and $n = n'$, then $m + n = m' + n'$ and $m \cdot n = m' \cdot n'$.

- (*closure*) For all $m, n \in \mathbb{N}$, the sum $m + n \in \mathbb{N}$ and the product $m \cdot n \in \mathbb{N}$.

As usual for multiplication, the symbol $\cdot$ for the operation is often elided so that multiplication is indicated by juxtaposition, as in mn (to mean $m \cdot n$).

Addition and multiplication of natural numbers are assumed to have the following properties.

Algebraic Properties of Natural Numbers D.1.1.

1. (associative law for addition) *For all $m, n, k \in \mathbb{N}$,*

$$(m + n) + k = m + (n + k).$$

2. (zero law) *There exists a* unique *element $0 \in \mathbb{N}$ such that, for every $n \in \mathbb{N}$,*

$$0 + n = n = n + 0.$$

3. (commutative law for addition) *For all $m, n \in \mathbb{N}$,*

$$m + n = n + m.$$

4. (associative law for multiplication) *For all $m, n, k \in \mathbb{N}$,*

$$(m \cdot n)k = m(n \cdot k).$$

5. (unity law) *There exists a* unique *element $1 \in \mathbb{N}$ such that $1 \neq 0$ and, for every $n \in \mathbb{N}$,*

$$1 \cdot n = n = n \cdot 1.$$

6. (commutative law for multiplication) *For all $m, n \in \mathbb{N}$,*

$$m \cdot n = n \cdot m.$$

7. (distributive laws)[1] *For all* $m, n, k \in \mathbb{N}$,

$$m(n + k) = (m \cdot n) + (m \cdot k),$$
$$(m + n)k = (m \cdot k) + (n \cdot k).$$

8. (cancellation laws) *For all* $m, n, k \in \mathbb{N}$,

$$m + k = n + k \implies m = n,$$
$$m \cdot k = n \cdot k \text{ and } k \neq 0 \implies m = n.$$

We continue to assume the legitimacy of mathematical induction.

Principle of Mathematical Induction D.1.2. *If I is a subset of $\mathbb{N}$ such that $0 \in I$ and, for each $n \in \mathbb{N}$, $n \in I \implies n + 1 \in I$, then $I = \mathbb{N}$.*

As usual, we denote $1 + 1$ by 2, $1 + 2$ by 3, etc. Because of the associative law for addition, for natural numbers m, n, and k, the sum $m + n + k$ can unambiguously be defined as both $(m + n) + k$ and $m + (n + k)$. For example, $3 = 1 + 1 + 1$.

The following additional properties of addition and multiplication of natural numbers are easy consequences of the ones above.

Proposition D.1.3.   *1. For all $n \in \mathbb{N}$, $n \cdot 0 = 0 = 0 \cdot n$.*

2. (no zero summands) *For all $m, n \in \mathbb{N}$,*

$$m + n = 0 \implies m = 0 \text{ and } n = 0.$$

3. (no zero divisors) *For all $m, n \in \mathbb{N}$,*

$$m \cdot n = 0 \implies m = 0 \text{ or } n = 0.$$

In $\mathbb{N}$ there is also an **order relation** $<$ that is assumed to have the following properties.

Ordering Properties of Natural Numbers D.1.4.

1. (transitive law) *For all $m, n, k \in \mathbb{N}$, if $m < n$ and $n < k$, then $m < k$.*

2. (trichotomy law) *For each $m, n \in \mathbb{N}$, exactly one of the following is true:*

$$m = n, \quad m < n, \quad m > n.$$

---

[1]The parentheses on the right-hand sides of the distributive laws are included to indicate the order of operations. We shall, of course, adopt the customary rule about the **order of precedence** of the operations: In the absence of parentheses indicating to the contrary, multiplication is done before addition. Then parentheses are no longer required on the right-hand sides.

3. (addition preserves order) *For all $m, n, k \in \mathbb{N}$, if $m < n$, then $m + k < n + k$.*

4. (multiplication preserves order) *For all $m, n, k \in \mathbb{N}$, if $m < n$ and $k \neq 0$, then $m \cdot k < n \cdot k$.*

5. $0 < 1$.

As with any order relation, so for the order relation $<$ in $\mathbb{N}$ we write $n > m$ to mean $m < n$, write $m \leq n$ to mean $m < n$ or $m = n$, and write $n \geq m$ to mean $m \leq n$.

In terms of the weak relation $\leq$ on $\mathbb{N}$ obtained from the strict relation $<$, the ordering properties may be restated as follows.

Proposition D.1.5. *1.* (reflexive law) *For each $n \in \mathbb{N}$, $n \leq n$.*

2. (antisymmetric law) *For all $m, n \in \mathbb{N}$, if $m \leq n$ and $n \leq m$, then $m = n$.*

3. (transitive law) *For all $m, n, k \in \mathbb{N}$, if $m \leq n$ and $n \leq k$, then $m \leq k$.*

4. (comparability) *For all $m, n \in \mathbb{N}$, $m \leq n$ or $n \leq m$.*

5. (addition preserves order) *For all $m, n, k \in \mathbb{N}$, if $m \leq n$, then $m + k \leq n + k$.*

6. (multiplication preserves order) *For all $m, n, k \in \mathbb{N}$, if $m \leq n$, then $m \cdot k \leq n \cdot k$.*

An easy induction, starting with the property $0 < 1$ for the base step, establishes the following proposition.

Proposition D.1.6. *For every $n \in \mathbb{N}$, if $n \neq 0$, then $n > 0$. In other words, $\mathbb{N}^* = \{ n \in \mathbb{N} : n > 0 \}$.*

Another easy induction establishes the following lemma.

Lemma D.1.7. *For each $n \in \mathbb{N}$ with $n \neq 0$, there exists $k \in \mathbb{N}$ such that $n = k + 1$.*

(This lemma, here a consequence of the assumed algebraic and ordering properties and the Principle of Mathematical Induction, is a form of one of the Peano Postulates—see Section D.2. In fact, all five of the Peano Postulates are consequences of the same assumed properties, as you are invited to verify.)

Lemma D.1.7, together with induction and the algebraic and ordering properties of $\mathbb{N}$ previously stated here, is all that is required to establish the following result, which was originally formulated as Lemma 1.4.1.

Lemma D.1.8 (Gap Lemma). *For each natural number m there is no natural number n with $m < n < m + 1$.*

With the aid of the Gap Lemma, Lemma D.1.7 can now be generalized as follows.

Proposition D.1.9. *For all $m, n \in \mathbb{N}$, if $m < n$, then there exists a unique $k \in \mathbb{N}^*$ such that $n = m + k$.*

Proof. Uniqueness follows from the cancellation law for addition [Algebraic Properties D.1.1, part 8]. For existence, we use induction on n.

Base step ($n = 0$). From Proposition D.1.6 and the trichotomy law, there is no natural number m with $m < 0$, and so the conclusion that $m < n$ implies $n = m + k$ for some k holds vacuously.

Inductive step. Let $n \in \mathbb{N}$ and assume that for every $m \in \mathbb{N}$, if $m < n$, then $n = m + k$ for some $k \in \mathbb{N}$. Let $m \in \mathbb{N}$ with $m < n + 1$. By the Gap Lemma, $m \le n$. If, on the one hand, $m = n$, then $m + 1 = n + 1$. If, on the other hand, $m < n$, then by the inductive assumption $m + k = n$ for some $k \in \mathbb{N}$ whence $m + (k + 1) = (m + k) + 1 = n + 1$. $\square$

If $m = n$, then of course 0 is the unique $k \in \mathbb{N}$ for which $m + k = n$. In view of the preceding proposition, it is therefore possible to define a partial operation of **subtraction** in $\mathbb{N}$: If $m, n \in \mathbb{N}$ and $m \le n$, define the **difference** $n - m$ to be the unique $k \in \mathbb{N}$ such that $m + k = n$.

Proposition D.1.10 (Distributive Laws for Subtraction). *Let $m, n \in \mathbb{N}$ with $m \le n$. Then for all $k \in \mathbb{N}$,*

$$k(n - m) = kn - km,$$
$$(n - m)k = nk - mk.$$

The proof of the **Ordinary Recursion Principle** (Theorem 1.2.9) outlined in Exercise B.4.2 relies only upon properties of the natural numbers stated above together with generalities about sets, relations, and functions. Hence we may now freely use recursion.

In particular, we may use recursion to define the operation of **exponentiation** on $\mathbb{N}$. For $a \in \mathbb{N}$, the powers a^n of a are defined recursively, as in Example 1.2.13, by

$$a^0 = 1,$$
$$a^{n+1} = a \cdot a^n \qquad (n \in \mathbb{N}).$$

In particular, 0^0 is 1, *not* 0.

The other familiar laws of exponents follow from this recursive definition. (Note that only nonnegative exponents are at issue here.)

Proposition D.1.11 (Laws of Exponents). *For all $a, b, m, n \in \mathbb{N}$*

1. $a^{m+n} = a^m a^n$.

2. $a^{mn} = (a^m)^n$.

3. $(ab)^n = a^n b^n$.

D.2　Natural Numbers: Peano Postulates

The properties of addition, multiplication, and ordering of natural numbers assumed above are not the most basic things one can assume about the natural numbers. In fact, all these properties—and, indeed, the very definitions of addition, multiplication, and ordering—can be derived from a few modest assumptions. These assumptions are embodied in the following five statements about the set $\mathbb{N}$, an element 0, and a function σ.

Peano Postulates D.2.1.

1. $0 \in \mathbb{N}$.

2. $\sigma \colon \mathbb{N} \to \mathbb{N}$ *is a function from $\mathbb{N}$ to $\mathbb{N}$.* [For each $n \in \mathbb{N}$, the element $\sigma(n)$ is called the **successor** of n. Thus this postulate states that each natural number has, in the set of natural numbers, exactly one successor.]

3. $0 \notin \operatorname{range}(\sigma)$, *that is, $0 \neq \sigma(n)$ for all $n \in \mathbb{N}$.* (In other words, 0 is not the successor of any natural number.)

4. *The function σ is one-to-one, that is, if $m \neq n$, then $\sigma(m) \neq \sigma(n)$.* (In other words, distinct natural numbers have distinct successors.)

5. *If $I \subset \mathbb{N}$ such that $0 \in I$ and $\sigma(n) \in I$ whenever $n \in I$, then $I = \mathbb{N}$.* (This is the zero base form of the Principle of Mathematical Induction.)

We define $1 = \sigma(0)$, $2 = \sigma(1)$, $3 = \sigma(2)$, etc. Moreover, we define the subset $\mathbb{N}^*$ of $\mathbb{N}$ by

$$\mathbb{N}^* = \{ n \in \mathbb{N} : n \neq 0 \}.$$

What the Peano Postulates denote by $\sigma(n)$ will turn out to be $n + 1$, once addition of natural numbers is defined (recursively) by means of the function σ.

The following property is sometimes included among the Peano Postulates but is actually an easy consequence of them.

Proposition D.2.2. $\mathbb{N}^* \subset \text{range}(\sigma)$

Proof. Apply Peano Postulate 5 (zero base induction) to the subset I of $\mathbb{N}$ defined by

$$I = \{ n \in \mathbb{N} : n = 0 \text{ or } n \in \text{range}(\sigma) \} = \{0\} \cup \text{range}(\sigma). \quad \square$$

According to Peano Postulate 3, the natural number 0 is not the successor of any natural number. According to this proposition, each other natural number is the successor of at least one natural number. Together, the proposition and Peano Postulate 4 say that *each natural number other than 0 is the successor of exactly one natural number.* Further, the proposition and Peano Postulate 3 together say that $\text{range}(\sigma) = \mathbb{N}^*$.

The **Principle of Mathematical Induction**, as originally stated in Chapter 1, used base 1, that is, concerned subsets of $\mathbb{N}^*$. This form of induction is readily deduced from Peano Postulate 5: Given a subset I of $\mathbb{N}^*$ such that $1 \in I$ and $(\forall n)(n \in I \implies \sigma(n) \in I)$, form the subset $J = \{ n \in \mathbb{N} : \sigma(n) \in I \}$, deduce that $J = \mathbb{N}$ from Peano Postulate 5, and conclude that $I = \mathbb{N}^*$.

The proof of the **Ordinary Recursion Principle** (Theorem 1.2.9) outlined in Exercise B.4.2 uses only the Peano Postulates and generalities about sets, relations, and functions. Hence the principle is still valid in the present context. Of course, until we formally define addition, the Ordinary Recursion Principle should be stated without using +, as follows.

> Given a set X, an element $z \in X$, and a map $G: X \to X$, there exists a unique sequence $(s_n)_{n \in \mathbb{N}}$ such that $s_0 = z$ and $s_{\sigma(n)} = G(s_n)$ for each $n \in \mathbb{N}$.

As indicated in Section 1.2, operations of **addition** and **multiplication** on $\mathbb{N}$ may be defined recursively and their basic properties established inductively. For each $m \in \mathbb{N}$ its sum $m + n$ with $n \in \mathbb{N}$ is defined recursively by

$$\begin{cases} m + 0 = m, \\ m + \sigma(n) = \sigma(m + n) \quad (n \in \mathbb{N}). \end{cases}$$

Since by definition $1 = \sigma(0)$, in particular

$$\boxed{m + 1 = \sigma(m)}$$

for all m. Then the recursive definition of addition takes the form

$$\begin{cases} m + 0 = m, \\ m + (n + 1) = (m + n) + 1 \quad (n \in \mathbb{N}). \end{cases}$$

This defines the operation of addition as a function from $\mathbb{N} \times \mathbb{N}$ to $\mathbb{N}$. The operation of multiplication on $\mathbb{N}$ is defined recursively in a similar way.

Properties D.1.1 are consequences of these recursive definitions. To prove this, of course, requires repeated use of induction (although not necessarily for every property). It is important to prove the properties in the right order, or you will find yourself going around in circles! Here is a suggested order:

- Associative law for addition.

- The part $0 + n = n$ of the zero law.

- Commutative law for addition.

- Cancellation law $m + k = n + k \implies m = n$ for addition.

- Uniqueness of zero.

- Distributive laws.

- Associative law for multiplication.

- The part $m \cdot 1 = m$ of the unity law; the part $1 \cdot n = n$ of the unity law.

- The special case $0 \cdot n = 0$ of the commutative law for multiplication.

- The commutative law for multiplication.

- The cancellation law $m \cdot k = n \cdot k, k \neq 0 \implies m = n$ for multiplication.

- Uniqueness of unity.

For $m, n \in \mathbb{N}$, let $m < n$ mean there exists some $k \in \mathbb{N}^*$ such that $m + k = n$. This defines the **order relation** $<$ in $\mathbb{N}$. As usual, define $m \leq n$ to mean $m < n$ or $m = n$; then $m \leq n$ if and only if there exists some $k \in \mathbb{N}$ such that $m + k = n$.

Immediately from the definition, $0 < 1$, which is part 5 of the Ordering Properties D.1.4. The other four ordering properties are readily proved. (For the trichotomy law, there are two things to establish for given m and n: at least one of $m = n$, $m < n$, and $n < m$ holds; and at most one of $m = n$, $m < n$, and $n < m$ holds.)

All the Algebraic Properties D.1.1 and Ordering Properties D.1.4 of $\mathbb{N}$ that were assumed in the first part of this section are thus actually consequences of the Peano Postulates. Hence so are the lemmas and propositions previously stated as following from these properties.

In particular, the **Gap Lemma**—for each natural number m there is no natural number n with $m < n < m + 1$—is a consequence of the Peano Postulates. Then, exactly as in Section 1.4, the **Well-Ordering Principle** (Theorem 1.4.4) and the **Principle of Strong Induction** (Theorem 1.4.11) also follow from the Peano Postulates.

The Peano Postulates are not the ultimate foundation for the set of natural numbers. After all, they do not answer the questions of what the set $\mathbb{N}$ is or what its members 0, 1, 2, etc., actually are. In a carefully developed axiomatic treatment of set theory, it is possible to so define $\mathbb{N}$ as to answer these questions and, in fact, to derive the Peano Postulates as consequences of the definition! What follows is a brief indication of how to do this.

For motivation, think of the natural number 0 as "emptiness," the number 1 as "oneness," the number 2 as "twoness," and so forth. More precisely, think of 0 as a set having no elements; there is only one such set, namely, the empty set $\varnothing$. Accordingly, define $0 = \varnothing$. Next, think of 1 as a particular set having a single element; the singleton $\{0\}$ is such a set.[2] Accordingly, define $1 = \{0\}$. Next, think of 2 as a particular set having exactly two elements; the set $\{0, 1\}$ is such a set. Accordingly, define $2 = \{0, 1\}$.

It is tempting to continue at this point by saying "define $3 = \{0, 1, 2\}$, etc." The "etc." seems to suggest some sort of recursion, but of course recursion at this point is out of the question because we are trying to define the set of natural numbers itself. So what can we do?

Observe that the just-suggested definitions of 0, 1, 2, 3 require

$$0 = \varnothing,$$
$$1 = \{0\} = 0 \cup \{0\},$$
$$2 = \{0, 1\} = \{0\} \cup \{1\} = 1 \cup \{1\},$$
$$3 = \{0, 1, 2\} = \{0, 1\} \cup \{2\} = 2 \cup \{2\}.$$

This suggests the following definition as a way of getting from one natural number to the next. For an arbitrary set x, its **successor** is the set denoted by x^+ and defined to be $x \cup \{x\}$. Thus the elements of x^+ are the elements of x together with x itself. [Eventually, the successor n^+ of a natural number n will turn out to be what we denoted by $\sigma(n)$ in the Peano Postulates.]

Particular natural numbers are going to be obtained by taking repeated successors starting with $0 = \varnothing$. But we still do not have an entire set of such numbers. To obtain this set, we make another definition. We say that a set I is **inductive** if

---

[2]Do not confuse the set $0 = \varnothing$, which has no elements, with the set $\{0\}$, which has a unique element, namely, 0. This is like the difference between an uncaged lion and a caged lion (compare page A4).

(a) $\emptyset \in I$; and

(b) for each $x \in I$, its successor $x^+ \in I$, too.

What is an example of an inductive set? The set $\mathbb{N}$ of natural numbers will be such an example, but we do not have this set defined yet. In fact, one of the fundamental axioms of set theory—the **Axiom of Infinity**—asserts the existence of at least one inductive set. We assume this axiom.

The **set of natural numbers** is denoted by $\mathbb{N}$ and is defined to be the intersection of the collection of all inductive sets—the set of all those n that belong to every inductive set. (Had we not assumed the Axiom of Infinity, then it could be that the collection of all inductive sets is empty. In that case, there would be difficulty in forming the intersection of this collection because then every object x in the universe of mathematics would, vacuously, belong to every inductive set.)

By definition, 0 belongs to every inductive set. Hence $0 \in \mathbb{N}$; this is Peano Postulate 1.

From its definition, evidently the set $\mathbb{N}$ itself is inductive. Then $n^+ \in \mathbb{N}$ whenever $n \in \mathbb{N}$. Hence the formula $\sigma(n) = n^+$ defines a function $\sigma: \mathbb{N} \to \mathbb{N}$. That there is such a function is stated in Peano Postulate 2.

By definition, $\mathbb{N}$ is the smallest inductive set—it is a subset of every inductive set. Then every inductive subset of $\mathbb{N}$ equals $\mathbb{N}$. In view of the definition just given for σ, this is the Principle of Mathematical Induction (zero base form)—Peano Postulate 5.

By definition, 0 is empty; for each $n \in \mathbb{N}$ we have $n \in n \cup \{n\} = n^+ = \sigma(n)$ whence $\sigma(n) \neq \emptyset$. Hence $0 \neq \sigma(n)$ for each $n \in \mathbb{N}$. This is Peano Postulate 3.

On the basis of our definition of $\mathbb{N}$, we have now established all the Peano Postulates except the fourth (that σ is one-to-one). To prove Peano Postulate 4 requires two preparatory lemmas. The first of the two is suggested by our tentative definitions of 0, 1, 2, 3 above. For example, the elements of 3 are 0, 1, and 2. Now $2 \subset 3$ because $3 = 2 \cup \{2\}$; $1 \subset 3$ because $1 \subset 2 = 1 \cup \{1\}$ and $2 \subset 3$; and $0 = \emptyset \subset 3$ because the empty set is a subset of every set. Thus each element of 3 is, in fact, a subset of 3.

Lemma D.2.3. *For every natural number n, each element of n is a subset of n.*

Proof. Use induction on n. The base step holds vacuously because $0 = \emptyset$ has no elements. For the inductive step, use the definition that $n^+ = n \cup \{n\}$. $\square$

Lemma D.2.4. *If $n \in \mathbb{N}$, then $n \notin n$. If $m, n \in \mathbb{N}$, then $m \notin n$ or $n \notin m$.*

Proof. For the first statement, use induction on n and apply the preceding lemma. For the second statement, assume that both $m \in n$ and $n \in m$; then $m \subset n$ and $n \subset m$ by the preceding lemma, whence $m = n$, which contradicts the first statement. □

We are now ready to complete the proof that all the Peano Postulates are consequences of our definition of $\mathbb{N}$.

Proposition D.2.5. *For all natural numbers m and n, if $\sigma(m) = \sigma(n)$, then $m = n$.*

Proof. Let $m, n \in \mathbb{N}$. Just suppose $\sigma(m) = \sigma(n)$ but $m \neq n$. Since $n \in \sigma(n) = \sigma(m) = m \cup \{m\}$, then $n \in m$ or $n = m$. Hence $n \in m$. Similarly, $m \in n$. This contradicts Lemma D.2.4. □

D.3 Integers: Working Assumptions

The **set of integers** is denoted by $\mathbb{Z}$. On this set we assume the existence of operations of **addition** and **multiplication**, denoted in the usual way. That addition and multiplication are operations on $\mathbb{Z}$ means, among other things, that the set $\mathbb{Z}$ is *closed* under these operations, that is, $m + n \in \mathbb{Z}$ and $m \cdot n \in \mathbb{Z}$ for all $m, n \in \mathbb{Z}$. Addition and multiplication are assumed to have the following properties.

Algebraic Properties of Integers D.3.1.

1. (associative law for addition) *For all $m, n, k \in \mathbb{Z}$,*

$$(m + n) + k = m + (n + k).$$

2. (zero law) *There exists a* unique *element $0 \in \mathbb{Z}$ such that, for every $n \in \mathbb{Z}$,*

$$0 + n = n = n + 0.$$

3. (negatives law) *For each $n \in \mathbb{Z}$, there is a unique element $-n \in \mathbb{Z}$ such that*

$$n + (-n) = 0 = (-n) + n.$$

4. (commutative law for addition) *For all $m, n \in \mathbb{Z}$,*

$$m + n = n + m.$$

5. (associative law for multiplication) *For all $m, n, k \in \mathbb{Z}$,*

$$(m \cdot n)k = m(n \cdot k).$$

6. (unity law) *There exists a* unique *element* $1 \in \mathbb{Z}$ *such that* $1 \neq 0$ *and, for every* $n \in \mathbb{Z}$,

$$1 \cdot n = n = n \cdot 1.$$

7. (commutative law for multiplication) *For all* $m, n \in \mathbb{Z}$,

$$m \cdot n = n \cdot m.$$

8. (distributive laws) *For all* $m, n, k \in \mathbb{Z}$,

$$m(n + k) = (m \cdot n) + (m \cdot k),$$
$$(m + n)k = (m \cdot k) + (n \cdot k).$$

9. (no zero divisors) *For all* $m, n \in \mathbb{Z}$,

$$m \cdot n = 0 \implies m = 0 \text{ or } n = 0.$$

There is some redundancy in the statements of the preceding properties. For example, from $1 \cdot n = n$ and the commutativity of multiplication, necessarily $n \cdot 1 = n$. The zero element and unity element assumed to exist are necessarily unique, as may be proved using others of these properties.

The operation of **subtraction** on $\mathbb{Z}$ is defined by $m - n = m + (-n)$ for all $m, n \in \mathbb{Z}$.

Further properties of addition and multiplication of integers are easy consequences of the ones above. Some of these same properties appear in Proposition 3.1.12 in the context of the real numbers and more general fields (where, unlike the situation for integers, nonzero elements have multiplicative inverses); the proofs there carry over to the context of integers, too.

Proposition D.3.2. *The following properties of integers are true.*

1. (cancellation laws) *For all* $m, n, k \in \mathbb{Z}$,

$$m + k = n + k \implies m = n,$$
$$m \cdot k = n \cdot k \text{ and } k \neq 0 \implies m = n.$$

2. *For all* $n \in \mathbb{Z}$, $-(-n) = n$.

3. *For all* $n \in \mathbb{Z}$, $n \cdot 0 = 0 = 0 \cdot n$.

4. *For all* $n \in \mathbb{Z}$, $(-1)n = -n$.

5. *For all* $m, n \in \mathbb{Z}$, $m(-n) = -(mn) = (-m)n$.

6. *For all $m, n \in \mathbb{Z}$, $(-m)(-n) = mn$.*

7. (distributive laws for subtraction) *For all $m, n, k \in \mathbb{Z}$,*

$$m(n - k) = mn - mk,$$
$$(m - n)k = mk - nk.$$

In $\mathbb{Z}$ there is also an **order relation** $<$ that is assumed to have the following properties.

Ordering Properties of Integers D.3.3.

1. (transitive law) *For all $m, n, k \in \mathbb{Z}$, if $m < n$ and $n < k$, then $m < k$.*

2. (trichotomy law) *For each $m, n \in \mathbb{Z}$, exactly one of the following is true:*

$$m = n, \qquad m < n, \qquad n < m.$$

3. (addition preserves order) *For all $m, n, k \in \mathbb{Z}$, if $m < n$, then $m + k < n + k$.*

4. (multiplication preserves order) *For all $m, n, k \in \mathbb{Z}$, if $m < n$ and $0 < k$, then $m \cdot k < n \cdot k$.*

As with any order relation, so for the order relation $<$ in $\mathbb{Z}$ we write $n > m$ to mean $m < n$, write $m \leq n$ to mean $m < n$ or $m = n$, and write $n \geq m$ to mean $m \leq n$. As usual, an element $n \in \mathbb{Z}$ is said to be **positive** if $n > 0$, **negative** if $n < 0$, and **nonnegative** if $n \geq 0$.

Here are some more ordering properties of the integers.

Proposition D.3.4. *For all $m, n, k \in \mathbb{Z}$:*

1. *If $m < n$, then $-n < -m$. In particular, $n > 0$ if and only if $-n < 0$.*

2. *If $m < n$ and $k < 0$, then $mk > nk$.*

3. *If $n \neq 0$, then $n \cdot n > 0$. In particular, $1 > 0$.*

In terms of the weak relation $\leq$ on $\mathbb{Z}$ obtained from the strict relation $<$, the ordering properties may be restated as follows.

Proposition D.3.5.   *1. (reflexive law) For each $n \in \mathbb{Z}$, $n \leq n$.*

2. (antisymmetric law) *For all $m, n \in \mathbb{Z}$, if $m \leq n$ and $n \leq m$, then $m = n$.*

3. (transitive law) *For all $m, n, k \in \mathbb{Z}$, if $m \le n$ and $n \le k$, then $m \le k$.*

4. (comparability) *For all $m, n \in \mathbb{Z}$, $m \le n$ or $n \le m$.*

5. (addition preserves order) *For all $m, n, k \in \mathbb{Z}$, if $m \le n$, then $m + k \le n + k$.*

6. (multiplication preserves order) *For all $m, n, k \in \mathbb{Z}$, if $m \le n$ and $k \ge 0$, then $m \cdot k \le n \cdot k$.*

When one starts with the Algebraic Properties D.3.1 and Ordering Properties D.3.3 formulated above as fundamental assumptions about the integers, there is no need to assume separately the corresponding properties of natural numbers formulated in Section D.1. In fact, the latter will be consequences of the former once $\mathbb{N}$ is defined as a subset of $\mathbb{Z}$.[3] The definition is

$$\mathbb{N} = \{\, n \in \mathbb{Z} : n \ge 0 \,\}.$$

At this point we do also assume the **Principle of Mathematical Induction** as formulated on page A61. Then all the consequences flowing from it are available for use in establishing further properties of integers and natural numbers.

If all you want to know about the integers is what basic properties it is fair to assume about them, then you may stop reading now or skip to Section D.5. If you want to know more, read the next section.

D.4   Integers: Construction

It is all well and good to stipulate that the integers have the properties enumerated above. But exactly what *is* the set of integers? The positive integers are easy enough to understand as counting numbers. Even the set of natural numbers, with its extra element zero, may be considered "natural." But what are the negative integers—really? Exercise 3.4.12 suggests one answer to this question, perhaps an unexpectedly complicated answer in that it involves an equivalence relation and equivalence classes (and it requires some cutting and pasting to ensure that the set of natural numbers is actually a subset of the set of integers).

Here we give a simpler answer to the question of what negative integers *really* are. This answer, like that based on equivalence relations, shows how to construct the set of integers from the natural numbers.

---

[3] However, if you want to *construct* the set of integers and its operations and order relation from the set of natural numbers—as in Section D.4—then of course you need to assume the fundamental properties of natural numbers formulated in Section D.1.

Naturally, we begin by assuming the properties of $\mathbb{N}$ enumerated in Section D.1.

What we are going to do is join to the set of natural numbers a new object of the form $-n$ for each natural number $n > 0$. But what does "object of the form $-n$" mean if you do not already know what the negative integers are? Our answer is that such an object can be thought of as an ordered pair $(-, n)$ of two things: an object $-$ and a natural number n. But what is this object $-$. Actually, it does not matter at all; we just have to select some convenient object to denote by $-$. Since we are assuming we already know what the set $\mathbb{N}$ is, we use the symbol $-$ to denote the set $\mathbb{N}$ itself!

Here, then, is our definition: For each $n \in \mathbb{N}^*$, let $-n$ denote the ordered pair $(-, n)$. Define the **set of integers** to be the set

$$\mathbb{Z} = \mathbb{N}^* \cup \{0\} \cup \{-n : n \in \mathbb{N}^*\},$$

in other words, $\mathbb{Z} = \mathbb{N} \cup \{-n : n \in \mathbb{N}^*\}$. Thus for each integer $n \in \mathbb{Z}$, exactly one of the following conditions holds:

$$n \in \mathbb{N}^*, \qquad n = 0, \qquad n = -k \text{ for a unique } k \in \mathbb{N}^*$$

For each $n \in \mathbb{Z} \setminus \mathbb{N}$, define $-n$ to be the unique $k \in \mathbb{N}^*$ such that $n = -k$. (If you find this confusing, you may wish temporarily to use some symbol other than $-$, say $\ominus$, when denoting elements of $\mathbb{Z} \setminus \mathbb{N}$. Then the preceding definition would be that $-n = k$ where $n = \ominus k$.)

It remains to define operations of addition and multiplication on $\mathbb{Z}$ and an order relation in $\mathbb{Z}$ in such a way that they extend the usual operations and order relation from $\mathbb{N}$ and satisfy the fundamental properties enumerated in Section D.3. That is easy: we force the definitions to follow the familiar rules of signs.

We start with the order relation. For $m, n \in \mathbb{Z}$, we define $m < n$ to be true[4] if and only if one of the following holds:

$$m \in \mathbb{N}, n \in \mathbb{N}, \text{and } m < n \text{ in } \mathbb{N},$$
$$m \in \mathbb{Z} \setminus \mathbb{N} \text{ and } n \in \mathbb{N},$$
$$m \in \mathbb{Z} \setminus \mathbb{N}, n \in \mathbb{Z} \setminus \mathbb{N}, \text{and } -m > -n \text{ in } \mathbb{N}.$$

As usual, we define $m > n$ to mean $n < m$, $m \leq n$ to mean $m < n$ or $m = n$, and $m \geq n$ to mean $n \leq m$. Then

$$m < n \iff -m > -n$$

for all $m, n \in \mathbb{Z}$. Moreover, for each $n \in \mathbb{Z}$,

$$n > 0 \iff n \in \mathbb{N}^* \iff -n < 0,$$
$$n < 0 \iff n \in \mathbb{Z} \setminus \mathbb{N} \iff -n > 0.$$

---

[4]In other words, the ordered pair (m, n) is an element of the relation $<$.

The transitive and trichotomy laws evidently hold for this relation $<$ in the set $\mathbb{Z}$.

Next, we define multiplication in $\mathbb{Z}$ (extending multiplication from $\mathbb{N}$ to $\mathbb{Z}$ is a bit easier than extending addition.) For $m, n \in \mathbb{Z}$, define their product $m \cdot n$ as follows:

$$m \cdot n = \begin{cases} \text{the product } m \cdot n \in \mathbb{N} & \text{if } m \geq 0, n \geq 0, \\ -[m \cdot (-n)] & \text{if } m \geq 0, n < 0, \\ -[(-m) \cdot n] & \text{if } m < 0, n \geq 0, \\ (-m) \cdot (-n) & \text{if } m < 0, n < 0. \end{cases}$$

(If you find this confusing, you may want to distinguish the multiplication assumed known on $\mathbb{N}$ from the multiplication being defined on $\mathbb{Z}$ by temporarily denoting the latter by $\odot$, say.)

Finally, we define addition in $\mathbb{Z}$. For $m, n \in \mathbb{Z}$, define their sum $m + n$ by

$$m + n = \begin{cases} \text{the sum } m + n \text{ in } \mathbb{N} & \text{if } m \geq 0, n \geq 0, \\ \text{the difference } m - (-n) \text{ in } \mathbb{N} & \text{if } m \geq 0, n < 0, m \geq -n, \\ -[(-n) - m] & \text{if } m \geq 0, n < 0, m < -n, \\ n + m \text{ as just defined} & \text{if } m < 0, n \geq 0, \\ -[(-m) + (-n)] & \text{if } m < 0, n < 0. \end{cases}$$

The elements 0 and 1 of $\mathbb{N}$ are the zero and unity, respectively, of $\mathbb{Z}$. For each $n \in \mathbb{Z}$, its negative is the element $-n$ as previously defined. Verifying all the algebraic and ordering properties of $\mathbb{Z}$ not already established is rather tedious; we leave it as an exercise.

D.5   Rational and Real Numbers

The properties assumed of rational and real numbers are discussed in detail in Chapter 3 and are therefore not repeated here. Briefly, each of the **set of rational numbers** $\mathbb{Q}$ and the **set of real numbers** $\mathbb{R}$, when equipped with appropriate operations of addition and multiplication, is a *field*. The properties of a field are enumerated in Section 3.1; they are the usual laws of algebra for manipulating addition, multiplication, subtraction, and division. When equipped with appropriate order relations, each of $\mathbb{Q}$ and $\mathbb{R}$ is an *ordered* field. The properties of an ordered field are enumerated in Section 3.2; they are the usual laws for manipulating inequalities.

Each of the ordered fields $\mathbb{Q}$ and $\mathbb{R}$ is *archimedean* in the sense that for each x, y in $\mathbb{Q}$ or $\mathbb{R}$ with $y > 0$, there is some natural number n

such that $ny > x$. In $\mathbb{Q}$, each element can be written (in more than one way) as a quotient m/n of integers m and n with $n \neq 0$. In $\mathbb{R}$, to the contrary, there are elements such as $\sqrt{2}$ that are **irrational**—not such a quotient of integers. (The irrationality of $\sqrt{2}$ is proved in Example 1.4.6 as well as in Example 2.2.20.) Such elements exist in $\mathbb{R}$ (their existence is proved in Proposition 3.3.15) because this archimedean ordered field is *complete* in the sense explained in Section 3.3.

A method for constructing the field of rational numbers from the set of integers is explained in Example 3.4.10. Two different methods for constructing the field of real numbers from the field of rational numbers are indicated in Exercises 3.3.18 and 3.5.46.

Index